# Food Packaging: The Smarter Way

Ashutosh Kumar Shukla
Editor

# Food Packaging: The Smarter Way

Springer

*Editor*
Ashutosh Kumar Shukla
Physics Department
Ewing Christian College
Prayagraj, Uttar Pradesh, India

ISBN 978-981-16-7195-1 ISBN 978-981-16-7196-8 (eBook)
https://doi.org/10.1007/978-981-16-7196-8



This Springer imprint is published by the registered company Springer Nature Singapore Pte Ltd.
The registered company address is: 152 Beach Road, #21-01/04 Gateway East, Singapore 189721, Singapore

*To my parents.*

# Preface

Food packaging is of interest to different stakeholders including producers, retailers and consumers in view of societal and environmental concerns. Right choice of packaging materials to improve packaging experience and shelf-life are therefore the areas that have attracted the scientific community. This volume is a collection of different chapters dealing with novel packaging materials and trends from pre- to post-packaging of food items. Initial chapters have been planned to acquaint the audience with food packaging industry and provide a link to latest research on smarter ways of food packaging including artificial intelligence applications. Biodegradable packaging solutions for fruits, vegetables and animal-based food products including dairy and fishery products have been covered in the following chapters. Edible films and coatings have been presented in a separate chapter highlighting the challenges and applications. Role of sensors to improve food packaging has also been presented in a separate chapter as well. Similarly, another chapter presents the applications of functional nanomaterials for food packaging. Rules and regulations related to the packaging have also been included to bring in the sense of completeness.

While going through the chapters, I learnt many things and expect the same for the intended audience including novice researchers. I thank the expert contributors from different laboratories/countries/disciplines making this volume truly interdisciplinary. Some of the authors accepted my request to review the individual chapters and provided their valuable comments. This helped me a lot to present the quality content before the readers.

I sincerely thank Dr. Naren Aggarwal, Editorial Director—Books, Asia, Medicine & Life Sciences, Springer for giving me the opportunity to present this book to the readers. I also thank Madhurima Kahali, Editor Books—Medicine & Life Sciences, Springer and Mr. Suraj Kumar, Production Editor (Books), Springer for their support during different stages of publication.

Prayagraj, Uttar Pradesh, India — Ashutosh Kumar Shukla

# Contents

# About the Editor

**Ashutosh Kumar Shukla** is an Associate Professor of Physics at Ewing Christian College, Prayagraj, a constituent college of the University of Allahabad, India. He has also served as Associate Professor in the Department of Pure and Applied Physics, Guru Ghasidas Vishwavidyalaya, Bilaspur, Chhattisgarh (2011–2013), on lien from his parent organization. His research interest is primarily based on the applications of Electron Spin Resonance Spectroscopy technique in different fields of societal interest including Condensed Matter Physics. He has been awarded the Bilateral Exchange fellowship and Visiting fellowship of Indian National Science Academy, New Delhi, in addition to research fellowships of Council of Science and Technology, Lucknow, and Council of Scientific and Industrial Research, New Delhi. Dr. Shukla is on the reviewer panel and editorial boards for international journals. With 20 years of experience of Physics teaching, he has to his credit journal publications, textbooks, and many edited volumes in different areas such as medicine, food science, and crude oil in collaboration with experts globally. Dr. Shukla has successfully completed research projects and presented his research in different international events. He is a member of the International EPR Society (IES) and web member of the International Society of Magnetic Resonance (ISMAR).

# Food Packaging Industry: An Introduction 1

Vilásia Guimarães Martins and Viviane Patrícia Romani

## 1.1 Introduction

The basic materials used in the food packaging industry are glass, metal, paper and paperboard, plastics, or their combinations in the same material. Innovations in the sector of food packaging are adding many different compounds and using various technologies to produce intelligent, active, sustainable, and biodegradable packaging.

There is a growing demand from consumers who desire safer, high-quality, extended shelf life, and less processed food products. In this case, the innovations in terms of packaging are remarkably interesting. Active compounds with antimicrobial and antioxidant properties, for example, can be incorporated in the packaging material. These compounds are released throughout the product's shelf life, and then, it is not necessary to add preservatives in the food product formulation. This type of material is called active packaging. Another technology highlighted in recent years is the intelligent packaging, which also contributes for food safety. Intelligent packaging informs the consumer through indicators and/or labels if the food product is appropriated to be consumed, even if it is within the expiration date described on the product packaging. In some cases, food products are not stored, transported, or even handled at the appropriate temperature, reducing their shelf life. Through intelligent packaging, consumers will know if the product was appropriately transported and handled, and thus, they will be sure about the quality of the product.

The world produces about 300 million tons of plastic waste each year, and so far, only 9% of this waste has been recycled (ONU 2019a). Also, each year at least eight

V. G. Martins (✉)
Federal University of Rio Grande, School of Chemistry and Food Engineering, Rio Grande, Brazil
e-mail: vilasiamartins@furg.br

V. P. Romani
Department of Food Engineering, State University of Midwest, Guarapuava, Brazil


A. K. Shukla (ed.), *Food Packaging: The Smarter Way*,
https://doi.org/10.1007/978-981-16-7196-8_1

million tons of plastic go to sea and oceans, often decomposing into small microplastics that end up in our food chain (ONU 2019b). Durability, resistance, lightness, flexibility, and low cost are some of the characteristics that make the plastic unique and indispensable in daily life (Heidbreder et al. 2019). However, the qualities that make the plastic a good material for consumer products can also make it extremely resistant to biodegradation and can remain in the environment for decades.

Given the above, a sector of the packaging industry that is expanding is the sustainable and biodegradable packaging. Research has been carried out all over the world trying to find solutions to make possible to replace mainly the synthetic plastic packaging by biodegradable packaging. The main raw materials used to produce biodegradable packaging are proteins, lipids, and carbohydrates, usually extracted from agroindustrial sources, and waste from food industries. Important findings are reported in the production of biodegradable materials for food packaging; however, their use is still limited due to the mechanical and barrier properties, which are inferior to those of synthetic polymers. Several techniques have been used by researchers to improve these properties, including the use of blends, chemical modifications of polymers, addition of reinforcement materials (e.g., nanoparticles and fibers), application of plasma, and UV light treatments.

In addition to the sustainable, renewable, and biodegradable materials being used in the development of new plastic packaging, it is also possible to observe this environmental concern for other materials as well. For example, the immense exchange that is taking place from glass, metal, and plastic packaging by cellulosic packaging also has this environmental appeal. Today is possible to find whiskey "bottles" in cardboard on the market, which would have been unthinkable a few years ago.

This chapter will briefly discuss information about the global food packaging market, innovations in the food packaging sector, and what are the challenges and perspectives for the packaging produced in research laboratories to reach the market.

## 1.2 Global Food Packaging Market and Sustainability

According to the World Packaging Organization (2020), the revenue of the global packaging industry is over USD 500 billion. Sustainability, convenience, efficiency, protection, and flexibility are the parameters associated with packaging solutions. For consumers, easy-to-use and sustainable packaging is becoming a priority now than ever before.

There are a lot of market analysis reports about the global food packaging market size at the Internet. Most of them project the future market of food packaging based on type of material, application, outlook, region, among others. The growth of the food packaging market is estimated based on some factors such as convenience, increase in the population, improvement of the shelf life, single serve packs, food delivery, and high-performance materials; all these parameters have a positive impact on the market. In the industries, the consumers' desire is what drives the vector of changes, and in the packaging industry, it would be no different. In addition to the improvement of the consumer experience when unpacking a product, its design and how it relates to sustainability also matter.

About 40% of all plastic produced in the world is used for food packaging, followed by building and construction industry (20%) (PlasticsEurope and EuPR 2015). Governments have been acting to reduce environmental impacts, establishing agreements with brands to reduce or change the use of raw materials, such as plastic. In the United Kingdom, for example, the government will introduce a tax on plastic packaging from 2022. Besides that, financial sanctions will be established to encourage the incorporation of recycled plastic in the production chain of these companies, creating an interesting precedent for their adhesion.

As mentioned by the Grand View Research (2020), the paper and paper-based material segment accounted for a revenue share of 31.9% of the total food packaging market in 2019. Growth of this segment is driven by high product adoption to substitute nonbiodegradable packing solutions. Innovations in design, ease of printability, and sustainability give paper packaging a competitive advantage over plastic and metal packaging solutions.

Regarding bioplastics, which represents a crucial way to reduce the environmental burden, European Bioplastics, Nova-Institute (2020), mentioned that the global production capacities of bioplastics including bio-based and biodegradable materials are 2.11 million tons in 2020. Bio-based/nonbiodegradable materials (PE, PET, PA, PP, PTT, and other) represent 41.9% of the bioplastics produced, and biodegradable materials (PBAT, PBS, PLA, PHA, starch blends, and other) represent 58.1% of the total. Going forward, the conscious consumption will be the meeting point for transformations in the global food industry. Implementing sustainable policies and philosophies in the manufacturing lines, from the product itself, through packaging and logistics, will be a necessity to survive in a modern market that is concerned with the planet.

## 1.3 Sustainable Food Packaging

Plastics are the most used materials in packaging applications in different fields. Commonly used polymers, such as polyvinylchloride (PVC), polyethylene terephthalate (PET), polypropylene (PP), polyethylene (PE), polyamide (PA), polystyrene (PS), and ethylene vinyl alcohol (EVOH), are cost-effective and have enough properties to protect packaged products since production to consumption (Geyer et al. 2017; Luzi et al. 2019). However, these fossil-based polymers are obtained from finite sources. In addition, the incorrect plastic disposal, which has a nonrecyclable or nonbiodegradable nature, is driving the growing need of sustainable alternatives, due to the environmental burden resulting from the high amount discarded (Ahmed et al. 2018).

Sustainable materials protect the environment since the raw material extraction to the final disposal, meaning that there is no damage to the nature. The food industry is one of the largest users of packaging (All4Pack 2018) and thus has a valuable responsibility in using sustainable packaging as well as motivating industries and consumers to adhere to the use of such materials. Then, the idea is to use renewable and abundant resources to produce biodegradable/compostable alternatives. In this context, it is important to consider that plastics produced by renewable resources are not certainly biodegradable or compostable. Likewise, biodegradable materials are not necessarily produced with renewable resources, as biodegradation is related to the chemical structure in the matrix instead of its origin (Lambert and Wagner 2017; Asgher et al. 2020). In summary, a sustainable packaging is a material obtained from renewable sources that in the end of its cycle is biodegraded or composted.

Bioplastics are a potential alternative that can contribute to the sustainable development in food packaging. Biopolymers or natural polymers can be generated by plants or microorganisms and/or extracted from food industry by-products. Some examples include carbohydrates (e.g., chitosan, starch, cellulose), proteins (e.g., keratin, gluten, collagen), polylactic acid (PLA) polyhydroxyalkanoates (PHAs), and exopolysaccharides (EPSs) (Benbettaïeb et al. 2016; Asgher et al. 2020), which are being widely explored in the development of films, wraps, and laminates for food packaging. Some examples of recent developments include an edible film packaging made from a milk protein for dairy, a box for pasta packaging prepared with wasted vegetables and fruits, and a plant-based paper cup (TrendHunter 2017).

The production of bio-based materials for food packaging is a multistep process. Generally, it starts with the breakage of the intermolecular linkages of the biopolymer, followed by the synthesis of a new molecular structure and finally the development of the three-dimensional network from the new linkages formed (Galić et al. 2011). The resulting material characteristics depend on the raw polymer structure and the processing conditions. Bioplastics can be produced by wet or dry processing (Blanco-Pascual et al. 2013). Wet processing, known as casting technique or continuous spreading, is based on the solubilization of the biopolymer in the solvent with the posterior solvent evaporation. The final polymer is influenced by the pH and temperature of the suspension, and the type of solvent (Khwaldia et al. 2004; Mellinas et al. 2016). Differently, the dry processing, including extrusion, or thermal processing, depends on the thermoplastic properties of the polymers, based on the theory of glass transition. This method consists of the conversion of the glassy structure into a semi-solid condition at a specific temperature, where the molecules are broken, and new linkages and bonds are formed (Khwaldia et al. 2004; Hernandez-Izquierdo and Krochta 2008).

Bioplastics can also be incorporated with various additives, for example, antioxidant and antimicrobial compounds, nutrients, and colorants. Agricultural by-products are a potential and cheap source of renewable additives. These additives reduce the possibility of microbial growth and lipid oxidation, and also have the capacity to increase the shelf life of products (Jafarzadeh et al. 2020). Studies reported the incorporation of various additives from renewable sources in different matrices, such as durian leaf waste to enhance antioxidant activity in gelatin films (Joanne Kam et al. 2018), jambolão skin extract in methylcellulose films to provide antioxidant and color-changing properties (da Silva Filipini et al. 2020), grape seed extract—carvacrol microcapsules in chitosan films to increase the shelf life of refrigerated Salmon (antimicrobial properties) (Alves et al. 2018), chitosan nanoparticles in starch films to inhibit bacteria in wrapped cherry tomatoes (Shapi'i et al. 2020), pomegranate peel particles into starch-based films as antimicrobial and reinforcing agent (Ali et al. 2019), kombucha tea in chitosan films to retard lipid oxidation and microbial growth in minced beef (Ashrafi et al. 2018), and pink pepper phenolic compound incorporation in starch/protein blends to inhibit apple browning (Romani et al. 2018). Bioplastics have strong potential as sustainable alternatives to synthetic materials because they are produced using renewable sources, a prerequisite as previously mentioned, and despite some challenges that their use still faces (discussed in the section ahead), generally they present fast biodegradation in soil and water.

## 1.4 Technology Developed in the Laboratories X Scale Up to Industries

Sustainable materials have been broadly studied and interesting results are being obtained, but the large-scale production for commercial uses is still limited and the commercial use is insignificant compared to the conventional plastics. There are different reasons that prevent the wide use of sustainable packaging, including (1) the material properties, which are not enough to protect the products and can compromise the food shelf life; (2) scale-up, since the process of production, development of the new technology, and financial capacity; (3) production and logistic for feedstocks and composting infrastructure; (4) regulation policies; and (5) consumer behavior (Rydz et al. 2018; Steenis et al. 2018; de la Caba et al. 2019).

New sustainable materials for packaging are generally produced using hydrophilic biopolymers, which result in the low physicochemical properties (such as mechanical and barrier performance) due to sensitivity to humidity. To overcome these limitations, researchers are searching for strategies for material improvement. Examples of strategies are the incorporation of reinforcing agents, blending different raw materials, and chemical, enzymatic, and physical methods to alter polymer chains (Bourtoom 2009). Promising results are being obtained; however, the challenge is the improvement of the overall performance of the polymer. Usually, the increase in the mechanical resistance does not come with the increase in barrier properties and vice versa. Indeed, the protection of foods requires the equilibrium of different material's characteristics to prevent the component oxidation and microbial spoilage, and thus, attention is necessary at the minimum performance necessary in the packaging to keep the quality of products (Martins et al. 2019). Additionally, the change in the properties of bioplastics during the interaction with the food material needs to be considered. Either, the compatibility of the polymers with the products can affect the food quality and gained minimal attention (Asgher et al. 2020).

Another important hurdle that affects the wide adoption of biopolymers use is the difficult scale-up of the production process. The production of biopolymers in laboratories is widely performed through the casting technique. It consists of pouring a suspension on plates (e.g., Petri dishes) controlling the mass of suspension to generate an uniform thickness, but variations are difficult to avoid. This technique is suitable to produce films up to 30 cm, not larger, and takes long time in the drying step making it useless at industrial scale. The tape casting, which consists of the spreading of suspension in larger supports or on continuous carrier tapes, is useful to produce larger films, but industries in general have well-established extrusion processes (De Moraes et al. 2013; Werner et al. 2017). To advance in the large-scale production of biopolymers, the proof of concept needs to be effectively produced in industries. Even though some biopolymers are not suitable for extrusion processes, as in the case of proteins due to their molecular architecture and spatial arrangement (Mensitieri et al. 2011), it is important to focus on strategies to adequate such polymers in a way to facilitate their processing by the industry technologies. Generally, industries adopt and license unique raw materials to produce plastics, and then, an industry adjustment would be necessary for the production of the biopolymers. In addition, besides the fact that academic research mostly stops at the proof-of-concept stage, patented technologies are frequently incomplete in scope to be adopted by industry (Nerkar and Shane 2007; Tolfree and Jackson 2008; Inns 2012), impeding the advance of such technologies. That is why, it is also important the pilot scale fabrication and characterization, otherwise it is possible to be financially inviable for industries to take high risks to adopt the sustainable technologies for packaging (Werner et al. 2017).

## 1.5 Final Remarks

To increase the use of biopolymers, besides the mentioned aspects, the resource efficiency and composting infrastructure need to be considered, as well as the regulation policies. Currently, raw materials used to produce biopolymers often compete with requirements for food-based products. The expansion of the first-generation bio-based polymer production can generate unsustainable demands (Babu et al. 2013). Furthermore, a composition structure is necessary for the suitable biopolymer disposal. As the definition of ASTM-D6400, a compostable material is a material that biologically degrades; therefore, just substances that can degrade biologically in a composting environment can be labeled as "compostable" (ASTM D 2004). In this sense, besides the processing of biopolymers, it is fundamental to the planning of resource efficiency, raw materials obtaining, and composting infrastructure in the end of the biopolymer life.

Also, the consumer purchase probability and willingness to pay for sustainable materials have a critical influence in the advance of new biopolymer technologies. The consumer acceptance of these new materials has also an important role. Firstly, the higher costs of biopolymer production result in superior market prices and consumers usually do not pay the price for sustainability. Also, traditional consumers can question the quality and/or safety of the new material due to the limited information about the advantages and the low familiarity with biopolymers (De Marchi et al. 2020). In addition, consumer insights related to sustainable designs help designers to develop coherent strategies to adopt new initiatives. The lack of these consumer insights also complicate the adoption of new sustainable alternatives (Steenis et al. 2018). Still regarding the attitudes to adopt sustainable materials, industries have an essential role as well. Some manufacturers tend to support the use of old technologies and prevent the adoption of more sustainable solutions due to the efforts needed in adjusting their processes (Keränen et al. 2020). Therefore, reorientation of existing industries toward sustainability demands more attention and consumers need to be informed regarding the importance of sustainable packaging and motivated to consume such products.

## References


Ahmed T, Shahid M, Azeem F et al (2018) Biodegradation of plastics: current scenario and future prospects for environmental safety. Environ Sci Pollut Res 25:7287–7298. https://doi.org/10.1007/s11356-018-1234-9

Ali A, Chen Y, Liu H et al (2019) Starch-based antimicrobial films functionalized by pomegranate peel. Int J Biol Macromol 129:1120–1126. https://doi.org/10.1016/j.ijbiomac.2018.09.068

All4Pack (2018) Market key figures, challenges & perspectives of worldwide packaging. Paris

Alves VLCD, Rico BPM, Cruz RMS et al (2018) Preparation and characterization of a chitosan film with grape seed extract-carvacrol microcapsules and its effect on the shelf-life of refrigerated Salmon (Salmo salar). LWT 89:525–534. https://doi.org/10.1016/j.lwt.2017.11.013

Asgher M, Qamar SA, Bilal M, Iqbal HMN (2020) Bio-based active food packaging materials: sustainable alternative to conventional petrochemical-based packaging materials. Food Res Int 137:109625. https://doi.org/10.1016/j.foodres.2020.109625

Ashrafi A, Jokar M, Mohammadi Nafchi A (2018) Preparation and characterization of biocomposite film based on chitosan and kombucha tea as active food packaging. Int J Biol Macromol 108: 444–454. https://doi.org/10.1016/j.ijbiomac.2017.12.028

ASTM D (2004) ASTM D6400-04: standard specification for compostable plastics. ASTM Int

Babu RP, O'Connor K, Seeram R (2013) Current progress on bio-based polymers and their future trends. Prog Biomater 2:8. https://doi.org/10.1186/2194-0517-2-8

Benbettaïeb N, Gay JP, Karbowiak T, Debeaufort F (2016) Tuning the functional properties of polysaccharide–protein bio-based edible films by chemical, enzymatic, and physical cross-linking. Compr Rev Food Sci Food Saf 15:739–752. https://doi.org/10.1111/1541-4337.12210

Blanco-Pascual N, Fernández-Martín F, Montero MP (2013) Effect of different protein extracts from Dosidicus gigas muscle co-products on edible films development. Food Hydrocoll 33: 118–131. https://doi.org/10.1016/j.foodhyd.2013.02.019

Bourtoom T (2009) Edible protein films: Properties enhancement. Int Food Res J 16:1–9

da Silva Filipini G, Romani VP, Guimarães Martins V (2020) Biodegradable and active-intelligent films based on methylcellulose and jambolão (Syzygium cumini) skins extract for food packaging. Food Hydrocoll 109:106139. https://doi.org/10.1016/j.foodhyd.2020.106139

de la Caba K, Guerrero P, Trung TS et al (2019) From seafood waste to active seafood packaging: an emerging opportunity of the circular economy. J Clean Prod 208:86–98. https://doi.org/10.1016/j.jclepro.2018.09.164

De Marchi E, Pigliafreddo S, Banterle A et al (2020) Plastic packaging goes sustainable: an analysis of consumer preferences for plastic water bottles. Environ Sci Pol 114:305–311. https://doi.org/10.1016/j.envsci.2020.08.014

De Moraes JO, Scheibe AS, Sereno A, Laurindo JB (2013) Scale-up of the production of cassava starch based films using tape-casting. J Food Eng 119:800–808

European Bioplastics, Nova-Institute (2020). www.european-bioplastics.org/market

Galić K, Ščetar M, Kurek M (2011) The benefits of processing and packaging. Trends Food Sci Technol 22:127–137. https://doi.org/10.1016/j.tifs.2010.04.001

Geyer R, Jambeck JR, Law KL (2017) Production, use, and fate of all plastics ever made. Sci Adv 3: 25–29. https://doi.org/10.1126/sciadv.1700782
Grand View Research (2020). https://www.grandviewresearch.com/industry-analysis/food-packaging-market
Heidbreder LM, Bablok I, Drews S, Menzel C (2019) Tackling the plastic problem: a review on perceptions, behaviors, and interventions. Sci Total Environ 668:1077–1093. https://doi.org/10.1016/j.scitotenv.2019.02.437
Hernandez-Izquierdo VM, Krochta JM (2008) Thermoplastic processing of proteins for film formation—a review. J Food Sci 73:R30–R39. https://doi.org/10.1111/j.1750-3841.2007.00636.x
Inns GR (2012) The packaging supply chain. In: Packaging technology. Elsevier, pp 10–23
Jafarzadeh S, Jafari SM, Salehabadi A et al (2020) Biodegradable green packaging with antimicrobial functions based on the bioactive compounds from tropical plants and their by-products. Trends Food Sci Technol 100:262–277. https://doi.org/10.1016/j.tifs.2020.04.017
Joanne Kam W-Y, Mirhosseini H, Abas F et al (2018) Antioxidant activity enhancement of biodegradable film as active packaging utilizing crude extract from durian leaf waste. Food Control 90:66–72. https://doi.org/10.1016/j.foodcont.2018.02.036
Keränen O, Komulainen H, Lehtimäki T, Ulkuniemi P (2020) Restructuring existing value networks to diffuse sustainable innovations in food packaging. Ind Mark Manag. https://doi.org/10.1016/j.indmarman.2020.10.011
Khwaldia K, Perez C, Banon S et al (2004) Milk proteins for edible films and coatings. Crit Rev Food Sci Nutr 44:239–251. https://doi.org/10.1080/10408690490464906
Lambert S, Wagner M (2017) Environmental performance of bio-based and biodegradable plastics: the road ahead. Chem Soc Rev 46:6855–6871. https://doi.org/10.1039/c7cs00149e
Luzi F, Torre L, Kenny J, Puglia D (2019) Bio- and fossil-based polymeric blends and nanocomposites for packaging: structure–property relationship. Materials (Basel) 12:471. https://doi.org/10.3390/ma12030471
Martins VG, Romani VP, Martins PC, Filipini GS (2019) Innovative packaging that saves food. In: Saving food. Elsevier, pp 171–202
Mellinas C, Valdés A, Ramos M et al (2016) Active edible films: current state and future trends. J Appl Polym Sci 133:42631. https://doi.org/10.1002/app.42631
Mensitieri G, Di Maio E, Buonocore GG et al (2011) Processing and shelf life issues of selected food packaging materials and structures from renewable resources. Trends Food Sci Technol 22: 72–80. https://doi.org/10.1016/j.tifs.2010.10.001
Nerkar A, Shane S (2007) Determinants of invention commercialization: an empirical examination of academically sourced inventions. Strateg Manag J 28:1155–1166. https://doi.org/10.1002/smj.643
ONU (2019a) Fungos podem ajudar no combate à poluição por plásticos, diz ONU Meio Ambiente. https://nacoesunidas.org/fungos-podem-ajudar-no-combate-a-poluicao-por-plasticos-diz-onu-meio-ambiente/. Accessed 25 May 2020
ONU (2019b) ONU Meio Ambiente aponta lacunas na reciclagem global de plástico. https://nacoesunidas.org/onu-meio-ambiente-aponta-lacunas-na-reciclagem-global-de-plastico/. Accessed 25 May 2020
PlasticsEurope, EuPC, EuPR, EPRO (2015) Plastics—the facts 2015. An analysis of European plastics production, demand and recovery for 2015
Romani VP, Prentice-Hernández C, Martins VG (2018) Pink pepper phenolic compounds incorporation in starch/protein blends and its potential to inhibit apple browning. Food Packag Shelf Life 15:151–158. https://doi.org/10.1016/j.fpsl.2018.01.003
Rydz J, Musioł M, Zawidlak-Węgrzyńska B, Sikorska W (2018) Present and future of biodegradable polymers for food packaging applications. In: Biopolymers for food design. Elsevier, pp 431–467

Shapi'i RA, Othman SH, Nordin N et al (2020) Antimicrobial properties of starch films incorporated with chitosan nanoparticles: in vitro and in vivo evaluation. Carbohydr Polym 230:115602. https://doi.org/10.1016/j.carbpol.2019.115602

Steenis ND, van der Lans IA, van Herpen E, van Trijp HCM (2018) Effects of sustainable design strategies on consumer preferences for redesigned packaging. J Clean Prod 205:854–865. https://doi.org/10.1016/j.jclepro.2018.09.137

Tolfree D, Jackson MJ (2008) Commercializing micro-nanotechnologies. CRC Press, Boca Raton

TrendHunter (2017) 25 examples of sustainable food packaging. https://www.trendhunter.com/slideshow/sustainable-food

Werner BG, Koontz JL, Goddard JM (2017) Hurdles to commercial translation of next generation active food packaging technologies. Curr Opin Food Sci 16:40–48. https://doi.org/10.1016/j.cofs.2017.07.007

World Packaging Organization (2020)

# Toward Smarter Food Packaging 2

Bambang Kuswandi

**Abstract**

Many food producers regularly know about new food packaging options, i.e., smart packaging, active packaging, intelligent packaging, and connected packaging. This chapter review shows that these types of packaging benefit specific types of food products, so producers can decide which ones most help them. Smart packaging is an umbrella term used to describe food packaging with enhanced functionality through technology. However, intelligent packaging that will be focused on in this chapter contains sensors to determine the condition (e.g., freshness or ripeness) of the food products. Another term used in smart packaging is active packaging, where it is used to interact with the packaged foods to enhance their condition, significantly to extend freshness or shelf life. In comparison, connected packaging allows consumers to interact with a food product through a label or code on the package that can be activated with a mobile device. Finally, the role of I.T. in smart packaging applications for food quality and safety monitoring is discussed.

**Keywords**

Smart packaging · Intelligent packaging · Sensors · Perishable foods · I.T

B. Kuswandi (✉)
Chemo and Biosensors Group, Faculty of Pharmacy, University of Jember, Indonesia, Jember, Indonesia
e-mail: b_kuswandi.farmasi@unej.ac.id


A. K. Shukla (ed.), *Food Packaging: The Smarter Way*,
https://doi.org/10.1007/978-981-16-7196-8_2

## 2.1 Introduction

There will be many food products that lose their nutritional values and properties fastly if they are not packaging properly. Food packaging plays a very important role in the food supply chain management system. Since it can be employed as a protective layer to avoid food contamination and maintain food quality and safety (Bambang Kuswandi et al. 2011), however, the classical food packaging system can only protect or isolate food product from the external condition and environment without informing food quality, such as freshness and safety, for users (Kuswandi and Moradi 2019). Physical, chemical, and microbial properties and sensory evaluation are traditional methods for the determination of food quality and safety. The drawbacks of these methods are laborious works, long procedure, high cost, requiring sample pretreatment, and destructions. Based on these methods, the detection of food quality and safety is difficult to be performed in fast and nondestructive methods. Therefore, smart food packaging that informs the consumer regarding the food quality and safety evaluation is needed. It drives by the recent trends and consumer needs as well as regulatory requirements on food products.

Food packaging also plays a crucial role in reducing waste and carbon dioxide ($CO_2$) caused by used packaging and their disposal (Kuswandi 2017). This is due to the fact that our society has produced 8.3 billion metric tons of waste, where it contains 76% of plastic, which is mainly not recycled (90.5%) for over the last 60 years (UN Environment Programme 2018). Furthermore, the current situation due to the pandemic of COVID-19 caused food scarcity, which has to pay further attention to the food waste problem. Even though all in the food sectors have a role play in avoiding and decreasing food waste, starting from food producers to consumers, the significance of domestic waste recommended that innovative packaging can be an important tool for preventing and reducing food waste. In this regard, packaging has become an important technology to assure quality and safety in the food supply chain, enhance shelf life of foods, fulfill consumer satisfaction, and prevent undesired consumer complaints (Kuswandi 2020). Since consumers are more concerned about food quality and safety and require more accurate food information on quality and safety, in this case, it needs to provide a simple and real-time monitoring method for food product quality and safety during food supply, distribution, storage, and display according to the consumer interests and rights. Although there are many major innovations in food packaging systems and materials, the basic food packaging principles are still the same. Among them, smart packaging systems offer a great approach in reducing or even preventing food wastes.

Smart packaging, especially intelligent, active, and connected packaging, could support the optimization of the food supply management system, such as enhancing food shelf life and food quality monitoring, and reducing food loss and waste. Active packaging system helps in increasing the food product shelf life, enhancing their condition, and significantly extending food quality by "interacting" with the food product, while an intelligent packaging system helps in informing the food product quality and safety during transportation and storage as well as a display by "monitoring" the food product condition (Kuswandi et al. 2011; Kuswandi 2017; Rai et al. 2019). Furthermore, the connected packaging allows consumers to interact with a food product via package label or code that can be accessed using a mobile device. This chapter discussed smart packaging that will be more focused on intelligent packaging development over active packaging since its "monitoring" function could enhance to be smarter with the help of ICT via Apps, artificial intelligence, or IoT (Internet of Things). In this case, connected packaging concepts are also discussed that allow consumers to interact with a food product through a label or code on the package that can be activated with a mobile device.

## 2.2 Intelligent Packaging

Based on the Commission of the European Communities, it is stated that intelligent food contact materials are defined as materials that allow monitoring the packaged food condition or the environment surrounding the food (Communities 2004). Hence, intelligent food packaging could be defined as a new packaging technology that integrates intelligent functions with food packaging systems. Intelligent packaging can detect, sense, trace, record, and communicate internal or external changes related to food products that associate with the formation of food quality and safety, and extended the packaging information functions during transport, storage, and display that promotes and enhances safety and quality of food producers or consumer (Kuswandi et al. 2011; Salinas et al. 2014a, b). The intelligent packaging functions are given in Fig. 2.1. Intelligent packaging performs the communication functions on the packaged food condition during distribution, storage, sales, and packaging waste disposal to food producers and consumers (Yam et al. 2005).

In order to allow real-time or online monitoring of a food product during distribution, storage, and display, numerous smart devices are employed, such as indicators (for monitoring food freshness, pH, leakage, integrity, time, and temperature), data carriers (bar codes, IoT), and sensors (gas chemical sensors and biosensor) (Ghaani et al. 2016; Kuswandi 2018; Müller and Schmid 2019). Intelligent packaging allows many applications in food freshness detection, spoilage, chemical, and microbial contaminant detection, traceability, authentication, etc. (Majid et al. 2018; Kalpana et al. 2019; Vanderroost et al. 2014; Popa et al. 2019). Some current example of various smart devices used and integrated into intelligent food packaging is shown in Table 2.1.

### 2.2.1 Freshness Indicator

A freshness indicator is a smart colorimetric device integrated into intelligent food packaging. This type of colorimetric indicator is divided into two types: direct colorimetric indicator and indirect colorimetric indicator as given in Fig. 2.2 (Kuswandi et al. 2011). The direct colorimetric indicator commonly works based on color development of indicator caused by the volatile gas released by food that

**Fig. 2.1** Intelligent packaging functions in food packaging

**Table 2.1** Some current examples of various smart devices integrated into intelligent food packaging

| Type of smart device | Sensitive material | Target analyte | Material used | Type of food | References |
|---|---|---|---|---|---|
| Indicators | Black carrot | pH | Bacterial cellulose | Pasteurized milk | Ebrahimi Tirtashi et al. (2019) |
| | Red cabbage | pH | Bacterial cellulose | Pasteurized milk | Kuswandi et al. (2020) |
| | Alizarin | pH | Chitosan | Fish | Ezati and Rhim (2020) |
| | Shikonin | pH | CMC<br>CNF | Fish fillet (mackerel) | Ezati et al. (2021) |
| Gas sensor | Oxygen sensor | $O_2$ | A blending of polyethylene and ethylene-vinyl acetate | Beef | Kelly et al. (2020) |
| | RFID | $O_2$ and $CO_2$ | Wheat gluten | Cheese | Saggin et al. (2019) |
| | RFID | $O_2$ and $CO_2$ | Cellulose | Vegetables | Eom et al. (2012) |

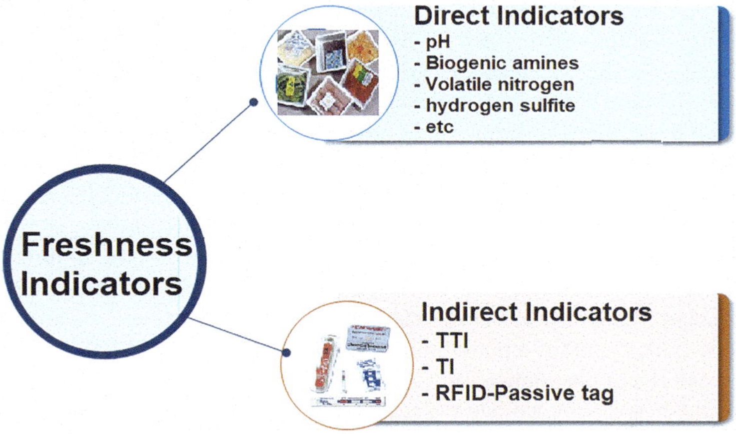


**Fig. 2.2** Types of freshness indicators

can be directly related to the food freshness. This colorimetric indicator allows for qualitative or semi-quantitative detection of food quality and safety as a result of physiological changes or microbial growth during food storage, which directly integrated into food packaging, where it can help consumers to judge food quality easily by the naked eye (Ghaani et al. 2016; Kuswandi 2018), while indirect colorimetric indicator works based on imitating or mimicking degradation of targeted food to be monitored, for example, TTI (time–temperature indicator) and T.I. (temperature indicator) (Kuswandi 2018; Kuswandi 2020). In other perspectives, these freshness indicators could also be classified into two types, i.e., the first one is an internal indicator, and the second, external indicator. The internal freshness indicators are placed inside the food package, and the external freshness indicators are physically placed outside the food package (Pavelková 2013). Both types of freshness indicators can inform consumers related to the presence or absence of specific compounds, or the concentration of a specific substance, related to the level of food freshness. Thus, a distinct characteristic of freshness indicator is the type of information they provide, either qualitative or semi-quantitative toward the user for freshness level.

Freshness indicators employed in food packaging inform a consumer on the issue of food quality and safety related to food degradation due to microbial activity or other food properties. Mostly, the information is presented by immediate color change variations, such as different color intensities or the dye diffusion inside the indicator area (Kalpana et al. 2019; Kuswandi 2020). Even though freshness indicators are simple devices but crucial in assuring food quality and safety, they allow a reduction in the food loss and the cost loss due to the food damage. Therefore, it is very useful for nondestructive testing of food freshness, especially for perishable foods and dairy products, such as fish, meat, milk, fruits, and vegetables. Therefore, in other senses, it is also called a spoilage indicator since it is also used to indicate food deterioration or spoilage. Even though intelligent food packaging has many benefits for perishable foods, however, its application in the food industry is still limited, and it needs further exploration and commercialization.

Commonly, freshness indicators are monitoring the packaged food quality by detecting metabolite compounds produced associated with the food freshness causing by natural metabolism or microbial growth (Kuswandi et al. 2011; Realini and Marcos 2014). Physicochemical changes during food distribution and storage are indicators for food freshness. Food metabolites change during storage, such as pH, lactic acid, carbon dioxide, ethanol, volatile nitrogen, biogenic amines, or hydrogen sulfite, are an indication of microbial activities inside foods, which, in turn, can be employed as indicators of food freshness (Arvanitoyannis and Stratakos 2012; Kuswandi et al. 2011). Freshness indicator monitors food freshness via this metabolite detection that has been reported in the literature (Kuswandi 2018), even though, their successful commercialization is still limited. For instance, FreshTag® and Toxin Guard™ are commercial freshness indicators that have been discontinued. The former was a colorimetric freshness indicator that detects the volatile amines produced in fish, based on the immobilization of antibodies into plastic films to detect food pathogens (Kuswandi et al. 2011).

### 2.2.2 Ripeness Indicators

Freshness indicators are mainly employed for freshness monitoring of protein-based food products, e.g., meat and seafood. Besides these, particularly, fruits and vegetables produce a large number of volatile compounds during ripeness and maturation that can be used to indicate ripeness and maturity. For the first time, in 2004, P-P Enterprise's supermarket in New Zealand released a new ripeness indicator, namely ripeSence (T.M.) for pear, as a new intelligent packaging system. It is the first intelligent label in the world that can indicate the fruit ripeness or maturity or even rotten. The ripeness indicator could detect fruit ripeness by determining the fruit aroma compounds released during fruit ripening. Volatile aldehydes, such as acetaldehyde produced during stone fruit maturation, such as apple, could be used to detect their maturity. An indicator label based on methyl red can detect the release of aldehyde from an apple. The ripeness indicator is fabricated by special ink printed directly on a paper medium. Under mature conditions, the indicator color developed from yellow to orange and then to red (Kim et al. 2018).

Ethylene produce can directly relate to fruit ripeness or vegetable maturity. A colorimetric ripeness indicator for color change upon reaction with ethylene has been developed. The indicator can be used to indicate the ripeness or maturity via the ethylene release during ripening of fruits and vegetables, such as apple (Lang and Hübert 2012) and kiwifruit (Hu et al. 2016). With the development of ripeness indicators, a variety release of aromatic substances during fruit and vegetable maturation or rotten can be employed as releases indicating fruit and vegetable ripeness.

pH could also be used as an indicator for fruit freshness since many metabolite changes during fruit ripeness or vegetable maturity relate to the pH change inside

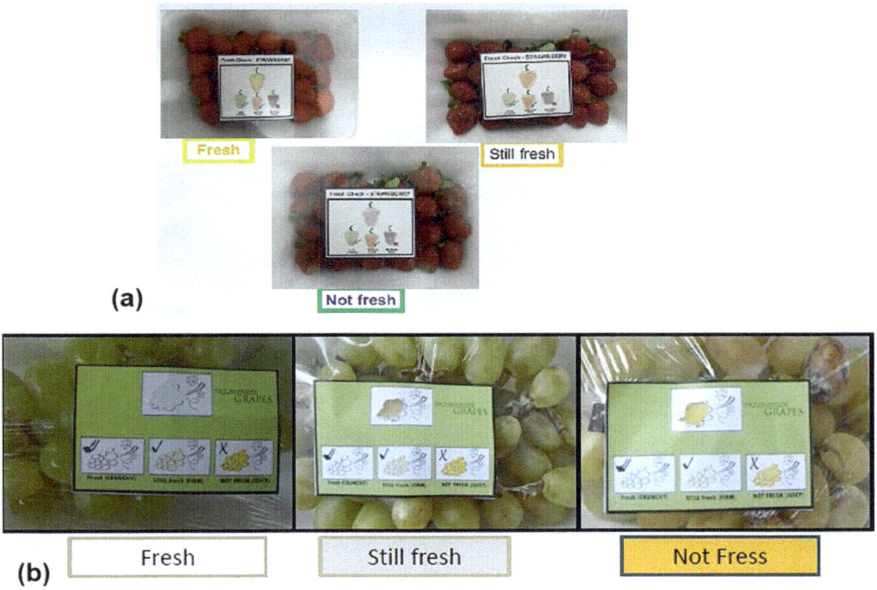


**Fig. 2.3** Ripeness indicators were used for monitoring fruit ripeness (freshness) for strawberries (**a**) and grapes (**b**)

food packaging (Kuswandi 2018). For example, a ripeness indicator was developed for monitoring guava (*Psidium guajava* L.) freshness based on immobilized bromophenol blue onto a cellulose membrane by absorption (Kuswandi et al. 2013). The ripeness color indicator developed from blue to green as the pH declined as a result of the guava over-ripening, where at this stage, the volatile organic compounds were released. This indicator was successfully applied to detect the guava ripeness at ambient temperature (28–30 °C). Furthermore, fruit ripeness indicators have been developed by the same group for strawberries (Kuswandi 2013) and grape (Kuswandi and Murdyaningsih 2017) using a similar principle based on pH change in fruits monitored as given in Fig. 2.3. Another indicator for vegetable maturity was developed based on a silver nanoparticle that has yellow color as a colorimetric sensor for the detection of organosulfur compounds produced when onion spoilage. The color developments were observed during 10 days, where the color developed with time, from yellow to orange, then pink, and lastly colorless when the onion spoilage.

### 2.2.3 Other Indicators and Sensors

Other indicators or often called colorimetric sensors since these smart devices can be classified as colorimetric chemical sensors or biosensors depend on the sensitive membrane or film used. One example of this type of indicator is integrity indicator, where it is informing regarding how long a packaged food has been opened or inform the gases presence inside the food package. Usually, this integrity indicator is employed with MAP (modified atmospheres in packaging), where the oxygen gas replaces other gases, such as nitrogen or carbon dioxide, which, in turn, enhances the shelf life of foods.

Ageless Eye, Tell-Tab, and $O_2$ Sense are examples of commercial integrity indicators that provide color changes upon oxygen exposure so that suitable for the naked eye detection in terms of oxygen concentrations at low concentration (0.5%). Another example is Food Sentinel System™ that has been discontinued already. It is a biosensor system for the food pathogen detection using a specific pathogen antibody immobilized onto the membrane creating a barcode part (Yam et al. 2005). Here, when the contaminating bacteria are present, it will produce a localized dark bar, which, in turn, causes the error or unreadable barcode while scanning when it is used in the supermarket.

## 2.3 Smart Indicator for Food Freshness

Based on the type of design construction, the food freshness indicator could be grouped into (1) single indicator; (2) dual indicator; and (3) indicator array (Kuswandi 2020). The single indicator is a freshness indicator that only used one indicator to detect, sense, trace, and inform the food freshness quality and safety inside the package. The dial indicator is the freshness indicator that used two indicators in monitoring and informing the food freshness quality and safety. At the same time, an indicator array is the multiple indicators that used several indicators in an array to create a pattern in detecting or sensing, and informing the food freshness quality and safety. Mostly, commercially available freshness indicators are a single indicator type, while dual indicator and indicator array are undergoing laboratory-scale developments. Figure 2.4 shows the ripeness indicator or sensors based on design, type, and application for food freshness.

The smart indicators integrated into food packaging can act as an important tool in sensing, detecting, monitoring, and tracing food freshness quality and safety.

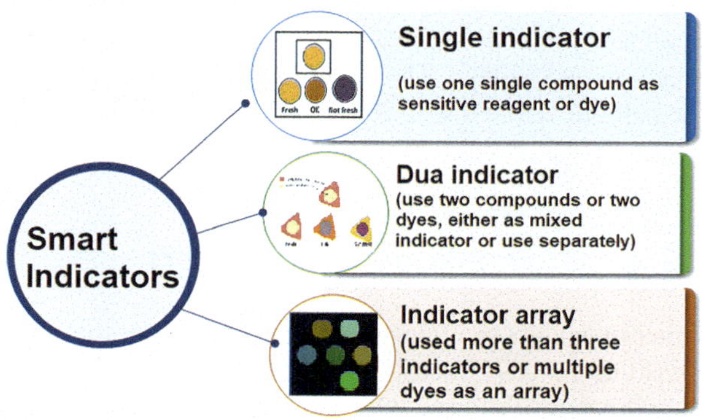


**Fig. 2.4** Types of smart indicators used in food packaging for freshness monitoring

These smart indicators or sensors are critically important when foods are stored outside their desired conditions, e.g., extreme hot or cold. Furthermore, in the case of processed foods, where they should not be frozen, a smart indicator could be necessary to monitor whether they had been stored in freezing conditions improperly or not. On the other hand, a smart indicator could also inform if the processed food is sensitive to heat where it had been exposed to hot conditions and the duration involved. Moreover, the detailed description of the smart indicators for food freshness quality and safety is described in the following subsections.

### 2.3.1 Single Indicator

Commonly, a single indicator or sensor is used in a single color indicator construction, and the choice of reagent dye has a great impact on its colorimetric indicator performance. In order to construct high sensitivity of the colorimetric indicator or sensor, the reagent dye should have high sensitivity as well. For instance, if a single indicator works based on pH change inside the headspace of food packaging, therefore, a large pH range of the reagent dye is needed, or it should be sensitive enough to pH change and easy to distinguish its color change related to the food freshness during storage (Kuswandi 2018). By using a single indicator, simple, fast, and sensitive detection of food freshness can be performed by using a noninvasive colorimetric method detected by the naked eye. For example, it is used for fish freshness detection, where it was observed to relate well with bacterial growth trends in whiting and codfish samples, which in turn allowing real-time monitoring of fish spoilage (Pacquit et al. 2007; Kuswandi et al. 2012a, b). This single indicator allows the high potential for creating "best-before" dates where it could make improvements in the food quality evaluation of food freshness active label.

Polyaniline (PANI) could make a distinctive color change with total volatile base nitrogen (TVBN), so it is often used as a single indicator in an intelligent packaging system. A single indicator was developed based on a similar working principle based on PANI film for milkfish sample (*Chanos chanos*) freshness monitoring in the fish package headspace (Kuswandi et al. 2012a, b). This single indicator showed color changes toward a variety of TVBN produced during fish deterioration as shown in Fig. 2.5. Here, the PANI film response presented as a color change is also related to microbial activities in fish samples (*Pseudomonas spp.* and TVC (total viable count)). The single indicator or sensor allows for the real-time spoilage detection of fish either at stable or fluctuating temperatures. Furthermore, a single biosensor was developed for xanthine (adenine nucleotide degradation product in animal tissue) detection (Arvanitoyannis and Stratakos 2012). In this work, the xanthine oxide was attached to the electrodes, e.g., silver, platinum, and pencil graphite electrode (Devi et al. 2013; Dolmacı et al. 2012; Realini and Marcos 2014). Another novel single indicator was developed based on curcumin for volatile amine (TVBN) detection in shrimp (Kuswandi et al. 2012a, b). The curcumin was absorbed onto the bacterial cellulose membrane to create a sensitive indicator and edible membrane for food applications. The indicator develops color from yellow to orange and finally to

**Fig. 2.5** Example of a single indicator used for fish freshness monitoring based on PANI

reddish orange for the indication of shrimp spoilage. Moreover, the indicator color change was correlated with bacterial activities in shrimp in ambient and chillier conditions.

A single indicator was constructed to monitor the guava ripeness. The single indicator was immobilized bromocresol blue on the cellulose membrane for acidic volatile compound detection. Here, when the indicator color changes from blue to green, the maturity of guava changes from ripeness to rotten (Kuswandi et al. 2013). It is often that ripeness and rotten fruit foods can be distinguished by only two color changes, since secondary ripeness fruits cannot be identified, such as medium condition. Therefore, the same group constructed a single indicator based on chlorophenol red as a ripeness indicator for the detection of grape maturity. Here, acid volatile organic compounds released during grape maturation create the indicator color development from white to beige and lastly to yellow that obviously show the ripeness, medium, and rotten grade of grape (Kuswandi and Nurfawaidi 2017).

### 2.3.2 Dual Indicator

Sometimes, a single indicator is difficult to distinguish the color transition in the process of color change, and therefore, mixture of two indicators could be employed to increase the sensitivity of color development end point. Dual indicator can be divided into three classes: First is to add an inert dye to the acid–base indicator, where the color does not change at the pH less than 7.0, and second is to mix the two acid–base indicators to accurately create the indicator color change in order to have a narrower range of color change by using the complementary effects of each indicator, and the last is using dual indicators simultaneously that used two indicators that referencing each other in detecting, monitoring, and informing the food freshness level, since the amount of TVBN and $CO_2$ in meat products is the key maker for food deterioration. Therefore, the mixed indicator color change was more obvious in the detection of TVBN at various levels of food freshness. A freshness indicator was developed based on bromothymol blue phenol red for skate fish (*Raja kenojei*) (Lee et al. 2016). Here, TVBN produced during the fish storage was increased in the fish package, causing the headspace pH to increase. As a result, the indicator changed from yellow to purple color when the fish product deteriorated.

The mixed dual indicator could also increase their sensitivity toward $CO_2$. The mixed dual indicator label of bromothymol blue and methyl red could accurately evaluate the skinless chicken freshness. The label works based on $CO_2$ detection as the main gas released by the microbial metabolism of skinless chicken. Here, when the mixed indicator shows green and orange color, it means freshness and spoilage, separately (Rukchon et al. 2014). A similar principle was developed using a mixed dye-based indicator in food spoilage for an effective shelf life detection by allowing dynamic freshness to be detected visually alongside the best-before date, which, in turn, reduces margins of error (Nopwinyuwong et al. 2010). Further applications of mixed dual indicator for food freshness indicator to other perishable food products are open up for future area of developments and commercialization.

The last type of dual indicator is dual indicators that used two indicators simultaneously as a label for food freshness monitoring. It was developed as a novel approach for food freshness monitoring, i.e., beef meat, and proposed to prevent the problem using a single indicator (Kuswandi and Nurfawaidi 2017). This is due to the fact that a single indicator is similar to traditional acid–base titration using a pH indicator dye, where it is often difficult to detect the onset of spoilage threshold, as it could be too late or too early if it is related to microbial activities on food products (Kuswandi et al. 2015). The dual indicator has several benefits as smart label, such as suitable for naked detection for the spoilage onset threshold, easily to be displayed and distinguished for each level of the food freshness as two color displays, more accurate food freshness determination, preventing false negative and positive for the level of freshness, more attractive due to two color tone as well as they referenced each other, as one indicator develops color a long side with other indicator color change. It was developed by using two pH dyes (methyl red and Bromocresol purple) to fabricate an on-package dual indicator label for the real-time detection of beef freshness (Kuswandi and Nurfawaidi 2017) (Fig. 2.6).

### 2.3.3 Indicator Array

Apart from single or dual indicator or colorimetric sensor, indicator or sensor array based on imaging approaches offers many benefits in nondestructive determination of food quality and safety. This imaging technique is using an indicator array or sensor array to capture a pattern of an "odor" fingerprint related to the food quality and safety (Chen et al. 2016; Morsy et al. 2016). Most of the developed indicator or sensor arrays are employed for monitoring chemical species in the food packaging headspace that are related to food freshness or spoilage of perishable foods, e.g., meat, seafood, fruits, and vegetables, which have high value. Since the cost of the sensor array is the most expensive and not simple compared to other smart indicators

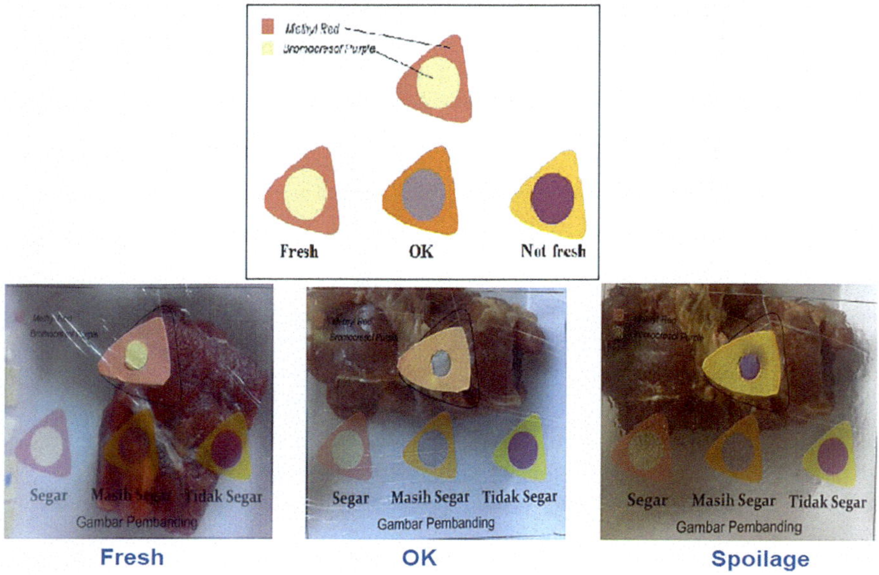


**Fig. 2.6** Example of dual indicator for beef freshness packaging used two pH indicators separately

or sensors, furthermore, sensor array needs additional tools, such as smartphone or mobile device coupled with the dedicated app that have signal processing and artificial intelligence inside that used in capturing odor pattern of food related to food freshness or spoilage.

An indicator or sensor array, i.e., the electronic nose, is a system that works based on the principle of imitating or mimicking the mammalian olfactory system in a device proposed to obtain measurements that allow the classification of aroma mixtures contained in the food odor. The array creates a pattern as a specific signal to each odor, savor, or flavor. For example, e-nose contains a chemical sensor or biosensor array that specifically enables the pattern recognition of simple or even complex odor, savor, or flavor (Gardner and Bartlett 1994; Vanderroost et al. 2014b). This system was used in the quality determination of packaged beef and fresh yellow fin tuna (Blixt and Borch 1999; Dobrucka and Cierpiszewski 2014). Furthermore, it is also employed for freshness detection of broiler chicken cuts (Rajamäki et al. 2006), for the freshness of boiled marinated turkey (Salinas et al. 2014a, b), for fresh pork sausage spoilage (Salinas et al. 2014a, b), and for squid spoilage (Zaragozá et al. 2015).

Currently, the indicator array-based pH indicators and a colorimetric dye, selective for thiols, were developed for the spoilage monitoring of various meats, i.e., chicken, beef, pork, and fish, which produced the different models of the mimicking degradation pathway. The spoilage monitoring for each type of meat then followed the array color evolution using multivariate analysis by three-way PCA (principal component analysis). Using this method, it found similar protein degradation pattern

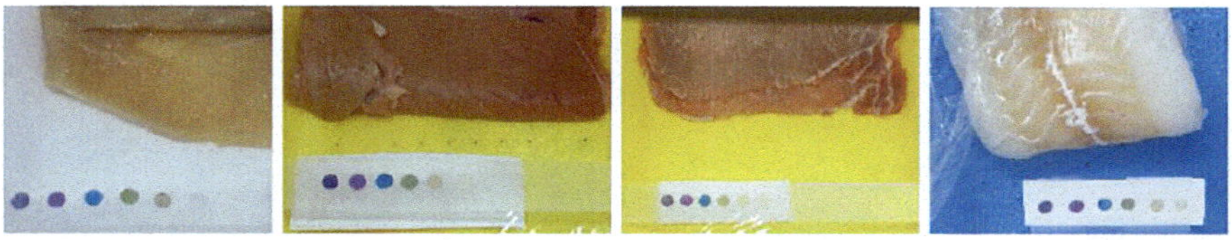

(a) (b) (c) (d)

**Fig. 2.7** Example of indicator array used in food packaging for freshness monitoring of numerous meats, (**a**) chicken, (**b**) beef, (**c**) pork, and (**d**) codfish (Magnaghi et al. 2020)

of meats, which can be used to monitor meat spoilage of different meats as shown in Fig. 2.7 (Magnaghi et al. 2020). Generally, the results were compared to the results of classical methods, such as sensory evaluation, microbiological assay, and head-space gas analysis. The indicator or sensor array allows for better distinctive of the food deterioration from fresh food and can be applied for different foods.

## 2.4 ICT for Smart Packaging

In the industrial era 4.0, ICT (Information and Communication Technology) plays a major role in our daily activities. Therefore, the use of ICT in terms of smartphones and other mobile devices has become an important part of our lifestyle. In this regard, the use of these gadgets to access and connect with many things has become obvious, including food packaging. Thus, the concept of "connected packaging" clearly will become another part of smart packaging. Of course, the term connected packaging is not only limited to food packaging but also limited to other applications in almost every kind of consumer product. It could be connected for medicines, cosmetics, category, apparel, etc. In this connected packaging, the applications could be connected with a special printed code on or within the packaged consumer products so that the consumers can activate this code with smartphones or other mobile devices to receive exclusive information regarding the content of products. In food packaging, the content could be online calorie and nutritional value, originality, halalness, etc. Hence, by applying this novel packaging concept, the smart packaging will become smarter as following applications (Fig. 2.8).

### 2.4.1 Artificial Intelligence

Recently, there is increasing use of artificial intelligence (AI) and machine learning (ML) for smart sensor system development for the problem-solving-based approach. For instance, online or real-time AI and ML or new algorithms are developed, and various methods are employed to integrate these algorithms in a smartphone or other mobile devices for sensor systems to allow for autocalibration measurement, enhancing linear range, etc. New clustering and classification techniques, learning methods,

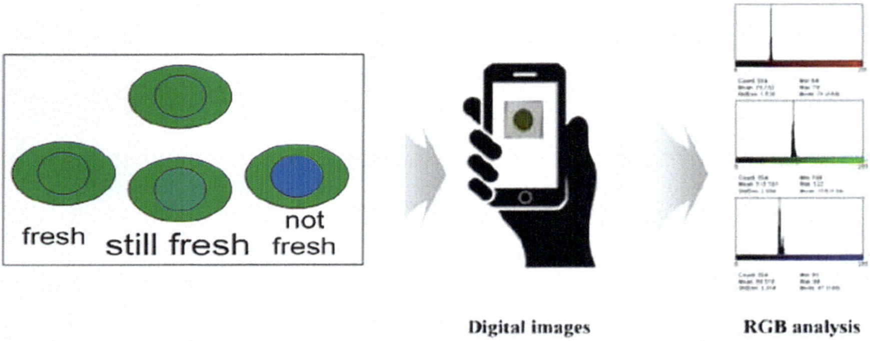


**Fig. 2.8** Example of smartphone-based analysis in intelligent food packaging for food freshness monitoring

and data quality methods are required, including distributed AI algorithms to make indicators or sensors to be more accurate in the target analyte detection. For instance, the integration of an ML algorithm to a sensor based on luminescence detection has been used for time-resolved luminescence data for the detection of $O_2$ in the sample (Dordević et al. 2018). This approach also has great potential to be integrated into the smart packaging system, including IoT, image processing, and artificial intelligence (Schaefer and Cheung 2018). For instance, this approach was used for the protein degradation pattern of meats and was applied to monitor meat spoilage of different meats (Magnaghi et al. 2020).

Another approach that can be applied in smart packaging used a neural network (NN). For example, in food freshness monitoring using an indicator or sensor array, a NN learns to correlate the quantities image sensed by a sensor array with the analyte target concentration. In this case, as the number of measurements is limited, typically, the NN training was done with a simulated training dataset. Besides, the temperature that needs to be counted in this array was kept constant but also could be predicted. The best performing network was optimized by employing different architectures and hyper-parameter tuning. Then, the trained network was used in the data experiment to check its applicability when used to other data. The error in the freshness-level prediction was caused by the synthetic training data that were calculated using an approximate model. By employing this approach for the training dataset, the experimental data would reduce the absolute error in the freshness-level measurements. This approach was applied for the measurement of oxygen concentration, and the results were comparable with many commercial oxygen sensors based on the traditional methods (Chu et al. 2016; Michelucci et al. 2019). Moreover, the AI and ML integration support other fields, e.g., IoT, big data. Therefore, in smart packaging development, these technologies would enhance the reliability of the sensing system. In this case, the data were used to help consumer, particularly when it allows the systems to monitor and to learn indicator or sensor responses that can be related to food freshness status. Therefore, the future challenges in smart packaging development are the novel approaches to use AI algorithms via smartphone or other mobile devices in the food freshness monitoring system. This work will cover new hybrid AI systems, novel ML models, etc., that could be applied to enhance indicator or sensor array performance in monitoring of food freshness status. Moreover, AI soon can be expected to play a more important role in improving machine learning, data mining, and decision support systems that are allowed to adjust processes and decision in real time and on the basis of big data (Vanderroost et al. 2014).

### 2.4.2 IoT

The IoT is a concept focused on serving a global infrastructure of interconnected networks for connecting objects to the cyber world. It allows for the tracking and control of devices integrated with indicators, sensors, and actuators (Schaefer and Cheung 2018). For instance, RFID tags equipped to the packaging objects will be easily tracked along their chain from producer to customer (Cui 2016). Therefore, by the integration of smart packaging in the IoT, the product lost or damaged during shipments and distribution could potentially be prevented and reduced significantly. In few years, not only smartphones and other gadgets will be part of the IoT, other appliances, e.g., food packaging, furniture, and cars, or even machinery and factories will also become part of the industrial IoT (IIoT) (Da Xu et al. 2014).

One of the most important areas in IoT is a further advancement in intelligent food packaging technologies. It is improved in monitoring and controlling the conditions of food packaging and managing online or real time. This advancement will affect food quality and safety and food waste reduction significantly, which would increase consumer health. A suitable and reliable IIoT infrastructure and the associated ICT equipped into both the food packaging and their supply chain, generally speaking, would create cyber-physical production and delivery networks within a company and across several companies. Moreover, the further advantages from these digitalization approaches of the food product life cycle and supply chain are easy to be managed, organized, and distributed.

## 2.5 Conclusions

The recent advances in smart packaging, particularly intelligent packaging, depends on the smart materials development and sensor technology, wherein some way allow for direct detection or indirect detection by mimicking the packaged food freshness condition to help a customer in judging the quality of food freshness and safety, as well as its shelf life and computability. For freshness indicator or sensor, it should be compatible with printing technology for mass production, low cost over the food value, simple, user-friendly, reliable, accurate, and reproducible in their range of operation, food contact safe, and environmentally friendly.

The integration of an indicator or sensor in food packaging has created innovative ways in smart packaging developments. These innovations have allowed increasing food quality, safety, shelf life, and usability, reducing food loss and waste. Today, most packaging advances have been paved by the consumer preferences and trends worldwide, such as ICT, including imaging technology, A.I., big data, and IoT, with their numerous mobile devices. In addition, some advances have derived from the emerging technology, such as nanotechnology, the sensing technology in nanosize. Undoubtedly, novel smarter indicator or sensor in smart packaging will be a rise in the year to come as the marriage between emerging these technologies, so that the smart packaging could inform the consumers smarter, not only by the color development but also by their smart mobile devices in terms of food preferences, quality, and safety. Furthermore, smart packaging could become even smarter in the near future by numerous functions, such as tracking, tracing, authentication, preferences, online calorie, and nutritional value, halalness, and food sustainability. These functions can be covered via the connected packaging, where it allows consumers to communicate with the food product via a mobile device using an active package label or code that contains many pieces of information regarding the food products.

**Acknowledgments** The author gratefully thanks the LP2M, University of Jember, for supporting this work.

## References

Arvanitoyannis IS, Stratakos AC (2012) Application of modified atmosphere packaging and active/smart technologies to red meat and poultry: a review. Food Bioprocess Technol 5:1423–1446. https://doi.org/10.1007/s11947-012-0803-z

Blixt Y, Borch E (1999) Using an electronic nose for determining the spoilage of vacuum-packaged beef. Int J Food Microbiol 46(2):123–134. https://doi.org/10.1016/S0168-1605(98)00192-5

Chen Q et al (2016) Nondestructively sensing of total viable count (TVC) in chicken using an artificial olfaction system based colorimetric sensor array. J Food Eng 168:259–266. https://doi.org/10.1016/j.jfoodeng.2015.08.003

Chu CS, Lin KZ, Tang YH (2016) A new optical sensor for sensing oxygen based on phase shift detection. Sensors Actuat B Chem 223:606–612. https://doi.org/10.1016/j.snb.2015.09.155

Communities (2004) Regulation (EC) No. 1935/2004 of the European Parliament and of the Council of 27 October 2004 on materials and articles intended to come into contact with food and repealing Directives 80/590/EEC and 89. 1935/2004

Cui X (2016) The internet of things. In: Ethical ripples of creativity and innovation. Palgrave Macmillan UK, London, pp 61–68. https://doi.org/10.1057/9781137505545_7

Da Xu L, He W, Li S (2014) Internet of things in industries: a survey. IEEE Trans Indus Inform 10 (4):2233–2243. https://doi.org/10.1109/TII.2014.2300753

Devi R et al (2013) Electrochemical biosensor based on gold coated iron nanoparticles/chitosan composite bound xanthine oxidase for detection of xanthine in fish meat. J Food Eng 115 (2):207–214. https://doi.org/10.1016/j.jfoodeng.2012.10.014

Dobrucka R, Cierpiszewski R (2014) Active and intelligent packaging food research and development—a review. Pol J Food Nutr Sci 64(1):7–15

Dolmacı N et al (2012) An amperometric biosensor for fish freshness detection from xanthine oxidase immobilized in polypyrrole-polyvinylsulphonate film. Artif Cells Blood Substit Immobil Biotechnol 40(4):275–279. https://doi.org/10.3109/10731199.2011.646410

Dordević N et al (2018) Machine learning for analysis of time-resolved luminescence data. ACS Photonics 5(12):4888–4895. https://doi.org/10.1021/acsphotonics.8b01047

Ebrahimi Tirtashi F et al (2019) Cellulose/chitosan pH-responsive indicator incorporated with carrot anthocyanins for intelligent food packaging. Int J Biol Macromol 136:920–926. https://doi.org/10.1016/j.ijbiomac.2019.06.148

Eom KH, et al (2012) The vegetable freshness monitoring system using RFID with oxygen and carbon dioxide sensor. Int J Distributed Sensor Networks, 2012. https://doi.org/10.1155/2012/472986

Ezati P, Rhim JW (2020) pH-responsive chitosan-based film incorporated with alizarin for intelligent packaging applications. Food Hydrocolloids 102:105629. https://doi.org/10.1016/j.foodhyd.2019.105629

Ezati P et al (2021) CMC and CNF-based intelligent pH-responsive color indicator films integrated with shikonin to monitor fish freshness. Food Control 126:108046. https://doi.org/10.1016/j.foodcont.2021.108046

Gardner JW, Bartlett PN (1994) A brief history of electronic noses. Sensors Actuat B Chem 18 (1–3):210–211. https://doi.org/10.1016/0925-4005(94)87085-3

Ghaani M et al (2016) An overview of the intelligent packaging technologies in the food sector. Trends Food Sci Technol 51:1–11. https://doi.org/10.1016/j.tifs.2016.02.008

Hu XG et al (2016) Nondestructive monitoring of Kiwi ripening process using colorimetric ethylene sensor. Bull Korean Chem Soc 37(5):759–762. https://doi.org/10.1002/bkcs.10745

Kalpana S et al (2019) Intelligent packaging: trends and applications in food systems. Trends Food Sci Technol 93(September):145–157. https://doi.org/10.1016/j.tifs.2019.09.008

Kelly CA et al (2020) Application of O2 sensor technology to monitor performance of industrial beef samples packaged on three different vacuum packaging machines. Sensors Actuat B Chem 304:127338. https://doi.org/10.1016/j.snb.2019.127338

Kim YH et al (2018) Non-destructive monitoring of apple ripeness using an aldehyde sensitive colorimetric sensor. Food Chem 267:149–156. https://doi.org/10.1016/j.foodchem.2018.02.110

Kuswandi B (2013) Simple and low-cost freshness indicator for strawberries packaging. Acta Manilana 61:147–159

Kuswandi B (2017) Environmental friendly food nano-packaging. Environ Chem Lett 15 (2):205–221. https://doi.org/10.1007/s10311-017-0613-7

Kuswandi B (2018) Freshness sensors for food packaging. In: Reference module in food science. Elsevier. https://doi.org/10.1016/b978-0-08-100596-5.21876-3

Kuswandi B (2020) Real-time quality assessment of fruits and vegetables: sensing approaches. In: Sensor-based quality assessment systems for fruits and vegetables. Apple Academic Press, pp 1–30. https://doi.org/10.1201/9781003084174-1

Kuswandi B, Moradi M (2019) Sensor trends in beverages packaging. Trends Beverage Packag 16: 279–302. https://doi.org/10.1016/b978-0-12-816683-3.00010-4

Kuswandi B, Murdyaningsih EA (2017) Simple on package indicator label for monitoring of grape ripening process using colorimetric pH sensor. J Food Measurement Characterization. https://doi.org/10.1007/s11694-017-9603-5

Kuswandi B, Nurfawaidi A (2017) On-package dual sensors label based on pH indicators for real-time monitoring of beef freshness. Food Control 82:91–100. https://doi.org/10.1016/j.foodcont.2017.06.028

Kuswandi B et al (2011) Smart packaging: sensors for monitoring of food quality and safety. Sens Instrumen Food Qual 5(3–4):137–146. https://doi.org/10.1007/s11694-011-9120-x

Kuswandi B, Jayus, Restyana A et al (2012a) A novel colorimetric food package label for fish spoilage based on polyaniline film. Food Control 25(1):18–189. https://doi.org/10.1016/j.foodcont.2011.10.008

Kuswandi B, Jayus, Larasati TS et al (2012b) Real-time monitoring of shrimp spoilage using on-package sticker sensor based on natural dye of curcumin. Food Anal Methods 5(4):881–889. https://doi.org/10.1007/s12161-011-9326-x

Kuswandi B et al (2013) Real time on-package freshness indicator for guavas packaging. J Food Measurement Characterization 7(1):29–39. https://doi.org/10.1007/s11694-013-9136-5
Kuswandi B et al (2015) Simple and low-cost on-package sticker sensor based on litmus paper for real-time monitoring of beef freshness. J Math Fundam Sci 47(3):236–251. https://doi.org/10.5614/j.math.fund.sci.2015.47.3.2
Kuswandi B et al (2020) Edible pH sensor based on immobilized red cabbage anthocyanins into bacterial cellulose membrane for intelligent food packaging. Packaging Technol Sci 33 (8):321–332. https://doi.org/10.1002/pts.2507
Lang C, Hübert T (2012) A colour ripeness indicator for apples. Food Bioprocess Technol 5 (8):3244–3249. https://doi.org/10.1007/s11947-011-0694-4
Lee GY, Lee S, Shin HS (2016) Evaluation of gas freshness indicator for determination of skate (Raja kenojei) quality during storage. Food Sci Biotechnol 25(5):1497–1500. https://doi.org/10.1007/s10068-016-0232-2
Magnaghi LR et al (2020) Colorimetric sensor array for monitoring, modelling and comparing spoilage processes of different meat and fish foods. Foods 9(5):684. https://doi.org/10.3390/foods9050684
Majid I et al (2018) Novel food packaging technologies: innovations and future prospective. J Saudi Soc Agric Sci 17:454–462. https://doi.org/10.1016/j.jssas.2016.11.003
Michelucci U, Baumgartner M, Venturini F (2019) Optical oxygen sensing with artificial intelligence. Sensors 19(4):777. https://doi.org/10.3390/s19040777
Morsy MK et al (2016) Development and validation of a colorimetric sensor array for fish spoilage monitoring. Food Control 60:346–352. https://doi.org/10.1016/j.foodcont.2015.07.038
Müller P, Schmid M (2019) Intelligent packaging in the food sector: a brief overview. Foods 8(1):–16. https://doi.org/10.3390/foods8010016
Nopwinyuwong A, Trevanich S, Suppakul P (2010) Development of a novel colorimetric indicator label for monitoring freshness of intermediate-moisture dessert spoilage. Talanta 81 (3):1126–1132. https://doi.org/10.1016/j.talanta.2010.02.008
Pacquit A et al (2007) Development of a smart packaging for the monitoring of fish spoilage. Food Chem 102(2):466–470. https://doi.org/10.1016/j.foodchem.2006.05.052
Pavelková A (2013) Time temperature indicators as devices intelligent packaging. LXI (1), pp 245–252
Popa A et al (2019) An intelligent IoT-based food quality monitoring approach using low-cost sensors. Symmetry 11(3):374. https://doi.org/10.3390/sym11030374
Rai M et al (2019) Smart nanopackaging for the enhancement of food shelf life. Environ Chem Lett 17(1):277–290. https://doi.org/10.1007/s10311-018-0794-8
Rajamäki T et al (2006) Application of an electronic nose for quality assessment of modified atmosphere packaged poultry meat. Food Control 17(1):5–13. https://doi.org/10.1016/j.foodcont.2004.08.002
Realini CE, Marcos B (2014) Active and intelligent packaging systems for a modern society. Meat Sci 98(3):404–419. https://doi.org/10.1016/j.meatsci.2014.06.031
Rukchon C et al (2014) Development of a food spoilage indicator for monitoring freshness of skinless chicken breast. Talanta 130:547–554. https://doi.org/10.1016/j.talanta.2014.07.048
Saggin B, Belaizi Y, Vena A, Sorli B, Guillard V, Dedieu I (2019) A flexible biopolymer based UHF RFID-sensor for food quality monitoring. In: Paper presented at the 2019 IEEE International Conference on RFID Technology and Applications (RFID-TA)
Salinas Y, Ros-Lis JV, Vivancos JL, Martínez-Mánez R et al (2014a) A chromogenic sensor array for boiled marinated Turkey freshness monitoring. Sensors Actuat B Chem 190:326–333
Salinas Y, Ros-Lis JV, Vivancos JL, Martínez-Máñez R et al (2014b) A novel colorimetric sensor array for monitoring fresh pork sausages spoilage. Food Control 35(1):166–176
Schaefer D, Cheung WM (2018) Smart packaging: opportunities and challenges. Procedia CIRP 72 (June):1022–1027. https://doi.org/10.1016/j.procir.2018.03.240

UN Environment Programme (2018) Single-use plastics: a roadmap for sustainability

Vanderroost M et al (2014) Intelligent food packaging: the next generation. Trends Food Sci Technol 39(1):47–62. https://doi.org/10.1016/j.tifs.2014.06.009

Yam KL, Takhistov PT, Miltz J (2005) Intelligent packaging: concepts and applications. J Food Sci 70(1):R1–R10. https://doi.org/10.1111/j.1365-2621.2005.tb09052.x

Zaragozá P et al (2015) Development of a colorimetric sensor array for squid spoilage assessment. Food Chem 175:315–321. https://doi.org/10.1016/j.foodchem.2014.11.156

# Novel Packaging Development, Assessment and Authentication Using Smart Technologies, Non-invasive Biometric Sensory Tools and Artificial Intelligence

# 3

Claudia Gonzalez Viejo, Paul A. Gurr, Frank R. Dunshea, and Sigfredo Fuentes

**Abstract**

Packaging creates the first impression from consumers when selecting commercial food or beverages. Different packaging components are important as they contain all areas of interest related to branding, shape, design and nutritional information, which could determine willingness to purchase and success of products in the market. However, traditional packaging acceptability assessments based on focus groups, acceptance and preference tests may be biased and subjective. Therefore, novel assessment methods have been developed based on more objective parameters, including non-invasive biometrics such as eye tracking, emotional responses from consumers and changes in physiological parameters, such as heart rate and body temperature. Emerging technologies have also been studied for packaging assessment, such as virtual/augmented reality and artificial intelligence tools, including computer vision and machine learning modelling. Furthermore, counterfeiting has been a major issue among commercial products, with food and beverages accounting for 10% counterfeited, including packaging and branding. This chapter focuses on the latest research on

C. Gonzalez Viejo (✉) · S. Fuentes
Digital Agriculture, Food and Wine Sciences Group, School of Agriculture and Food, Faculty of Veterinary and Agricultural Sciences, The University of Melbourne, Parkville, VIC, Australia
e-mail: cgonzalez2@unimelb.edu.au

P. A. Gurr
Polymer Science Group, Department of Chemical Engineering, Faculty of Engineering and Information Technology, The University of Melbourne, Parkville, VIC, Australia

F. R. Dunshea
Digital Agriculture, Food and Wine Sciences Group, School of Agriculture and Food, Faculty of Veterinary and Agricultural Sciences, The University of Melbourne, Parkville, VIC, Australia

Faculty of Biological Sciences, The University of Leeds, Leeds, UK


A. K. Shukla (ed.), *Food Packaging: The Smarter Way*,
https://doi.org/10.1007/978-981-16-7196-8_3

intelligent and digital technologies for packaging development, assessing consumer acceptability towards packaging and authentication using new and emerging digital technologies.

## 3.1 Introduction

It is well known that 95% of new food and beverage products are deemed to fail in the market without proper evaluation in terms of consumer acceptability and liking assessments (Buss 2018; Fuentes et al. 2020; Gonzalez Viejo and Fuentes 2020). The latter is especially relevant for food and beverage packaging since it is the first contact with the product and creates the first impression among consumers. Traditionally, the design and information added to packaging have been determined by food companies hiring visual design companies to present several alternatives. Then, the best packaging design is subjectively chosen by an executive group from the food company, without any other objective measures being considered.

When including sensory science in the decision-making process for effective packaging, food companies can use either focus groups conformed by professionals in the food and beverage area or by selecting potential consumers in the respective age group the food and beverage products are targeted. More in-depth studies can be conducted using consumer testing of 3D packaging prototypes or 2D and 3D image rendering on screens. These techniques can give a more objective sense of why consumers prefer different packaging features, depending on the number of questions included in the test.

However, all these techniques are either subjective, require trained professionals or rely on the conscious responses of panellists according to several pre-set questions. Recent research has determined that human sensory perception and emotional response towards images, food and beverages are mostly determined by the response of the autonomic nervous system (ANS) (Fuentes et al. 2020, 2021a; Gonzalez Viejo et al. 2019a, b, 2020, 2021; Biju et al. 2021; Gunaratne et al. 2019a). Hence, the conscious responses to preference or packaging appeal may only render partial information for decision-making.

New and emerging technologies have been applied in recent years in an effort to tap into missing information, such as the responses of the ANS to packaging design from consumers and how different features of the packaging elicit different emotional and physiological responses that may be linked to liking and preference.

The packaging design is also important to avoid anti-counterfeiting, which is a major problem within the industry and may affect consumers' confidence when selecting a food product. It has been shown that, when incorporating anti-counterfeiting elements using different technologies from intelligent or smart packaging, consumers' trust and loyalty are reinforced (Francis 2019).

This chapter focuses mainly on these new and emerging technologies applied to packaging assessment. This includes biometrics for physiological and emotional response assessment, eye trackers, virtual and augmented reality, artificial intelligence (AI), the development of intelligent or smart food packaging as innovative additions and anti-counterfeiting solutions, and consumer acceptance assessments for these technologies.

## 3.2 Biometrics in Packaging Assessment

Typical assessment of consumers' responses towards packaging/labels consists of the use of acceptance tests using hedonic and just about right scales, and ranking or paired preference tests. Other scales such as 100- or 15-cm non-structured scales and check-all-that-apply (CATA) tests have also been used to assess emotional responses from consumers. However, all those methods are self-reported, which means that participants go through a thinking process for decision-making to provide their answers. This leads to subjective responses that may be affected by different types of bias, such as suggestion effect, distraction error, halo effect and cultural factors (Kemp et al. 2011; Meilgaard et al. 2006).

The use of biometrics in sensory analysis has become more popular, especially in the last 5 years, to assess consumers' responses from the ANS and their emotions (Fuentes et al. 2021a; Gonzalez Viejo et al. 2019b, c). The ANS controls the body's involuntary function, such as heart rate, body temperature, pupil dilation, respiration, skin conductance and blood pressure. Once a stimulus is received, the brain is able to integrate the information and elicit an involuntary response to the environment, which, depending on the type of stimulus, may trigger either the sympathetic (stress) or parasympathetic (resting) nervous system (Donato 2018; Silverthorn 1998). On the other hand, emotional responses depend on the release of different hormones, which are released depending on the stimuli that have been received. Some of these hormones include adrenaline and cortisol, which trigger negative emotions, and dopamine, oxytocin, and serotonin that elicit positive emotions; these are the subconscious responses. After receiving motivation, this affects the behaviour, leading to the thinking process and ending in decision-making (Smith 2015). In previous studies, these responses have been reported to be different according to consumers' cultural background, such as Asians vs Westerners. This is because the reactions to stimuli also depend on participants' previous experiences and familiarity with the products (Fuentes et al. 2020; Torrico et al. 2018a; Gonzalez Viejo et al. 2018a; Gunaratne et al. 2019b).

Most of the published studies use contact sensors placed either on the chest, extremities or head to measure the physiological responses; however, this creates awareness of the participants about being monitored and affects their ANS responses as possible creation of a stressful environment (Gonzalez Viejo et al. 2018b; Frelih et al. 2017). Therefore, some researchers have developed non-invasive techniques using video analysis to reduce bias (Gonzalez Viejo et al. 2018b, 2019b; Gonzalez Viejo Duran 2020; Mohamed et al. 2020; Jain et al. 2016). On the other hand, for the assessment of subconscious emotional responses, some commercial software has been developed such as FaceReader™ (Noldus Information Technology, Wageningen, Netherlands), Affectiva (Affectiva, Boston, MA, USA), iMotions

(iMotions, Copenhagen, Denmark) and MorphCast® (MorphCast, Florence, Italy). Furthermore, commercial eye trackers such as Tobii (Tobii, Danderyd Municipality, Sweden) and Gazepoint (Gazepoint, Vancouver, BC, Canada) have been used to assess gaze parameters such as duration and number of fixations when evaluating packaging/labels.

### 3.2.1 Physiological Assessment

The most common physiological responses measured while assessing packaging/labels are heart rate and skin conductance (Table 3.1). Among the latest studies using these technologies is the research from Liao et al. (2015), who assessed different elements in packaging such as the image, colours and typefaces. In this test, they assessed the self-reported responses and skin conductance of 120 participants while evaluating 12 samples. The authors used two sensors (Procomp2, ThoughtTech, Montreal, Quebec, Canada) attached to the middle and index fingers of the left hand of participants to measure skin conductance. The main findings consisted of significant differences ($p < 0.05$) in skin conductance between samples with a negative image and without any image. Cuesta et al. (2018) assessed consumers' physiological response based on skin conductance towards four types of packaging with different colours and areas of interest (AOI). The authors did not report the method or sensors used to assess skin conductance, but it was claimed that a similar skin conductance response was obtained for all samples. However, no statistical analysis was reported. Rodriguez-Escudero et al. (2019) evaluated four different label layouts (changes in size and location) of smoothies' packaging and recorded skin conductance of 42 participants using electrodes (Bitbrain Technologies, Zaragoza, Spain) attached to the fingers of the left hand. The authors converted the skin conductance results to two parameters: activation and impact, and found significant differences between samples for both variables at $p < 0.05$ and $p < 0.01$. Modica et al. (2018) conducted a study using chocolate and rice packaging to assess consumers' heart rate and skin conductance responses towards different brands (major vs private label), hedonic value (comfort vs daily food) and familiarity (foreign vs local products). The authors used a pulse oximeter placed on the thumb of participants to measure heart rate and Shimmer electrodes (Shimmer Sensing, Dublin, Ireland) placed on the non-dominant hand for skin conductance. However, no results were reported for these parameters.

Schulte-Holierhoek et al. (2017) evaluated milk packaging with different colours to assess heart rate and skin conductance responses from consumers of different nationalities. The authors used the Vrije University Ambulatory Monitoring System (VU-AMS, ver. 3.9) (Willemsen et al. 1996). Heart rate was measured using the electrocardiogram method with seven electrodes (Kendall™ ECG Electrodes; H98SG; Cardinal Health, Dublin, OH, USA) attached to the chest, while for skin conductance, electrodes attached to the index and middle fingers of the non-dominant hand were used. For heart rate, significant differences ($p < 0.05$) were found between responses from participants according to their nationality. Skin

**Table 3.1** Summary of recent studies using biometrics in the assessment of consumer responses towards food packaging

| Food packaging | Assessment elements | Biometrics | Method | Number of consumers | Number of samples | Time of image display | Main findings | References |
|---|---|---|---|---|---|---|---|---|
| *Physiological responses* | | | | | | | | |
| Chocolates | Images<br>Typefaces<br>Colours | SC | Attached sensors<br>Fingers | 120 | 12 | 6 s | Significant differences ($p < 0.05$) between packaging with negative image and no image | Liao et al. (2015) |
| Juice | Colours<br>Areas of interest | SC | NR | 35 | 4 | 10 s | Similar results for all samples | Cuesta et al. (2018) |
| Smoothies | Label layout | SC | Attached sensors<br>Fingers | 42 | 4 | 2 s | Significant differences ($p < 0.05$; $p < 0.10$) between samples with different layout | Rodríguez-Escudero et al. (2019) |
| Chocolate<br>Rice | Brand (major vs private)<br>Familiarity (local vs foreign)<br>Hedonic value (comfort vs daily food) | SC<br>HR | Attached sensors:<br>– Palm<br>– Thumb | 32 | 8 | 15 s visual<br>15 s interaction | NR | Modica et al. (2018) |
| Milk | Colours | SC<br>HR | Attached sensors<br>– Fingers<br>– Chest | 98 | 4 | 10 s | At ($p < 0.05$):<br>SC higher in red vs blue packaging<br>SC lower in low vs high popularity<br>HR significant differences based on nationality of consumers | Schulte-Holierhoek et al. (2017) |

(continued)

**Table 3.1** (continued)

| Food packaging | Assessment elements | Biometrics | Method | Number of consumers | Number of samples | Time of image display | Main findings | References |
|---|---|---|---|---|---|---|---|---|
| Healthy product | Virtual vs physical packaging | SC<br>HR | Attached sensors<br>– Palm<br>– Finger | 83 | 6 | 11 s | SC and HR correlations ($p < 0.01$; $r > 0.90$) between virtual and physical assessment | Vila-López et al. (2021) |
| Healthy product | Colours and design/message | SC | Attached sensors<br>– Palm | 83 | 6 | 11 s | SC ratio significantly lower in blue packaging | Vila-López and Küster-Boluda (2019) |
| *Emotional responses* | | | | | | | | |
| Juice | Colours<br>Areas of interest | Emotions | NR | 35 | 4 | 10 s | Packaging with yellow colour higher in positive emotions | Cuesta et al. (2018) |
| Chocolates | Images<br>Typefaces<br>Colours | Emotions (valence) | Attached sensors face | 120 | 12 | 6 s | Significant differences ($p < 0.05$) between packaging with positive image and no image | Liao et al. (2015) |
| Yoghurt | Recyclable logos (recyclable, non-recyclable, no logo) | Emotions | FaceReader™ | 89 | 6 | 6 s | More positive emotions towards samples with recyclable logo | Songa et al. (2019) |
| Snacks | Spatial representations (foreground vs background) | Emotions | FaceReader™ | 40 | Study 1: 4<br>Study 2: 4 | 60 s | Foreground representation elicited higher happiness, while background had higher sadness | Vergura and Luceri (2018) |

| | | | | | | | | |
|---|---|---|---|---|---|---|---|---|
| Olive oil | Health claims | Emotions (arousal) | FaceReader™ | 80 | 4 | 60 s | Interaction between perceived happiness and arousal | Pichierri et al. (2021) |
| Milk | Materials/shapes/colours | Emotions | FaceReader™ | 50 | 6 | 30 s | Significant differences ($p < 0.05$) between emotions for each sample | Clark et al. (2021) |
| Chocolate | AOI<br>Familiarity (novel vs familiar packaging) | Emotions | FaceReader™ | 60 | 12 | 10 s | Significant differences ($p < 0.05$) between samples in sadness for novel packaging<br>More positive emotions elicited by the brand name AOI | Gunaratne et al. (2019a) |
| Coffee | Packaging concepts (bold, every day, fun, premium, classic, natural) | Emotions | Computer app based on Affectiva | 69 | 6 | 15 s | Positive emotions related to everyday concept<br>Valence related to premium concept<br>Negative emotions related to bold concept | Gonzalez Viejo et al. (2021) |
| | | | | *Eye tracking* | | | | |
| Juice | Colours<br>AOI | Fixation path<br>Fixation time | Tobii ET | 35 | 4 | 10 s | Assessment starts in the centre of packaging, then top (logo) and finish in the bottom (text) | Cuesta et al. (2018) |
| Smoothies | Label layout (AOI) | Fixations | Tobii ET | 42 | 4 | 2 s | The "cold press" claim and product name had high fixations regardless of the position | Rodríguez-Escudero et al. (2019) |
| Yoghurt | Recyclable logos | Pupil diameter<br>Eye movement | SMI-RED250 ET | 89 | 6 | 6 s | Larger pupil diameter with samples with recyclable logo | Songa et al. (2019) |

(continued)

**Table 3.1** (continued)

| Food packaging | Assessment elements | Biometrics | Method | Number of consumers | Number of samples | Time of image display | Main findings | References |
|---|---|---|---|---|---|---|---|---|
| Chocolate | AOI<br>Familiarity<br>(Novel vs familiar packaging) | Fixation duration<br>Number of fixations | Gazepoint ET | 60 | 12 | 10 s | Higher number of fixations in the image and brand of familiar samples<br>Higher number of fixations in country of origin, ingredients and manufacturer of novel packaging | Gunaratne et al. (2019a) |
| Baby formula (thermochromic label) | Colour transitions in different location (brand plus Koala image and snowflake image) | Fixation duration<br>Number of fixations | Eye tribe tracker | 40 | 3 | 7 s | Brand plus Koala image had significantly ($p < 0.05$) higher fixation time and number of fixations than other samples | Torrico et al. (2018b) |
| Olive oil dressing | Label<br>Packaging type | Eye fixations | RealEye system<br>Real-time video analysis | 15 | 6 | NR | Higher fixations in product name, image and country of origin<br>In transparent bottles, fixations are also high in the neck and cap of bottle | Fazio et al. (2020) |
| Wine | Bottle design | Fixation time<br>Time to first fixation | Tobii ET | 37 | 8 | 7 s | Classic wine bottles had higher general visibility | Merdian et al. (2020) |

| | | | | | | | | |
|---|---|---|---|---|---|---|---|---|
| Yoghurt | Nutrition claims | Fixation time<br>Fixation count | Tobii ET | 100 | 251 labels<br>6 nutrition claims | 15 s | Higher fixation time in low-sugar claim | Ballco et al. (2019) |
| Apples | Organic labels | Fixation time<br>Fixation count<br>Number of visits | Tobii Pro Glasses 2 ET | 73 | 6 | NR | Visual attention related to preference and product choice | Meyerding and Merz (2018) |
| White wine | AOI | Fixation time<br>Visual attention | SMI-RED250 ET | 44 | 7 | 10 s | Highest visual attention to award star rating<br>Lowest visual attention to sweetness, vintage and country of origin | Mokrý et al. (2016) |
| Organic food | AOI | Fixation time<br>Visual attention | SMI-RED250 ET | Group 1: 88<br>Group 2: 59 | Group 1: 9<br>Group 2: 9 | NR | Higher visual attention to brand, image and organic claims | Drexler et al. (2018) |
| Gluten-free cookies | Gluten-free claims | Dwell time<br>Revisits<br>Fixation count<br>Average fixation duration<br>Visual attention | SMI ET glasses 2 wireless | 67 | 5 | 7 s | Significantly ($p < 0.05$) higher attention to verbal (text) than non-verbal (graphic) organic claims | Sielicka-Różyńska et al. (2021) |

(continued)

**Table 3.1** (continued)

| Food packaging | Assessment elements | Biometrics | Method | Number of consumers | Number of samples | Time of image display | Main findings | References |
|---|---|---|---|---|---|---|---|---|
| Soda | Warning sugar labels | Fixation time<br>Fixation count | SMI Mobile remote ET | 180 | 2 | 10 s | Significantly ($p < 0.01$) lower dwell time in non-warning AOIs in samples with warning sign | Popova et al. (2019) |
| Potato chips | Colour and text | Time to first fixation | SMI-RED250 ET | 24 | 8 | 15 s | Participants initially fixate on the colour | Huang et al. (2021a) |

*NR* not reported, *SC* skin conductance, *HR* heart rate, *ET* eye tracker, *AOI* areas of interest conductance was significantly higher when evaluating red rather than blue packaging, while it was lower when assessing packaging with low popularity. On the other hand, Vila-Lopez et al. (Vila-López and Küster-Boluda 2019; Vila-López et al. 2021) published two papers using the same low-fat food packaging samples with different colours (black, blue and red) and design/message (added a small drawing). In both publications, the authors measured skin conductance of 83 participants using a PowerLab/16SP electrode (AD Instruments, Sydney, NSW, Australia) attached to the palm of the non-dominant hand; however, for the second publication, heart rate was included, and it was measured using a PowerLab/16SP electrode (AD Instruments, Sydney, NSW, Australia) attached to the finger. In the first paper, the authors reported that the skin conductance was significantly ($p < 0.05$) lower when evaluating blue packaging samples. In their second paper, they compared results when evaluating the packaging virtually and physically and found that for heart rate and skin conductance there was a positive and significant ($p < 0.01$) correlation ($r = 0.90$).

As highlighted by previous research, there is an opportunity to conduct research and develop new methods in consumer evaluation towards food packaging using non-invasive biometrics to assess physiological responses to avoid the bias generated by contact sensors. There are novel techniques developed to measure heart rate and body/skin temperature using computer vision to assess food and beverages (images, videos and tasting). The heart rate method consists of recording videos of participants using the BioSensory Application (App; The University of Melbourne, Parkville, VIC, Australia) while evaluating the products. The videos are analysed using computer vision algorithms developed in MATLAB® (MathWorks Inc., Natick, MA, USA) using the photoplethysmography (PPG) method based on the luminosity changes in the forehead and cheeks produced by the blood flow (Gonzalez Viejo et al. 2018b; Fuentes et al. 2018). On the other hand, the contactless body/skin temperature method consists of using a thermal infrared camera integrated with the BioSensory App that can capture thermal images to assess consumers' temperature changes when evaluating the samples. These are then automatically analysed using MATLAB® to obtain the maximum temperature of the eye section (Gonzalez Viejo et al. 2018c, 2019a).

### 3.2.2 Emotional Assessment

Recent studies have used the subconscious assessment of emotions from consumers when evaluating food packaging/labels (Table 3.1). While the traditional technique uses sensors attached to specific muscles in the face to assess facial movements, more recently, the most common method is the non-invasive video analysis using commercial software. The software used to analyse videos for emotions works by automatic face detection and recognition of features such as eyebrows, mouth, eyes and nose. Once these are identified, the software can track the features identifying the micro- and macro-movements, through the duration of the video, using a specific model or algorithm developed by each software, such as an active appearance model in FaceReader™ and histogram of an oriented gradient in Affectiva. Then, these are translated into emotions such as happy, sad, joy, contempt, angry and scared using machine learning models based on artificial neural networks (ANNs) in FaceReader™ and support vector machine (SVM) in Affectiva. This non-invasive automatic process is convenient to assess emotional responses towards stimuli such as food packaging/labels, as consumers are not aware of the video recording throughout the study, making them have more natural/spontaneous reactions when evaluating the stimuli and, therefore, reduces bias.

Authors such as Liao et al. (2015) still use the traditional method of attaching two electrodes to the muscles above the left eyebrow using the Procomp2 sensors; however, this technique only measures the valence of emotions. This study found that chocolate packaging with a positive image elicited significantly higher valence than those with no image; the authors also found that the results from this analysis were different from those self-reported for emotions. As mentioned before, this may be due to consumers' awareness of being monitored with the attached sensors. Therefore, authors such as Songa et al. (2019) conducted a study with yoghurt packaging to assess consumers' emotional responses to recyclable logos (recyclable, non-recyclable, no logo) included in labels. This was conducted using FaceReader™ 5, which is capable of analysing facial expressions and translate them into emotions such as happy, sad, angry and surprised. The authors found that packaging with the recyclable logo elicited more positive emotions. Similarly, Vergura and Luceri (2018) evaluated spatial representations of packaging from different snacks (biscuits, focaccia, crackers and cake) by placing the product representation in the foreground and background to assess consumers' emotional responses. Authors also used the FaceReader™ 5 to analyse the videos from consumers while assessing the samples, and it was found that there were significantly ($p < 0.05$) more positive emotions (happiness) for the foreground and higher sadness for the background representation.

Pichierri et al. (2021) evaluated four samples of olive oil packaging with different health claims and assessed consumers' emotions analysed using FaceReader™. However, the authors only focused on the arousal value and discarded any other parameters provided by the software. They found an interaction between the perceived healthiness and arousal. Clark et al. (2021) also used FaceReader™ to assess emotional responses from consumers, but with a different objective, as they displayed milk packaging samples with different materials/colours and no label. They did not find significant ($p < 0.05$) differences between samples for any emotion, but there were significant differences between emotions for each sample. Gunaratne et al. (2019a) used the BioSensory App to gather self-reported responses and videos from consumers when evaluating chocolate packaging (six familiar and six novel). The authors also focused on assessing consumers' emotional responses towards different AOI (brand, ingredients, manufacturer, nutrition facts, net content and bar code). The videos from consumers were analysed using FaceReader™ and found significant differences ($p < 0.05$) between samples from the novel packaging for sadness. The authors also found that more positive emotions were elicited by the brand AOI of novel packaging. On the other hand, Gonzalez

Viejo et al. (2021) evaluated six different coffee pods packaging designed with different concepts (bold, fun, premium, classic, natural, everyday) based on the TNS NeedScope model™ (NeedScope International, Auckland, New Zealand). The authors used a novel virtual method to conduct the sensory session using Zoom (Zoom Video Communications, Inc., San Jose, CA, USA) and Google Forms (Google, LLC, Mountain View, CA, USA) to display the questionnaire and record consumers' reactions during the test. These videos were analysed using a computer application developed by the Digital Agriculture Food and Wine Group from the University of Melbourne, Australia, based on the Affectiva (Affectiva, Boston, MA, USA) software development kit (SDK). Results showed that positive emotions were related to the everyday packaging concept, valence with the premium sample and more negative emotions were associated with the bold packaging.

### 3.2.3 Eye Tracking

Eye tracking is the most used biometric method to assess consumers' subconscious responses towards food packaging/labels (Table 3.1). This is, generally, a non-invasive or contactless method, except for the alternative of eye tracking glasses, which are often used as a portable method when the test is conducted in an external location. Eye trackers are composed of camera-based sensors with an integrated infrared light that can detect the eyes and pupils. Once these are detected, it can follow and measure the gaze movements and positioning, and pupil size (Gonzalez Viejo et al. 2018c, 2019b; Torrico et al. 2018b). Using specific software connected to the eye tracker, different parameters such as fixation duration, the number of fixations, time to the first fixation and pupil dilation can be determined (Table 3.1). Fixation duration refers to the time spent looking at a specific location or AOI and is related to the importance that consumers give to the specific elements in the sample (Torrico et al. 2018b; Merdian et al. 2020). The number of fixations is used to assess how many times the participants fixate in a specific AOI during the test and is related to the importance that consumers give to the AOI (Torrico et al. 2018b; Meyerding and Merz 2018). Besides, the time to first the fixation is the time that participants take to fixate in a specific AOI for the first time (Merdian et al. 2020). On the other hand, pupil dilation refers to the pupil size and its changes during the evaluation of the samples; when it is more dilated, it has been associated with negative stimuli or emotions (Oliva and Anikin 2018; Shechner et al. 2017). The eye trackers can also obtain qualitative data such as heat maps and fixation maps to assess the areas of higher visual focus and the path or order in which the sample was evaluated.

The most used eye trackers among the recent publications are the Tobii (Cuesta et al. 2018; Rodríguez-Escudero et al. 2019; Merdian et al. 2020; Meyerding and Merz 2018; Ballco et al. 2019), and SMI eye trackers (SensoMotoric Instruments, Teltow, Germany) (Songa et al. 2019; Mokrý et al. 2016; Drexler et al. 2018; Sielicka-Różyńska et al. 2021; Popova et al. 2019; Huang et al. 2021a), and, to a lesser extent, Gazepoint (Gunaratne et al. 2019a), The EyeTribe Tracker (The

EyeTribe©, Copenhagen, S. Denmark) (Torrico et al. 2018b) and the RealEye System (RealEye, Poznań, Poland). The latter consists of an online software program capable of tracking eye movements in real time using a computer-integrated webcam (Fazio et al. 2020). The number of consumers recruited for the tests in the recent publications varied from 15 to 180, while the number of samples ranged from 2 to 251, and the time used to display the stimulus is within 2–15 s (Table 3.1). However, the experiment must be carefully designed to avoid any possible bias in the results; in this context, using 251 samples for a test (Ballco et al. 2019) is too large and may cause consumers' fatigue. Furthermore, displaying the stimulus for 2 s (Rodríguez-Escudero et al. 2019) may not be enough time for the participants to evaluate the sample. The studies mentioned in Table 3.1 were mainly conducted in packaging belonging to (1) beverages (Cuesta et al. 2018; Rodríguez-Escudero et al. 2019; Merdian et al. 2020; Mokrý et al. 2016; Popova et al. 2019), (2) snacks (Gunaratne et al. 2019a; Sielicka-Różyńska et al. 2021; Huang et al. 2021a), (3) yoghurt (Songa et al. 2019; Ballco et al. 2019), (4) organic food (Meyerding and Merz 2018; Drexler et al. 2018) and others such as baby formula (Torrico et al. 2018b) and olive oil dressing (Fazio et al. 2020). Furthermore, these publications focus on the assessment of mainly four elements such as (1) claims (Songa et al. 2019; Meyerding and Merz 2018; Ballco et al. 2019; Sielicka-Różyńska et al. 2021; Popova et al. 2019), (2) AOI or layout (Gunaratne et al. 2019a; Cuesta et al. 2018; Rodríguez-Escudero et al. 2019; Mokrý et al. 2016; Drexler et al. 2018), (3) colours (Cuesta et al. 2018; Torrico et al. 2018b; Huang et al. 2021a) and (4) design (Merdian et al. 2020; Fazio et al. 2020). More recently, Fuentes et al. (2021b) presented a novel and integrated method that allows assessing subconscious emotional and eye tracking responses towards each AOI from packaging and labels using non-invasive biometrics. This method eases the interactive and real-time evaluation and re-design only of the specific AOIs that require any modifications, rather than the entire label. This method has the capability of integrating other ANS responses such as heart rate, blood pressure and skin temperature to obtain more responses from consumers.

## 3.3 Virtual and Augmented Reality

Two main types of reality, virtual and augmented, have been developed to either modify the environment, create scenarios or incorporate virtual elements in real-life environments or objects. Virtual reality (VR) consists of a 3D environment or scenario presented through a headset covering the eyes to isolate the subject from the real situation. On the other hand, augmented reality (AR) combines virtual or digital 2D or 3D images or animations overlayed in the real objects or environment. This is often through special glasses such as the Microsoft HoloLens (Microsoft Corporation, Redmond, WA, USA) or devices such as smartphones (Fuentes et al. 2021a; Djurdjevic et al. 2019; Crofton et al. 2019). The third type of reality, named mixed reality, has been developed, consisting of a combination of a VR environment displayed using an AR device that allows mixing both the real and virtual environments with interaction (Fuentes et al. 2021a). However, it has not been used much in food packaging evaluation.

Using VR technology, stores have been created to experience the store environment without going to the physical place. This environment allows customers to virtually walk through the store aisles, browse the products through the shelves, see their position and use a controller to pick the products, read the label and add them to the cart for purchasing (Lombart et al. 2020; Siegrist et al. 2019). In VR stores, the packaging acquires higher importance because consumers cannot touch, assess freshness, look at the expiry date or smell the product; therefore, the visual aspects become the sole driver of purchase decision-making. In this context, VR is a very convenient tool for product development. It allows researchers and developers to conduct consumer tests to assess the acceptability of the prototypes and easily modify the packaging characteristics according to consumers' feedback (Lombart et al. 2020). Virtual reality may also help assess whether the success or failure of a product may be due to its packaging or its position on the shelves in a virtual store without the need to take consumers to the supermarket for testing.

Some researchers have explored whether the physical and VR store experiences are similar to assess if conducting VR tests is reliable. Siegrist et al. (2019) conducted a test with 68 participants in a physical and VR store with 33 cereal products and instructed consumers to walk through the store and pick the products to read the nutritional label. The authors found no significant differences ($p > 0.05$) between the VR and physical environments in consumers' browsing behaviour and the number of times they looked at the nutritional labels. Similarly, Xu et al. (2021) assessed the behaviour of 98 consumers in physical and VR environments when looking for cereals asking them to rate their perceived healthiness. Participants were also asked to walk through the aisles and interact with the product using a controller. Results from both environments were highly correlated ($r = 0.91$) when rating the cereals according to the perceived healthiness, but also found that participants took longer evaluating the products in VR. Pizzi et al. (2019) evaluated 95 confectionery products displayed on supermarket shelves in physical and VR environments. In this study, 50 consumers participated in the study, and results also showed that the sample assessment and consumers' behaviour in both settings are comparable. On the other hand, Huang et al. (2021b) conducted a session with 80 participants to evaluate eight potato chip samples. The study aimed to assess consumers' behaviour when browsing for potato chip packaging with flavour–colour congruency and incongruency using a VR headset to walk through the store aisle and look at the products on the shelves. It was found that consumers were less efficient locating the incongruent flavour–colour packaging.

Augmented reality has been integrated into food packaging by incorporating an element in the product, such as an image or quick response (QR) code that can be read with any smartphone camera. This is then able to display a virtual object, image, animation or game overlayed with the real product so that consumers can interact with it in real time. This is used at purchasing or at home to provide more information about the product such as nutritional facts, cooking instructions, freshness of the product, country or region of origin, display animated characters, and educational content or games (Djurdjevic et al. 2019; Yang 2019). These technologies applied to packaging have also been used as anti-counterfeiting solutions as consumers can assess their authenticity by scanning the packaging before purchasing them. Some examples of products in the market with AR integrated into their packaging are Kellogg's cereals, which display mini-games and quizzes as educational tools for kids when using a smartphone. Likewise, Nesquik cereals display content for entertainment and learning when pointing the smartphone camera to the packaging. On the other hand, in 2016, McDonald's launched an AR app in the UK to display an advent calendar when pointing the camera at their products; this displayed different games, filters, vouchers and animations every day (Konopelko 2019). In Australia, McDonald's released an AR monopoly game that could be played with each product purchase, and consumers could win different prizes (Tran 2020). To test the usefulness of AR in food packaging, Sonderegger et al. (2019) conducted a test with 84 participants to compare AR and static information in a smartphone when consumers assessed five food products' nutritional and environmental information. The authors found that the participants learned more about the product when using the AR version.

## 3.4 Intelligent Food Packaging

The term smart or intelligent is interchangeable when it comes to packaging technology. This refers to the use of embedded sensor technologies in the packaging (Francis 2019). Food packaging materials play several important roles in transport and delivery to the consumer and are predominantly made of plastics. The advantages of plastic over other materials are its ability to block pathogens, moisture and gases depending on the type of food being delivered. Active packaging involves the ability for the product to actively deter degradation and prolong the lifetime of the foodstuff. However, the consumer is often unaware of these technologies, despite the important role they play in fighting against food spoilage, counterfeiting and supply chain tracing.

Covert intelligent technologies designed to manage the supply chain include anti-counterfeiting technologies such as invisible UV inks, DNA tracking or small compounds known as taggants concealed within the packaging for detection by specialized instruments. The Australian plastic banknote, first released in 1988, is one good example that incorporated many security devices, some overt including holograms, intaglio features, microprinting and fluorescent inks, while others remain covert (Australia TRBo 2021). In its first release, its acceptance by the public was 48%, with 26% disliking the new technology, while today, with the gradual decline in cash, the public has come to accept this diminishing form of currency (Prime and Solomon 2010).

### 3.4.1 Data Loggers

Intelligent technologies designed to inform the supply chain from the producers to the retailers often use relatively expensive data loggers. Here devices are included with a bundle of products to track such things as the location, temperature and humidity over time and can be communicated at all stages of the supply chain through near-infrared (NIR)/radio frequency identification (RFID), ultra-high frequency (UHF) light-emitting diode (LED) or simple universal serial bus (USB) data storage.

The majority of intelligent or smart packaging, on the other hand, are deliberately designed to inform the consumer of the quality of the food contained within, guide them in the best storage for consumption temperature, identify its authenticity or whether the packaging has been improperly tampered with or adulterated. Overt intelligent food packaging is designed to be obvious and requires little interaction to understand the information triggered or contained within the device. In addition, the removal or absence of these technologies on a package can indicate a tampered or substituted product or that it is counterfeited.

### 3.4.2 QR Codes

A common form of communicating product information is the addition of a QR code, which can be linked to online information about the product, promotions, advertising or use-by-date information. It could also be used for product recall in a safety breach simply by changing the online information linked to a separate set of QR codes.

Consumers have become familiar with this form of intelligent technology. Generally, consumers accept the presence of these technologies in their day-to-day life, with QR codes being routinely used for things such as movie tickets and sporting event entry being scanned directly from their phones. Although consumer awareness of QR codes has gradually increased since their initial applications, the use of this technology has had a slower increase. Consumer surveys on the use of QR codes on the packaging have seen a gradual increase in their awareness and acceptance. However, in one survey, it was shown that, depending on the type of content the consumer could access, this might be a reason not to use the technology in the future (Cox and Shiffler 2014). In a 2019 survey of tourists visiting Serbia, the impact of QR codes linked to hospitality venues was studied. The majority of the respondents were satisfied with the technology and the information it garnered; however, the overall satisfaction of the destination was more likely due to the venue itself and other local factors (Vuksanović et al. 2021). In a targeted survey with 240 young Indian adults, the majority were aware and used QR codes, while 46% scanned for fun and 40% scanned for information (Chandramouli Rajaiah 2017). However, the nature of the product would be expected to be a large factor in its use.

With the appearance of COVID-19 worldwide and the advantages gained by social distancing, QR codes have been successfully used for contact tracing (Chen

2021; Faggiano and Carugo 2020; Wu et al. 2021; Li and Guo 2020). This allows the health authorities to track those who have contracted the virus and the people they have come in contact with. In Australia, for example, this has led to people routinely registering with a QR code at every venue they frequent. In the light of social distancing rules, restaurants have used QR codes to allow for contactless ordering of food while remaining seated and distanced from other patrons (Patil and Karekar 2019). According to Rogers' adoption model of new products (Rogers 1976), the uptake is led by innovators, early adopters, followed by the early and late majority and, finally, laggards. With new technologies such as intelligent packaging, gender and age affect the category of adoption for each group. In a study on the perception of active and intelligent packaging of European Consumers, Tiekstra et al. (Tiekstra et al. 2021) chose the generation Y respondents to validate their findings since they represent an important population segment. Advertisers recognizing that they are economically valuable, that abandoned their existing methods to this cohort and are expected to become richer over time represent an important potential market. This concurs with the findings by Erika et al. (2020), who determined using the Kano model a fear of packaging innovation by older responders, although they had the highest need for such technology. Careful selection of consumer groups can be used to either enhance and target specific innovator groups or enhance and improve trust and understanding among the majority and laggards of the population.

### 3.4.3 Chromic Sensors and Indicators

The recent development of chromic (colour changing) sensors and indicators for the food industry has produced sensors placed outside the packaging, which actively monitor and display various food attributes. These sensors can also be used to identify substituted products, in the case of irreversible sensors, or where removed or absent may be an indication of counterfeiting. The area has grown to include several scientific approaches, including colour changing sensors triggered by changes in pH (Alizadeh-Sani et al. 2020; Liu et al. 2019) (halochromism), temperature (Liu et al. 2020a) (thermochromism), humidity (Moustafa et al. 2021) (hydrochromism), mechanical forces (Qiu et al. 2019) (mechanochromism), light (photochromism), bacterial growth (Weston et al. 2021a) (biochromism) or the passage of time (chronochromism) (Zhang et al. 2013). Of these technologies, a range of commercial examples has been used with great success. The most notable of these was the inclusion of a thermochromic sensor onto the outside of the Coors beer can that changed colour when chilled to the correct consumption temperature. The consumer acceptance of this product increased sales of this product by 3% in 1 year attributed to the intelligent technology alone (Alsever 2009). A summary of some commercial examples of intelligent technology sensors for the food industry is listed in Table 3.2. The most prolific of these sensors on the market is the freshness/integrity indicators, thermochromic inks and time–temperature indicators (TTIs). These devices are preferred to be used on impermeable containers such as glass or metal to stop the migration of inks or chemicals to the food within or contain a suitable barrier where paper and plastic are insufficient. For freshness indicators that

**Table 3.2** Commercial examples of Intelligent Packaging Technologies

| | Trigger | Commercial example | Company |
|---|---|---|---|
| Freshness/ integrity indicators | Degradation gases | Fresh Tag | COX Technologies |
| | | ripeSense | ripeSense™ ad ort Research |
| | | Food Fresh | Vanprob |
| | Oxygen gas | O2xyDot® | OxySense |
| | | Ageless Eye® | Mitsubishi Gas Chemical Inc. |
| | | $O_2$Sense™ | FreshPoint Lab |
| | | Tell-Tab | IMPAK |
| | | O2 Sense™ | FreshPoint Lab |
| | Pathogens | Food Sentinel® | SIRA Technologies |
| | | SensorQ® | DSM NV and Food Quality Sensor |
| | Polymer degradation | Mimica Touch® | Mimica Lab |
| | Toxin antibodies | Toxin Guard™ | Toxin Alert |
| Time-temperature indicators | Enzymatic degradation | FreshTag™ | VITSAB International AB |
| | | Checkpoint® | VITSAB |
| | Polymerization | Fresh-Check® | Temptime Corp. |
| | | HEATmarker® | Temptime Corp. |
| | Diffusion | WarmMark | ShockWatch |
| | | Monitor Mark™ | 3M Company |
| | | Novas® | Insignia Technologies Ltd |
| | | Tempix® | Tempix AB |
| | | Timestrip® Plus™ | Timestrip Plc |
| | Photochromic | ONVU™ | FreshPoint and Ciba |
| | | Lay's Chips | Chromatic Technologies (CTI) |
| | Fluorescent | Glow in the Dark Products | Chromatic Technologies (CTI) |
| | Biological | TRACEO® | Cryolog |
| Thermochromic inks | Temperature | Coors Beer | Chromatic Technologies (CTI) |
| | | Thermax® | LCR Hallcrest |
| | | Paint, Hypercolor T-Shirts | Matsui Int. Comp., Inc. |
| | | Colour-Therm | Colour-Therm |

(continued)

**Table 3.2** (continued)

| | Trigger | Commercial example | Company |
|---|---|---|---|
| Temperature data logger | Radiofrequency identification technologies | CS8304 | Convergence Systems |
| | | Intelligent box | Mondi Pic |
| | | PakSense Express | PakSense |
| | | K1-2: ESCREC014 | Cryopak (tiptemp.com) |
| Time-temperature data logger | Ultra-high-frequency LED | TempTRIP | TempTRIP LLC |

are triggered by the product themselves, such as those relying on pathogen, toxin, gas or oxygen release, alternative non-toxic sensors are required that avoid the safety issues associated with their components (Liu et al. 2019, 2020a, b; Seeboth et al. 2013). Many critical reviews of intelligent food packaging in the food industry have been reported covering both the commercial and research-based technologies currently being developed. These reviews go into great detail about the mechanism of these devices and their advantages and disadvantages, including their safety, cost and accuracy (Firouz et al. 2021; Fang et al. 2017; Biesuz and Magnaghi 2021; Weston et al. 2021b).

### 3.4.4 Consumer Acceptance of Intelligent Packaging

A systematic review of consumer perceptions of smart packaging was reported by Young et al. in 2020, including 28 studies (Young et al. 2020). The models used in these studies included the Siegrist risk/benefit model (five studies on nanopackaging), Kano model (five studies from one group), value-driven model (one study), the relationship between neophobia and acceptance of food technology (one study on nanopackaging), random utility theory (two studies on nanopackaging) and the theory of planned behaviour (two studies on meat products), while 13 studies did not cite any theories or models. Familiarity is generally low for active and intelligent packaging, which is not countered by educational communication. Well-known brands' awareness can help reduce the risk perception and increase acceptance. Five studies were cited, which identified that the provision of information to consumers about the packaging technology would increase trust, improve attitude and reduce the perceived risk. They concluded that the acceptance of intelligent and active packaging is not a well-researched area and should include more longitudinal studies, a broader geographical spread, specific examples and "real" applications and consumer responses to specific food groups. This was confirmed by Li et al. (2020), who showed that, when presented with product-specific applications, the approval of intelligent packaging increased. They also reiterated that surveys should be specific about the type of product group, for example, cheese and specific types of intelligent packaging; TTIs should be conducted to lead to "rich detail and industry-relevant results".

Regional surveys of consumers on the perception of intelligent packaging have included geographical surveys from China (Li et al. 2020), Slovakia (Erika et al. 2020; Loucanova et al. 2019), Latvia (Kocetkovs et al. 2019), Italy, Europe (Tiekstra et al. 2021; Pennanen et al. 2015) and Turkey (Aday and Yener 2015) with varying findings.

A survey conducted in the different regions of Lativa in 2019, with 865 respondents, determined how familiar they were with the terms intelligent and smart packaging (Kocetkovs et al. 2019). The results showed that the majority of respondents had insufficient knowledge or understanding about smart packaging. However, it did show that there has been a decrease in the percentage of respondents not willing to pay more for innovative packaging, from 29% in 2017 to 6% in 2019. The report suggested that if customers were introduced to these new technologies through trust, comfort and satisfaction, there would be an improvement in perception and an increase in willingness to use smart packaging. These findings were similar to that of Slovak consumers that, in 2019, the results indicated they had a low level of awareness of intelligent packaging overall (Loucanova et al. 2019) and as an ecological innovation in the context of the bioeconomy (Erika et al. 2020).

In an earlier publication in 2013, consumers were tested for their acceptance of "Innovative packaging". The terms intelligent and active packaging were not used yet; the questions specifically targeted each category. In that study, 265 Turkish respondents were asked 24 multiple-choice questions. Here, homogeneity analysis was used to determine, among other findings, that intelligent packaging was preferred over active packaging as consumers want to be able to visually monitor product quality themselves using either freshness, microbial growth or toxin risk as well as shelf life conditions such as temperature sensors/indicators (Aday and Yener 2015). The respondents indicated that education through advertising would be the most effective way to increase innovative packaging acceptability. It was noted that increased awareness of innovative packaging in warmer climates such as Italy, Spain and Turkey might be due to their increased concern regarding microbial spoilage.

In a quantitative survey of intelligent food packaging using the intercept method in Beijing, 371 consumers were tested for their acceptance of the new technology (Li et al. 2020). In this study, a high percentage of respondents (81%; $n = 181$) had a qualification in science and technology, although no significant association was measured for education, employment, gender or age. Overall, acceptance of novel packaging was high (56%), with levels increasing when presented with product-specific applications.

A 2020 study of European generation Y respondents ($n = 1249$) used a 7-point scale, where 1 represented "not important" and 7 "very important" to determine the impact of various active and intelligent packaging features" (Tiekstra et al. 2021). Overall, it was clearly determined that packaging was important (score of 5.3) to the consumer with the main properties of protection (6.0), sustainability (5.8), economy, i.e. low cost or good value (5.8), information (5.5), especially to guarantee authenticity and detailed content, convenience (5.2), portability and storage (4.8) and promotion (4.0). Of the intelligent packaging products tested, respondents reported that they would buy packaging that indicated that it had been maintained at the correct temperature (5.9), with a higher response for women (5.9) over men (5.7). This would represent sensors such as the already commercially present freshness indicators and TTIs in Table 3.2. Sensors that showed the ideal consumption temperature, such as CTI's Coors beer label, were the next highest-ranked (5.5) and would include reversible sensors such as the thermochromic devices and inks. Products that brought with it more interaction with the product ranked next in importance (4.7) with a significant preference by women (4.8) over men (4.5) once again. Packaging that provided engagement and leisure (e.g., QR codes, VR and conductive inks) was less appreciated (4.0), while the last appreciated were those that emitted smell (3.8), light or sound (2.7). More than half (51%) of the consumers were willing to pay less than 10% more for intelligent packaging, while almost more a third would pay between 10 and 50% more (28%).

In a 2021 publication, Cammarelle et al. (2021) surveyed 260 Italian consumers using a modified theory of planned behaviour (TPB) model focusing on reducing household food waste. They found that the respondents were more willing to purchase intelligent rather than active packaging. The survey identified the intelligent packaging as freshness indicators (colour change sensors), gas sensors (emitted from the food), leak indicators, temperature indicators (ready to drink/eat) and TTIs. The questionnaire measured several parameters, including attitudes towards food waste, subjective norms, perceived behavioural control, awareness, shopping, planning and leftover reuse routines. The coefficient for willingness to purchase intelligent overactive packaging was higher for all measures aside from the attitude (favourable/unfavourable evaluation of the behaviour), with significant preference of intention to reduce household waste through willingness to purchase intelligent (0.81) overactive packaging (0.68).

The main considerations when developing intelligent packaging and understanding consumer acceptance include improved education and awareness, visual communication of safety authenticity and quality, increased studies specific to individual technologies and food groups, geographical and demographic considerations, and economic and ecological impacts in an atmosphere of openness and trust.

Most of the studies reported were in the form of written or oral questionnaires. Considering non-invasive biometrics, physiological measurement, virtual reality and AI, greater insights into consumer-targeted intelligent packaging could be achieved. Improvements to the application of such technologies could lead to full acceptance of this new and emerging approach to food packaging.

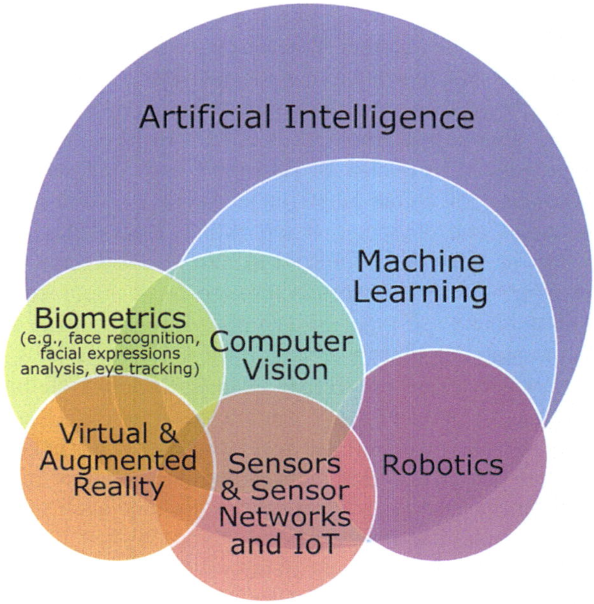


**Fig. 3.1** Venn diagram depicting the artificial intelligence sub-disciplines and their relationship within each other. Diagram adapted from Gonzalez Viejo et al. (2019b). IoT: Internet of Things

## 3.5 Artificial Intelligence in Food Packaging

The implementation of AI has gained increasing popularity in the last decade, especially in the food and beverage sciences (Fuentes et al. 2021a; Gonzalez Viejo et al. 2019b). It is important to note that AI does not involve only machine learning modelling, which is a discipline within AI, but also involve the integration with other disciplines, such as sensor technology, sensor networks and Internet of Things (IoT), robotics, biometrics, VR and AR, computer vision using imaging and multispectral cameras for smart or intelligent assessments (Fig. 3.1).

Different AI modelling strategies have been implemented for food and beverage applications, including packaging assessment, through the use of supervised machine learning and deep learning depending on the data availability and the objectives and targets of the models to be developed (Gonzalez Viejo et al. 2019b). Once the AI models are developed, they can be deployed using the respective sensor technology that is readily available, which can be cost-effective. One of the main principles to consider for efficient AI model deployment is the parsimony requirement of models, which states that the inputs of the models need to be simple enough and easier to acquire compared to the targets (Robbins n.d.).

Recent research is of special relevance, which implemented non-invasive biometrics for emotional and physiological response assessment from panellists in sensory tests. An interesting application of AI was proposed by the Digital Agriculture, Food and Wine research group at the University of Melbourne, using the automated integration of eye trackers and emotional response from participants assessed through computer vision (Fuentes et al. 2019). Through this methodology, it was possible to synchronize the assessment of different components of packaging labels, or AOI and the emotional response elicited by each (Fuentes et al. 2021b). Many other critical applications of AI in the food sector have been identified, such as improvement of quality of packaging and food safety (Paul et al. 2021), and the assessment of coffee labels under social isolation using online video conference resources and computer vision, which makes these studies relevant in times of pandemic (Gonzalez Viejo et al. 2021). Also, data collection and AI modelling have been facilitated recently through the development of a specialized BioSensory App to obtain self-reported data and video and infrared thermal imagery to process them using computer vision algorithms (Fuentes et al. 2018). Although it has not been widely explored, AI may be integrated into intelligent packaging technologies as more efficient anti-counterfeiting strategies. This is a promising application of AI, which is being explored and developed, especially in the most recent years (Schaefer and Cheung 2018). This includes the integration of machine learning to intelligent packaging sensors as support systems for food quality and safety (Sohail et al. 2018). For example, this may predict the product shelf life in real time according to the storage and transportation conditions (Loisel et al. 2021).

## 3.6 Conclusion

The implementation of new and emerging technologies for intelligent packaging assessment and smart feature integration shows promising results for the food and beverage industries. Implementing AI for packaging assessment of consumer acceptability by integrating sensor technology and machine learning could increase objectivity for the analysis of packaging for new products, which can support the decision-making process more efficiently. More research should focus on the efficient deployment of AI models and sensor integration since most recent developments are only up to the stage of model development. The latter will secure the proper application of AI in smart or intelligent packaging to benefit the industries and consumers through higher information, transparency, provenance, traceability and combat counterfeiting.

**Acknowledgements** Cited research published by the co-authors was partially funded by the Australian Government through the Australian Research Council [Grant Number IH120100053] "Unlocking the Food Value Chain: Australian industry transformation for ASEAN markets".

## References

Aday MS, Yener U (2015) Assessing consumers' adoption of active and intelligent packaging. Br Food J 117(1):157–177

Alizadeh-Sani M, Mohammadian E, Rhim J-W, Jafari SM (2020) pH-sensitive (halochromic) smart packaging films based on natural food colorants for the monitoring of food quality and safety. Trends Food Sci Technol 105:93–144

Alsever J (2009) Behind Coors' color-changing beer cans. CNN Money (online). money.cnn.com: CNN

Australia TRBo (2021) Banknotes: list of security features

Ballco P, de Magistris T, Caputo V (2019) Consumer preferences for nutritional claims: an exploration of attention and choice based on an eye-tracking choice experiment. Food Res Int 116:37–48

Biesuz R, Magnaghi LR (2021) Role of biogenic amines in protein foods sensing: myths and evidence. Meat and nutrition: IntechOpen

Biju S, Fuentes S, Gonzalez Viejo C, Torrico DD, Inayat S, Gupta D (2021) Silicon supplementation improves the nutritional and sensory characteristics of lentil seeds obtained from drought-stressed plants. J Sci Food Agric 101:1454–1466

Buss D (2018) Food companies get smart about artificial intelligence. Food Technol 72:26–41

Cammarelle A, Viscecchia R, Bimbo F (2021) Intention to purchase active and intelligent packaging to reduce household food waste: evidence from Italian consumers. Sustainability 13(8):4486

Chandramouli Rajaiah S (2017) QR codes in marketing: an empirical study on awareness and opinions regarding QR code usage among young indian adults. Int J Market Business Commun 6(3):11–18

Chen AT-Y (2021) How fragmentation can undermine the public health response to Covid-19. Interactions 28(2):64–69

Clark EA, Duncan SE, Hamilton LM, Bell MA, Lahne J, Gallagher DL et al (2021) Characterizing consumer emotional response to milk packaging guides packaging material selection. Food Qual Prefer 87:103984

Cox S, Shiffler R (2014) Extent of QR code adoption by consumers. Int J Business Human Technol 4(6):1–4

Crofton EC, Botinestean C, Fenelon M, Gallagher E (2019) Potential applications for virtual and augmented reality technologies in sensory science. Innovative Food Sci Emerg Technol 56: 102178

Cuesta U, Niño JI, Martínez-Martínez L, editors (2018) Neuromarketing: analysis of packaging using Gsr, eye-tracking and facial expression. Paper presented at the European conference on media, communication & film

Djurdjevic S, Novakovic D, Dedijer S, Kasikovic N, Zeljkovic Z, editors (2019) Development of augmented reality application for interactive smart materials. MATEC web of conferences; 2019. EDP Sciences

Donato A (2018) Peripheral nervous system. The University of Queensland. https://qbi.uq.edu.au/brain/brain-anatomy/peripheral-nervous-system/autonomic-nervous-system

Drexler D, Fiala J, Havlíčková A, Potůčková A, Souček M (2018) The effect of organic food labels on consumer attention. J Food Prod Mark 24(4):441–455

Erika L, Jan P, Martina N, Dopico A, Daniela H (2020) Perception of intelligent packaging in the context of bioeconomy. Stud U Vasile Gol Ec 30(3):77–89

Faggiano A, Carugo S (2020) Can the implementation of electronic surveys with quick response (QR) codes be useful in the COVID-19 era? Int J Epidemiol 49(5):1732–1733

Fang Z, Zhao Y, Warner RD, Johnson SK (2017) Active and intelligent packaging in meat industry. Trends Food Sci Technol 61:60–71

Fazio M, Reitano A, Loizzo MR, editors (2020) Consumer preferences for new products: eye tracking experiment on labels and packaging for olive oil based dressing. Multidisciplinary Digital Publishing Institute Proceedings

Firouz MS, Mohi-Alden K, Omid M (2021) A critical review on intelligent and active packaging in the food industry: research and development. Food Res Int 141:110113

Francis CM (2019) Counterfeiting and design-driven deterrents: trust perceptions of anti-counterfeiting strategies for food and product packaging. University of Melbourne

Frelih NG, Podlesek A, Babič J, Geršak G (2017) Evaluation of psychological effects on human postural stability. Measurement 98:186–191
Fuentes S, Gonzalez Viejo C, Torrico D, Dunshea F (2018) Development of a biosensory computer application to assess physiological and emotional responses from sensory panelists. Sensors 18(9):2958
Fuentes S, Gonzalez Viejo C, Torrico DD, Dunshea FR (2021) Digital integration and automated assessment of eye-tracking and emotional response data using the BioSensory App to maximize packaging label analysis. Sensors 21(22)
Fuentes S, Gonzalez Viejo C, Torrico DD, FRD (2019) Evaluation of labels and their elements (areas of interest) using sensory and eye-tracking techniques. In: 13th Pangborn sensory science symposium; 29 July 2019; Edinburgh, Scotland
Fuentes S, Wong YY, Gonzalez Viejo C (2020) Non-invasive biometrics and machine learning modeling to obtain sensory and emotional responses from panelists during entomophagy. Foods 9(7):903
Fuentes S, Tongson E, Gonzalez Viejo C (2021a) Novel digital technologies implemented in sensory science and consumer perception. Curr Opin Food Sci 41:99–106
Fuentes S, Gonzalez Viejo C, Torrico DD, Dunshea FR (2021b) Digital integration and automated assessment of eyetracking and emotional response data using the BioSensory App to maximize packaging label analysis. Sensors 21(22)
Gonzalez Viejo Duran C (2020) The effect of bubble formation within carbonated drinks on the brewage foamability, bubble dynamics and sensory perception by consumers. The University of Melbourne
Gonzalez Viejo C, Fuentes S (2020) Beer aroma and quality traits assessment using artificial intelligence. Fermentation 6(2):56
Gonzalez Viejo C, Fuentes S, Torrico D, Howell K, Dunshea F (2018a) Assessment of beer quality based on foamability and chemical composition using computer vision algorithms, near infrared spectroscopy and machine learning algorithms. J Sci Food Agric 98(2):618–627
Gonzalez Viejo C, Fuentes S, Torrico D, Dunshea F (2018b) Non-contact heart rate and blood pressure estimations from video analysis and machine learning modelling applied to food sensory responses: a case study for chocolate. Sensors 18(6):1802
Gonzalez Viejo C, Fuentes S, Howell K, Torrico D, Dunshea FR (2018c) Robotics and computer vision techniques combined with non-invasive consumer biometrics to assess quality traits from beer foamability using machine learning: a potential for artificial intelligence applications. Food Control 92:72–79
Gonzalez Viejo C, Fuentes S, Howell K, Torrico D, Dunshea F (2019a) Integration of non-invasive biometrics with sensory analysis techniques to assess acceptability of beer by consumers. Physiol Behav 200:139–147
Gonzalez Viejo C, Torrico D, Dunshea F, Fuentes S (2019b) Emerging technologies based on artificial intelligence to assess the quality and consumer preference of beverages. Beverages 5(4):62
Gonzalez Viejo C, Torrico DD, Dunshea FR, Fuentes S (2019c) Bubbles, foam formation, stability and consumer perception of carbonated drinks: a review of current, new and emerging technologies for rapid assessment and control. Foods. 8(12):596
Gonzalez Viejo C, Villarreal-Lara R, Torrico DD, Rodríguez-Velazco YG, Escobedo-Avellaneda Z, Ramos-Parra PA et al (2020) Beer and consumer response using biometrics: associations assessment of beer compounds and elicited emotions. Foods. 9(6):821
Gonzalez Viejo C, Zhang H, Khamly A, Xing Y, Fuentes S (2021) Coffee label assessment using sensory and biometric analysis of self-isolating panelists through videoconference. Beverages. 7(1):5
Gunaratne NM, Fuentes S, Gunaratne TM, Torrico DD, Ashman H, Francis C et al (2019a) Consumer acceptability, eye fixation, and physiological responses: a study of novel and familiar chocolate packaging designs using eye-tracking devices. Foods. 8(7):253

Gunaratne TM, Viejo CG, Fuentes S, Torrico DD, Gunaratne NM, Ashman H et al (2019b) Development of emotion lexicons to describe chocolate using the check-all-that-apply (CATA) methodology across Asian and Western groups. Food Res Int 115:526–534

Huang J, Peng Y, Wan X (2021a) The color-flavor incongruency effect in visual search for food labels: an eye-tracking study. Food Qual Prefer 88:104078

Huang J, Zhao P, Wan X (2021b) From brain variations to individual differences in the color–flavor incongruency effect: a combined virtual reality and resting-state fMRI study. J Bus Res 123: 604–612

Jain M, Deb S, Subramanyam A, editors (2016) Face video based touchless blood pressure and heart rate estimation. In: Multimedia signal processing (MMSP), 2016 IEEE 18th international workshop on IEEE

Kemp S, Hollowood T, Hort J (2011) Sensory evaluation: a practical handbook. Wiley, Hoboken

Kocetkovs V, Muizniece-Brasava S, Kirse-Ozolina A (2019) Consumer awareness and attitudes towards active and intelligent packaging Systems in the Latvian Market. Foodbalt, pp 222–6

Konopelko M (2019) Augmented reality packaging in food & beverages industry. Saimaa University of Applied Sciences

Li J, Guo X (2020) COVID-19 contact-tracing apps: a survey on the global deployment and challenges. arXiv preprint arXiv:200503599

Li T, Lloyd K, Birch J, Wu XM, Mirosa M, Liao XJ (2020) A quantitative survey of consumer perceptions of smart food packaging in China. Food Sci Nutr 8(8):3977–3988

Liao LX, Corsi AM, Chrysochou P, Lockshin L (2015) Emotional responses towards food packaging: a joint application of self-report and physiological measures of emotion. Food Qual Prefer 42:48–55

Liu BX, Ranji-Burachaloo H, Gurr PA, Goudeli E, Qiao GG (2019) A non-toxic reversible thermochromic binary system via pi-pi stacking of sulfonephthaleins. J Mater Chem C 7(30): 9335–9345

Liu BX, Mazo AR, Gurr PA, Qiao GG (2020a) Reversible non-toxic Thermochromic microcapsules. ACS Appl Mater Interfaces 12(8):9782–9789

Liu BX, Gurr PA, Qiao GG (2020b) Irreversible spoilage sensors for protein-based food. ACS Sensors 5(9):2903–2908

Loisel J, Duret S, Cornuéjols A, Cagnon D, Tardet M, Derens-Bertheau E et al (2021) Cold chain break detection and analysis: can machine learning help? Trends Food Sci Technol 112:391–399

Lombart C, Millan E, Normand J-M, Verhulst A, Labbé-Pinlon B, Moreau G (2020) Effects of physical, non-immersive virtual, and immersive virtual store environments on consumers' perceptions and purchase behavior. Comput Hum Behav 110:106374

Loucanova E, Parobek J, Nosalova M, Dopico A (2019) The perception of respondents of intelligent packaging in Slovakia as ecological innovations. Stud U Vasile Gol Ec. 29(1):58–67

Meilgaard MC, Carr BT, Civille GV (2006) Sensory evaluation techniques, 4th edn. Taylor & Francis, Routledge

Merdian P, Piroth P, Rueger-Muck E, Raab G (2020) Looking behind eye-catching design: an eye-tracking study on wine bottle design preference. Int J Wine Bus Res

Meyerding SG, Merz N (2018) Consumer preferences for organic labels in Germany using the example of apples–combining choice-based conjoint analysis and eye-tracking measurements. J Clean Prod 181:772–783

Modica E, Cartocci G, Rossi D, Martinez Levy AC, Cherubino P, Maglione AG, et al. (2018) Neurophysiological responses to different product experiences. Comput Intell Neurosci 2018

Mohamed MB, Yoshizawa M, Sugita N, Yamaki S, Ichiji K (2020) Non contact monitoring of heart rate responses to taste stimuli using a video camera. Indonesian J Electric Eng Comp Sci 18(1): 293–300

Mokrý S, Birčiaková N, Slováčková T, Stávková J, Nagyová Ľ (2016) Perception of wine labels by generation Z: eye-tracking experiment. Potravinarstvo Slovak J Food Sci 10(1):524–531

Moustafa H, Morsy M, Ateia MA, Abdel-Haleem FM (2021) Ultrafast response humidity sensors based on polyvinyl chloride/graphene oxide nanocomposites for intelligent food packaging. Sensors Actuators A Phys 331:112918
Oliva M, Anikin A (2018) Pupil dilation reflects the time course of emotion recognition in human vocalizations. Sci Rep 8(1):1–10
Patil K, Karekar N (2019) Restaurant automation system using QR codes
Paul A, Gaur S, Ahamed M (2021) Artificial intelligence: importance in food chain industry. SSRN 3855252
Pennanen K, Focas C, Kumpusalo-Sanna V, Keskitalo-Vuokko K, Matullat I, Ellouze M et al (2015) European consumers' perceptions of time-temperature indicators in food packaging. Packag Technol Sci 28(4):303–323
Pichierri M, Peluso AM, Pino G, Guido G (2021) Health claims' text clarity, perceived healthiness of extra-virgin olive oil, and arousal: an experiment using facereader. Trends Food Sci Technol 116:1186–1194
Pizzi G, Scarpi D, Pichierri M, Vannucci V (2019) Virtual reality, real reactions?: comparing consumers' perceptions and shopping orientation across physical and virtual-reality retail stores. Comput Hum Behav 96:1–12
Popova L, Nonnemaker J, Taylor N, Bradfield B, Kim A (2019) Warning labels on sugar-sweetened beverages: an eye tracking approach. Am J Health Behav 43(2):406–419
Prime EL, Solomon DH (2010) Australia's plastic banknotes: fighting counterfeit currency. Angew Chem Int Edit 49(22):3726–3736
Qiu WL, Gurr PA, Qiao GG (2019) Color-switchable polar polymeric materials. ACS Appl Mater Interfaces 11(32):29268–29275
Robbins B (n.d.) Machine learning: how black is this beautiful black box: towards Data Science Inc. https://towardsdatascience.com/machine-learning-how-black-is-this-black-box-f11e4031fdf
Rodríguez-Escudero AI, Carbonell P, Moreno-Albaladejo P (2019) The conjoint effect of front-label claims' surface size and distance-to-center on customers' visual attention and emotional response. J Appl Packag Res 11(3):4
Rogers EM (1976) New product adoption and diffusion. J Consum Res 2(4):290–301
Schaefer D, Cheung WM (2018) Smart packaging: opportunities and challenges. Procedia CIRP 72: 1022–1027
Schulte-Holierhoek A, Verastegui-Tena L, Goedegebure RP, Fiszman BP, Smeets PA (2017) Sensory expectation, perception, and autonomic nervous system responses to package colours and product popularity. Food Qual Prefer 62:60–70
Seeboth A, Lötzsch D, Ruhmann R (2013) First example of a non-toxic thermochromic polymer material–based on a novel mechanism. J Mater Chem C 1(16):2811–2816
Shechner T, Jarcho JM, Wong S, Leibenluft E, Pine DS, Nelson EE (2017) Threats, rewards, and attention deployment in anxious youth and adults: an eye tracking study. Biol Psychol 122:121–129
Siegrist M, Ung C-Y, Zank M, Marinello M, Kunz A, Hartmann C et al (2019) Consumers' food selection behaviors in three-dimensional (3D) virtual reality. Food Res Int 117:50–59
Sielicka-Różyńska M, Jerzyk E, Gluza N (2021) Consumer perception of packaging: an eye-tracking study of gluten-free cookies. Int J Consum Stud 45(1):14–27
Silverthorn DU (1998) Human physiology: an integrated approach. Prentice Hall, Hoboken
Smith TW (2015) The book of human emotions: an encyclopedia of feeling from anger to wanderlust, Profile Books
Sohail M, Sun D-W, Zhu Z (2018) Recent developments in intelligent packaging for enhancing food quality and safety. Crit Rev Food Sci Nutr 58(15):2650–2662
Sonderegger A, Ribes D, Henchoz N, Groves E, editors (2019) Food talks: visual and interaction principles for representing environmental and nutritional food information in augmented reality. In: 2019 IEEE international symposium on mixed and augmented reality adjunct (ISMAR-Adjunct). IEEE

Songa G, Slabbinck H, Vermeir I, Russo V (2019) How do implicit/explicit attitudes and emotional reactions to sustainable logo relate? A neurophysiological study. Food Qual Prefer 71:485–496
Tiekstra S, Dopico-Parada A, Koivula H, Lahti J, Buntinx M (2021) Holistic approach to a successful market implementation of active and intelligent food packaging. Foods 10(2):465
Torrico DD, Fuentes S, Gonzalez Viejo C, Ashman H, Gunaratne NM, Gunaratne TM et al (2018a) Images and chocolate stimuli affect physiological and affective responses of consumers: a cross-cultural study. Food Qual Prefer 65:60–71
Torrico DD, Fuentes S, Viejo CG, Ashman H, Gurr PA, Dunshea FR (2018b) Analysis of thermochromic label elements and colour transitions using sensory acceptability and eye tracking techniques. LWT Food Sci Technol 89:475–481
Tran C (2020) McDonald's Australia brings back its popular monopoly game—and there's a huge $532MILLION worth of prizes up for grabs: Daily Mail Australia; 2020 [cited 2021 7 July]. https://www.dailymail.co.uk/femail/article-8688161/McDonalds-Australia-brings-popular-Monopoly-Game.html
Vergura DT, Luceri B (2018) Product packaging and consumers' emotional response. Does spatial representation influence product evaluation and choice? J Consumer Market
Vila-López N, Küster-Boluda I (2019) Consumers' physiological and verbal responses towards product packages: could these responses anticipate product choices? Physiol Behav 200:166–173
Vila-López N, Kuster-Boluda I, Alacreu-Crespo A (2021) Designing a low-fat food packaging: comparing consumers' responses in virtual and physical shopping environments. Foods 10(2): 211
Vuksanović N, Bajrami DD, Petrović MD, Grigorieva EM (2021) QR codes as a tool for receiving feedback about guests' satisfaction at destinations. J Place Manag Dev 14(1):19–31
Weston M, Kuchel RP, Chandrawati R (2021a) Digital analysis of polydiacetylene quality tags for contactless monitoring of milk. Anal Chim Acta 1148:238190
Weston M, Geng S, Chandrawati R (2021b) Food sensors: challenges and opportunities. Adv Mater Technol 6(5):2001242
Willemsen GH, DeGeus EJ, Klaver CH, VanDoornen LJ, Carrofl D (1996) Ambulatory monitoring of the impedance cardiogram. Psychophysiology 33(2):184–193
Wu J, Xie X, Yang L, Xu X, Cai Y, Wang T et al (2021) Mobile health technology combats COVID-19 in China. J Infect 82(1):159–198
Xu C, Demir-Kaymaz Y, Hartmann C, Menozzi M, Siegrist M (2021) The comparability of consumers' behavior in virtual reality and real life: a validation study of virtual reality based on a ranking task. Food Qual Prefer 87:104071
Yang S (2019) An experiment on the effect of augmented reality on user experience in digitized shopping
Young E, Mirosa M, Bremer P (2020) A systematic review of consumer perceptions of smart packaging technologies for food. Front Sustain Food Syst 4, 63
Zhang C, Yin A-X, Jiang R, Rong J, Dong L, Zhao T et al (2013) Time–temperature indicator for perishable products based on kinetically programmable Ag overgrowth on Au nanorods. ACS Nano 7(5):4561–4568

# Biodegradable Packaging Materials and Techniques to Improve Their Performance

4

Vilásia Guimarães Martins, Simone Canabarro Palezi,
Gisele Fernanda Alves-Silva, and Luan Gustavo Santos

**Abstract**

The most used materials for the development of food packaging are petroleum-based polymers. Despite the advantages that synthetic polymers have, the low rates of recycling, coupled with their nonbiodegradability and nonrenewable nature, drive the concerns regarding environmental pollution. Alternatives to replace synthetic polymers include different types of biomolecules, and these molecules might be provided by the metabolism of microorganisms, from biomass, by-products from food industries, and chemical synthesis of bio-derived monomers. Even with many sources being investigated as raw materials for food packaging development, the large use of these bio-based materials is still limited by different reasons. Many strategies can be used to improve the performance of these materials, such as blends, reinforcing agents, cold plasma, UV light, and chemical and enzymatic methods. Thus, an opportunity remains in the search to improve the overall performance of biodegradable materials in order to overcome the limitations regarding their packaging performance for the commercial use.

**Keywords**

Biopolymers · Bioplastics · Sustainable packaging

V. G. Martins (✉) · S. C. Palezi · G. F. Alves-Silva · L. G. Santos
Laboratory of Food Technology, School of Chemistry and Food Engineering, Federal University of Rio Grande, Rio Grande, Brazil


A. K. Shukla (ed.), *Food Packaging: The Smarter Way*,
https://doi.org/10.1007/978-981-16-7196-8_4

## 4.1 Introduction

Packaging has a primordial role in the maintenance of the quality of the food products from their production to consumption. The four main functions of food packaging are containment, protection, convenience, and communication. Among the most widely used materials are petroleum-based polymers; however, the wide use of these nonrenewable sources of polymers coupled with the low recycling rates has been causing their accumulation in nature leading to serious environmental problems. In this way, the search for alternatives from renewable and nonpolluting sources is of utmost importance in order to replace these synthetic materials.

The most widely used molecules for the development of sustainable polymers for food packaging materials include polysaccharides, proteins, and lipids. The films developed with macromolecules from agricultural sources have different characteristics regarding their mechanical, color, and barrier properties. Although a wide variety of raw materials have been explored to develop these materials, several factors still limited the advancement of these technologies for commercial use. Among them, the hydrophilic behavior of many proteins and polysaccharides is responsible for the low performance of the developed films compared to the synthetic polymers, avoiding these materials to perform the basic functions of protection of the packaged products (Wihodo and Moraru 2013; Benbettaïeb et al. 2016b).

Polylactic acid (PLA) is a polymer that is characterized as biocompatible and biodegradable thermoplastic aliphatic polyester (Mochizuki 2009), and it is obtained industrially through the polymerization of lactic acid or the polymerization with opening of the lactide ring (Avérous 2008). PLA can be produced from renewable nonfossil natural resources by fermentation of polysaccharides such as sugar or extracted from corn, potatoes, cane molasses, and beets (Murariu and Dubois 2016). One of the major disadvantages of using PLA is its higher cost and deficiency in some properties, such as low thermal resistance and lower barrier properties, limiting its commercial applicability (Swaroop and Shukla 2018).

Polyhydroxyalkanoates (PHAS) are polyesters of natural origin, and these biopolymers are produced by microbial fermentation by a wide variety of bacteria. PHB has thermoplastic properties, which allow them to be molded or transformed into films for different applications. The high crystallinity of polyhydroxybutyrate (PHB) makes the films of these biopolymers very fragile (Ghanbarzadeh and Almasi 2013). Therefore, it is important to search for alternatives to improve the polymer network formed by macromolecules from agricultural sources, microorganisms, and bio-derived monomers.

Different strategies can be used to improve the performance of these materials, such as the mixing of different raw materials to combine their properties, incorporation of reinforcing agents, use of chemical and enzymatic methods for modification of polymer chains, and physical methods that alter the properties of the polymeric networks (Bourtoom 2009). Chemical treatments using acid or alkaline agents or crosslinking agents have shown promising results in relation to mechanical and film barrier performance (Shah et al. 2016; Benbettaïeb et al. 2016b). The use of physical methods such as cold plasma and UV light technologies has the advantages of not requiring chemical reagents, also these techniques are nongeneration waste, and the treatment application has a great uniformity. These technologies are known to alter the properties of polymers by generating radicals that initiate reactions in their structure, such as the formation of crosslinking, and/or the breaking and degradation of chains (etching) (Morent et al. 2011; Wihodo and Moraru 2013). Due to these characteristics, the study of cold plasma and UV light to modify the properties of films developed with molecules from agricultural sources is promising since they are nonpolluting technologies being used for the improvement of polymers from renewable sources.

This chapter aims to present a review of biodegradable polymers used in the development of food packaging, as well as the techniques that can improve the performance of these biodegradable materials in relation to physical, mechanical, and barrier properties.

## 4.2 Biopolymers

In view of the environmental impacts generated by the disposal of the use of nonbiodegradable packaging, the search for alternative materials for the production of biodegradable packaging is necessary and has been intensified and the latest years (Cazón et al. 2017; Thakur et al. 2018). In this context, biodegradable and/or bio-based polymers, also called biopolymers, are alternative sources that have been widely studied for this replacement since they are raw materials with the ability to decompose in nature in a short period of time after disuse (Gouveia et al. 2019; Nešić et al. 2020). Biopolymers are polymers obtained from natural sources, which can be extracted from plants or marine organisms, such as polysaccharides and proteins, or produced by microorganisms (Paixão et al. 2019), such as microbial polyesters (PHAs) or microbial polysaccharides (Nešić et al. 2020), and also could be chemically synthesized from biological material (Rogovina et al. 2019). Figure 4.1 shows the most used sources of biopolymers for food packaging production.

It is worth noting that in order to obtain biopolymeric films of proteins and polysaccharides, in most cases it is necessary to incorporate plasticizer additives to modify and improve properties, since the use of polymer alone is not sufficient to form rigid or elastic films with good barrier properties (Han 2014). Reinforcement materials such as composite materials or bioactive compounds can also improve the physical–chemical properties of the polymeric matrix (Zafar et al. 2016; Kumar et al. 2019).

The addition of plasticizer has the function of reducing the strain stress and hardness and increase the flexibility and resistance to fracture the biopolymeric films. However, the modification of the properties will be related to the combination of the polymer and the plasticizer used (Vieira et al. 2011; Espitia et al. 2014). According to Espitia et al. (2014), among the most used plasticizers are polyols (glycerol and sorbitol), lipids (monoglycerides, phospholipids, and surfactants), and sugars (glucose and sucrose).

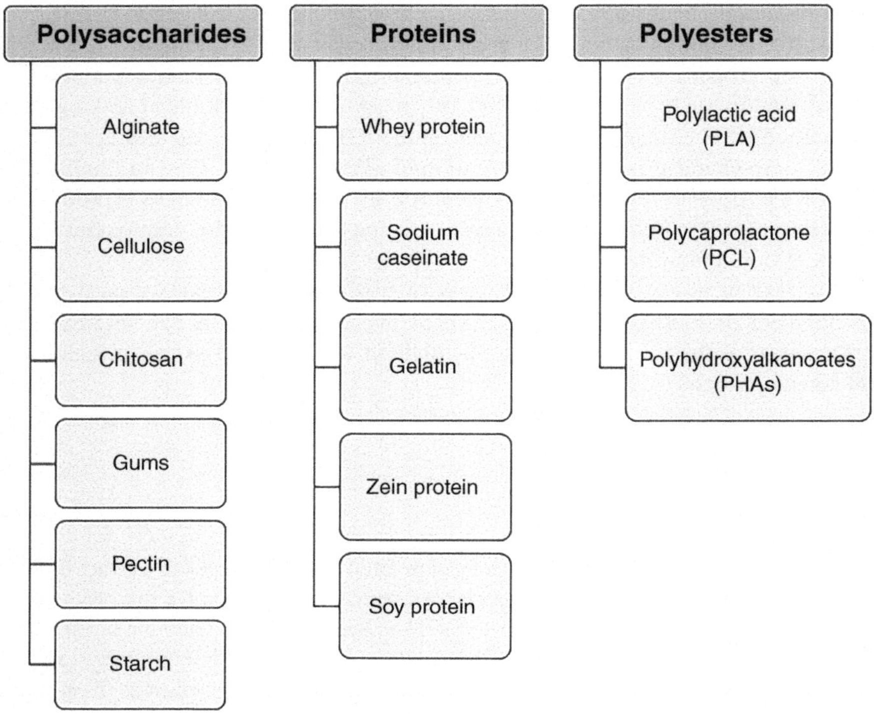


**Fig. 4.1** Types of biopolymers used for the preparation of food packaging

### 4.2.1 Polysaccharide-Based Films

Polysaccharides are promising sources for the development of packaging for food products, since they are a group of polymeric carbohydrates with desirable characteristics such as biocompatibility, biodegradability, and nontoxicity to live organisms (Chopin et al. 2014; Nešić et al. 2020). This biopolymer is a very abundant raw material, which could be originated from plants and marine biomass (Cazón et al. 2017; Nešić et al. 2020), bacteria, or fungi (Delattre et al. 2007).

In general, polysaccharide-based films have low barrier properties to water vapor and poor mechanical stability, due to the hydrophilic nature of most of these polymers (Cazón et al. 2017) However, polysaccharide packaging has good barrier properties to oxygen and carbon dioxide when exposed to low or moderate relative humidity (Nešić et al. 2020). The packages can be directed to products compatible with their properties; for example, some polysaccharides, e.g., alginate and carrageenan, are hydrophilic and hygroscopic and can be applied on the food surface to absorb water and slowing the loss of moisture (Bourtoom 2008; Cazón et al. 2017). Although there are some limitations for the preparation of biodegradable packaging, it is possible to improve the natural properties of polysaccharides through chemical modifications, in order to obtain molecular structures of interest and functionalized (Majid et al. 2018) The desired properties in polysaccharide films can also be obtained by combining them with other polymers and incorporating nanocomposites and/or bioactive compounds in the system (Hussain et al. 2017; Nešić et al. 2020).

Among the most used polysaccharides for the development of films are alginate, starch, cellulose, gums, pectin, and chitosan. In the next topic, the sources, desirable and undesirable characteristics, and some studies using different polysaccharides in the development of food packaging will be addressed.

#### 4.2.1.1 Alginate

Alginate is a polysaccharide obtained from the alkaline treatment of alginic acid, extracted from cell walls of brown seaweed (*Phaeophyceae*), also can be obtained from bacterial sources (Aziz et al. 2018). The algae species most used for alginate extraction are *Ascophyllum nodosum, Fucus* species, *Laminaria* species, *Macrocystis pyrifera,* and *Ecklonia* (Hay et al. 2013; SenturkParreidt et al. 2018). On the other hand, the extraction from a bacterial source occurs through the species of *Pseudomonas* and *Azotobacter* (Hay et al. 2013).

Due to the high hygroscopic, water-soluble properties, ease gel formation at room temperature, and emulsification capacity, alginate has shown great potential for the production of biodegradable films (Fig. 4.2). The films produced with alginate presented low permeability to oil and oxygen, in addition to having good flexibility, gloss, and absence of flavor and odor (Pawar and Edgar 2011; Paixão et al. 2019). According to Cazón et al. (2017), alginate films offer a low moisture barrier, and their hygroscopicity makes possible a delay dehydration of the foods in which they are applied. However, it can be used to mix polymers to get films with higher barrier properties, and despite the characteristics, alginate salts can also be modified by the process of crosslinking and/or adding lipids in the formulation, thus increasing the barrier to humidity (Benavides et al. 2012; SenturkParreidt et al. 2018).

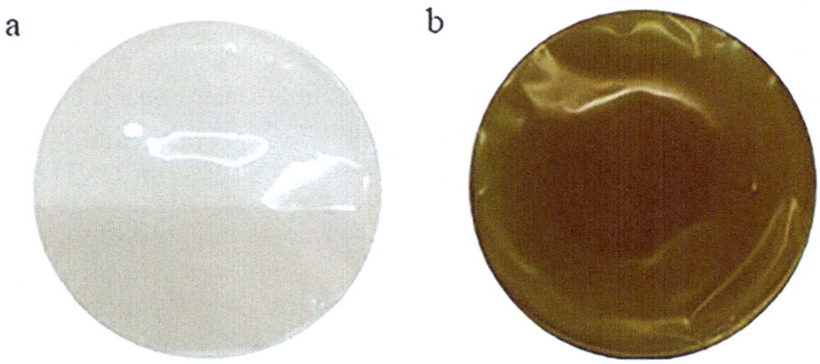


**Fig. 4.2** Sodium alginate-based films (**a**) and alginate based-film incorporated with *Caryocarbrasiliense* extract (**b**)

#### 4.2.1.2 Cellulose

Cellulose is the most abundant biopolymer in the world, being present mainly in plants due to cell development and biogenesis of cellulose in vegetables (Moon et al. 2011). However, this biopolymer can be found in fungi, algae, marine organisms, invertebrates, and gram-negative bacteria (Jõgi and Bhat 2020). It is one of the first biopolymers to be applied in the development of biodegradable packaging. Many packaging based on cellulose is used, as paper, cardboard box, wood box, among others.

A single cellulose fiber has Young's modulus values above 114 GPa, crystallinity about 89%, and a high degree of polymerization, with a thermal degradation temperature above 230 °C (Srivastava et al. 2018). Cellulose is commonly used in the paper and textile industries, generating materials of high quality and resistance, being applied as coatings and multilayer packaging for various types of food. However, due to the hydrophilic character, low solubility, and high crystallinity, the formation of films based on native cellulose becomes unfeasible because of the difficulty of the process and its high costs (Nechita and Iana-Roman 2020). Therefore, cellulose derivatives such as carboxymethylcellulose (CMC), methylcellulose (MC), ethyl cellulose (EC), hydroxypropyl and hydroxyethyl cellulose (HPC and HEC), and hydroxypropyl methylcellulose (HPMC) could be used to produce biodegradable films and membranes (Thakur and Thakur 2016).

In addition, due to the previously mentioned properties of high strength, crystallinity, and biodegradability, cellulose fibers can be reduced to nanocellulose (by mechanical or acid hydrolysis method) and applied as nanocomposites in the development of food packaging (Vestena et al. 2016; Hafemann et al. 2019). The incorporation of nanocellulose particles in filmogenic solutions can promote good emulsion and viscosity characteristics of fluids and, due to the hydroxyl groups present, nanoparticles can assist in the interaction between polymeric networks and lipophilic compounds, promoting the development of films more homogeneous and with greater resistance, malleability, and desirable barriers to UV light and gases (Mu et al. 2019).

#### 4.2.1.3 Chitosan

Chitosan is a copolymer, obtained from the deacetylation of chitin (Brasselet et al. 2019). Chitin is a naturally occurring polysaccharide that is among one of the most abundant biopolymers; however, its application is limited because it is not soluble in water or other common organic solvents, requiring its alkaline deacetylation in a solid state, which can also be obtained by fermentation (Jayakumar et al. 2010). After deacetylation, chitosan is obtained, and this polysaccharide is composed of two different monomers N-acetylglucosamine and glucosamine linked by β-(1 → 4)–glycosidic bonds (Younes and Rinaudo 2015).

According to Fernandez-Saiz (2011), chitosan is a polymer that presents bioavailability, biocompatibility, bio-adhesiveness, and biodegradability, is nontoxic, and can be ingested, in addition to intrinsic antimicrobial properties against fungi and bacteria (Laroche et al. 2018; Brasselet et al. 2019). Due to the physical and chemical properties of chitosan, several studies have demonstrated the potential of this biopolymer for the development of food packaging (Souza et al. 2017, 2019; Nešić et al. 2020). In films, chitosan has selective permeability to gases such as $O_2$ and $CO_2$ and good mechanical properties. However, the chitosan presents a poor water barrier due to its high sensitivity to water (Fernandez-Saiz 2011; Priyadarshi et al. 2018; Bouissil et al. 2019).

More recent studies have explored the development of active packaging using chitosan-based films incorporated with antioxidant and antimicrobial compounds as a strategy to improve the properties of chitosan films (Souza et al. 2017; Nešić et al. 2020). Bonilla et al. (2018), analyzing chitosan films (C) incorporated with clove oil (CO) and ginger oil (GO), observed that the incorporation of essential oil produced films with greater antioxidant activity, being CO > GO > C. Both films incorporated with essential oil showed better UV-Vis light barrier properties. The addition of clove oil also increased the properties of tensile strength and elastic modulus, compared to pure chitosan, showing that essential oils not only function as a bioactive compound, but also function as a reinforcing agent.

Studies also show that chitosan can be applied in the production of intelligent packaging, taking as an example the study of Wu et al. (2019), which used chitosan for the elaboration of intelligent packaging based on the incorporation of anthocyanins extracted from black rice bran, in which the response was characterized by pH sensitivity, higher antioxidant properties, and UV barrier effect.

Chitosan films can be applied to different products, as fruits, cuts of meat, biscuits, and butter (Drevinskas et al. 2017; Mostafavi and Zaeim 2020). In meat products, can be cited the study of Ruiz-Navajas et al. (2015), where chitosan films incorporated with nanocomposites of montmorillonite and rosemary essential oil were able to reduce the oxidation and color change of chicken meat.

#### 4.2.1.4 Gums

According to Salehi and Kashaninejad (2015), the term "gum" is applied to refer to a group of naturally occurring polysaccharides that have the capacity to form gels, viscous solutions, or emulsion stabilizers, also can be used as texture modifiers, thickeners, gelling agents, dietary fiber, and on the development of films and coatings (Kurt and Kahyaoglu 2014; Dick et al. 2015). Another nomenclature used is "hydrocolloids," referring to water-soluble gums (Salehi and Kashaninejad 2015). In general, the gums can be defined as molecules of high molecular weight that can produce gel when combined with appropriate solvents (Quiroga 2015).

The gums can be extracted from terrestrial or marine vegetables, products of microorganism biosynthesis, and chemical modification of natural polysaccharides (Quiroga 2015). Several studies have demonstrated the potential of the application of gums in the development of biodegradable food packaging, and some of them are shown in Table 4.1.

Among the polysaccharides capable of forming films, the gums are the class of mannans (consisting of glucomannan), present high molecular weight polymers, and have a strong interaction with water (Kurt and Kahyaoglu 2014).

Several studies have used agar as a raw material to produce films (Abdul Khalil et al. 2017; Contardi et al. 2019; Xu et al. 2019). In general, agar-based films are

**Table 4.1** List of gums more common used in the films production

| Gums | Source | References |
|---|---|---|
| Agar | Red seaweed of the class *Rhodophyceae* (*Gelidium* sp. and *Gracilaria* sp.) | Mostafavi and Zaeim (2020); Sousa and Gonçalves (2015) |
| Arabic | Exudate from Acacia Senegal and other species of the family *Leguminosae.* | Murmu and Mishra (2018); Murmu and Mishra (2017) |
| Carrageenan | Red seaweed of theRhodophyceae family: *Chondrus crispus* and *Gigantinamamillosa* | Rhim and Wang (2014); Siah et al. (2015); Sedayu et al. (2019) |
| Gellan | Secreted by the bacteria *Sphingomonas elodea* | Kim et al. (2015); Sapper et al. (2018) |
| Guar | *Cyamopsis tetragonolobus* seeds | Saurabh et al. (2015); Liu et al. (2020) |
| Locust | Ceratonia siliquo of the family *Leguminosae* | Barak and Mudgil (2014); Liu et al. (2020) |
| Unconventional | Chia seed | Barak and Mudgil (2014); Dick et al. (2015) |
| | Salep glucomannan of roots or tubers | Kurt and Kahyaoglu (2014) |
| | Basil seed (*OcimumbasilicumL.*) | Khazaei et al. (2014) |

relatively brittle, showing low elasticity and thermal stability, high solubility, and water vapor permeability (Arham et al. 2016). On the other hand, agar films have a high rate of shrinkage, are transparent, and heat-sealable, as well as biodegradable. However, worth to remember the film formation capacity of the agar is based on the gelling property and is closely linked to the origin of the source, the method of production of the film, and the components incorporated in the matrix (Sousa and Gonçalves 2015; Garrido et al. 2016; Huang et al. 2020).

Agar and carrageen are natural hydrophilic polymers extracted from red algae (Gonsalves et al. 2011; Siah et al. 2015), and films obtained from these polymers have limited use because it is a highly hydrophilic compound, but have a good barrier for fats and oils (Tavassoli-Kafrani et al. 2016). Capable of forming highly viscous solutions even at low levels of concentration, the xanthan gum has an excellent ability to form films (Mohsin et al. 2018; Murmu and Mishra 2018). However, the xanthan films also have poor mechanical properties (Kurt et al. 2017; Hao et al. 2018; Wu et al. 2018b). As well as other polysaccharides, the properties of these polymers can be improved by mixing other polymers (Hou et al. 2019) or by incorporation of bioactive compounds (Rhim and Wang 2014; Soni et al. 2016).

Cashew bark is a nonconventional gum source, and it is a nonallergenic, biocompatible, and biodegradable polymer obtained from the gummy exudate extracted from *Anacardium occidentale* (Nayak et al. n.d.). Films made only of cashew gum have poor mechanical properties and low stability; for this reason, some studies have analyzed the mixture of polymers with this matrix (Oliveira et al. 2018; Cruz et al. 2019).

Another nonconventional gum is the gum obtained from chia seeds, and the mucilage extracted from chia has the potential to produce edible films (Muñoz et al. 2012; Barak and Mudgil 2014), coatings (Dick et al. 2015), and soluble films (Fernandes et al. 2020). The mucilage is mainly composed of xylose, glucose, and methyl glucuronic acid that form a branched polysaccharide of high molecular weight with excellent gelling properties in aqueous solution, even in very low concentrations (Barak and Mudgil 2014).

Other sources are studied in order to evaluate the potential for the elaboration of biopolymeric packaging, such as *salep glucomannan* extracted from tuberculosis roots (Kurt and Kahyaoglu 2014), basil (*Ocimum basilicum L.*) seed gum (Khazaei et al. 2014), *Lepidium perfoliatum* seed gum (Seyedi et al. 2014), Brea gum extracted from *Circadian praecox* (Spotti et al. 2016), and tara gum (Wu et al. 2018b; Liu et al. 2020).

#### 4.2.1.5 Pectin

Pectin is one of the main structural polysaccharides of superior plant cells, and pectin is found mainly in citrus fruits and apples, and can also be extracted from by-products of the food industry, such as fruit and greenery marcs. It is an important biopolymer in the food industry due to its gelling properties, generally formed by water-soluble pectinic acids with varying levels of ester methyl, with the ability to form gels when combined with sugars and acids in appropriate proportions (Narasimman and Sethuraman 2016; Martău et al. 2019; Coman et al. 2020).

Chemically, pectin is constituted by poly $\alpha$1–4-galacturonic acids, can be classified according to its degree of esterification with methanol as high methoxy pectin (DE $<$50%) or low methoxy pectin (DE $>$50%) (Espitia et al. 2014). However, the source and conditions of extraction can affect the molecular weight of the pectin and the degree of esterification (DE) (Gouveia et al. 2019).

In the food industries, pectin is applied mainly as a stabilizer, thickener, and encapsulant (Rodriguez-Garcia et al. 2016). However, it is widely studied in different fields such as agriculture and medicine (Gupta et al. 2014; Smith et al. 2016). In addition to the ability to form gels, pectin has antimicrobial and antiviral properties, which can be directed toward the development of active films and coatings (Gouveia et al. 2019).

Generally, pectin films are produced by casting technique (Bátori et al. 2017), but other methods such as thermocompression or extrusion have been studied due to their simplicity and their ability to produce films without polymer solubilization (Gouveia et al. 2019).

Pectin films have high-water solubility and poor mechanical properties, limiting their application as food packaging (Nesic and Seslija 2017; Gouveia et al. 2019). In order to improve these properties Makaremi et al. (2019) incorporated clay nanoparticles into a pectin matrix, obtaining pectin films with better mechanical performance, thermal stability and water vapor barrier properties.

Another strategy to improve the properties of pectin films is the incorporation of polyols. In the study made by Norcino et al. (2020), biodegradable pectin films were elaborated with incorporation of copaiba oil nanoemulsions in order to study the interactions of the polymers and how they influenced the mechanical and morphological properties of the films. It was observed that the incorporation of copaiba oil resulted in a gradual decrease in the modulus of elasticity and tensile strength, while increasing elongation art break, it was also observed an inhibition of microbial growth after the addition of copaiba oil, thus obtaining active packaging.

#### 4.2.1.6 Starch

Starch is a polymer with great potential for manufacturing biodegradable packaging because it is a low cost and could be obtained from different plant sources (Dai et al. 2019). However, the properties of starch films are dependent on the source of the polysaccharide, and they are commonly extracted from rice, potatoes, cassava (Luchese et al. 2017b), corn, and wheat (Luchese et al. 2018; Sivakanthan et al. 2020). However, alternative sources for obtaining starch are studied, such as lentils, beans (Joshi et al. 2013), peas, jackfruit, mango, and yams (Li et al. 2019; Rodrigues et al. 2020).

Chemically, starch consists of a crystalline linear amylose polymer (poly-α-1,4-D-glucopyranoside) and highly branched amylopectin formed by a large number of D-glucose units linked by α-1,4 and α-1,6-glycoside bonds (Sivakanthan et al. 2020). When starch is solubilized in an aqueous solution, and subsequently heated to a certain temperature, the starch granules absorb a large amount of water, and swell or even collapse, releasing the amylose, occurring the process of gelatinization and increased viscosity of the solution, which is used in the development of edible coatings and biodegradable films (Dai et al. 2019; Li et al. 2019; Ruan et al. 2019).

In general, starch-based films are great alternatives for food packaging due to the transparent, tasteless, and odorless, being materials with good oxygen barrier properties (Sapper et al. 2018). On the other hand, starch film has limited use, due to the high sensitivity to water and retrogradation phenomena that influence the barrier and mechanical properties of these materials (Cano et al. 2014; Kumar et al. 2019).

Although the limitations inherent starch films, the properties can be improved through physical and chemical modifications of the polymer matrix (Nechita and Iana-Roman 2020). The starch modification consists of the alteration of the native characteristics of the starch by the introduction of new functional groups in the molecules of the polysaccharide, changing the molecular size of the polymer and the properties of the particles (Liu et al. 2018; Kumar et al. 2019). Physical modification generally consists of combinations of polymeric matrix or addition of nanoparticles (Kumar et al. 2019), while chemical modification consists of structural modification as the crosslinking process.

In the study of Dai et al. (2019) was evaluated the physicochemical properties of native cassava starch and films of cassava starch modified by different methods. It has been observed that modified starches have better properties than native starch. When comparing the modification methods, they observed that the starch modified by crosslinking had better mechanical properties and water vapor barrier in relation to the esterified starch and the oxidized starch.

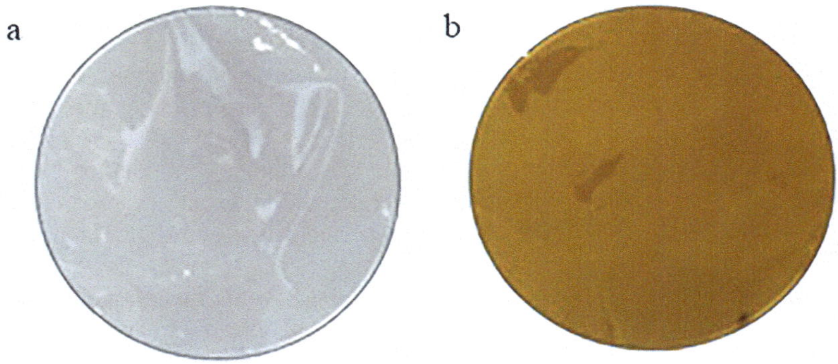


**Fig. 4.3** Cassava starch films plasticized with glycerol (**a**) and active cassava starch-based film incorporated with Annatto extract (*Bixa orellana*) (**b**)

Luchese et al. (2018) evaluated the properties of films made with different conventional sources of starch (wheat, potatoes, cassava, and corn), and it was observed that cassava starch presented more hydrophobic surfaces than corn and wheat starch films. Higher water vapor permeability values and lower mechanical properties values were also found for wheat starch films. While films of cassava starch showed higher values of elongation at break and lower permeability to water vapor.

Acosta et al. (2015) propose mixing starch with other biopolymers to improve the physical and functional properties of films, in addition to the incorporation of hydrophobic substances or antimicrobial compounds to produce active films (Fig. 4.3). Evangelho et al. (2019) incorporated orange essential oil in corn starch. The films incorporated with essential oil showed greater antibacterial activity against *Staphylococcus aureus* and *Listeria monocytogenes*, obtaining active packaging. The addition of oil also influenced the properties of the films, causing a reduction in tensile strength and elongation, increasing the moisture content, water solubility, and water vapor permeability.

Menzel et al. (2019) evaluated potato starch films incorporated with sunflower peel extract for preparation of active packaging. The addition of sunflower peel extract proved to be a valuable source of a natural antioxidant extract, and the incorporation of up to 6% of active compounds altered the mechanical properties, generating less elastic module and more durable films, and attributing this alteration mainly to the interactions of the phenolic compounds of the extracts with the polymeric starch network.

In general, it was observed that several sources of starches have potential for the development of biodegradable packaging; however, many of these matrices require chemical and or physical modifications to obtain packaging with more applicable properties.

## 4.3 Protein-Based Films

Proteins are among the most used natural biopolymers to produce films, and it is due to the presence of several polar and nonpolar amino acids that confer specific structural and functional variations (Scudeler et al. 2020). Proteins are heteropolymer composed of more than one hundred amino acids (monomers) linked through peptide bonds. Each of these amino acids contains a central carbon linked to a carboxyl group, a hydrogen, an amino group, and a R group (Nur Hanani et al. 2014).

When applied to packaging development, proteins have promising properties due to their ability to form three-dimensional networks stabilized and strengthened by hydrogen bonds, hydrophobic interactions, and disulfide bonds, allowing the creation of intermolecular bonds and cohesive matrices (Benbettaïeb et al. 2016a). However, the hydrophilic nature of proteins affects mechanical and barrier performance, which are essential properties for food packaging materials (Benbettaïeb et al. 2016c).

Protein biopolymers from animal and vegetable origin are of great interest for film production due to their relatively low cost, inherent biodegradability, and high availability as by-products from food and agriculture industries (Janjarasskul and Krochta 2010; Reddy and Yang 2013). Another advantage is that protein polymers can be processed by various methods, such as solvent evaporation or thermomechanical processes to produce films, developing materials with excellent oxygen barrier properties and adequate mechanical properties (Guerrero et al. 2010; Song et al. 2011).

There are several sources of proteins that have the ability to produce films. Some of them are whey proteins, sodium caseinate, gelatin (bovine, swine, and fish), corn, and soy proteins (Huntrakul et al. 2020).

### 4.3.1 Whey Protein

Whey proteins have been of great interest; currently, the world production of whey is about $1.8–1.9 \times 10^8$ tons/year. Whey proteins are responsible for 20% of the total milk protein, being a co-product of cheese production, and it is recovered through ultrafiltration or diafiltration. The main constituents of whey protein are α-lactalbumin, β-lactoglobulin, immunoglobulin, and bovine serum albumin (Smithers 2008; Tsai and Weng 2019).

According to Zhang et al. (2020b), whey proteins include whey protein isolate (WPI), whey protein concentrate (WPC), and whey protein hydrolysates. WPI has values above 90% of protein, and the main components are α-lactalbumin and β-lactoglobulin. The whey protein concentrate (WPC) is separated from the whey by ultrafiltration, and has strong nutritional properties and film formation capacity, which is transparent and elastic, making it a biodegradable and renewable source of protein of high biological value (Smithers 2008).

Protein-based films have no aroma and flavor and present good transparency, allowing a wide range of applications. Despite the promising use for packaging, the films produced have moderate mechanical characteristics, but poor water resistance due to their hydrophilic fractions. The high-water vapor permeability limits their application as packaging material when compared to synthetic polymers (Oymaci and Altinkaya 2016; Yayli et al. 2017).

GökkayaErdem et al. 2019 investigated the effect of the addition of sunflower oil on the formation of edible biocomposite films based on whey protein. The authors found that the water vapor permeability values decreased regardless of the added oil concentration, and their barrier properties were improved. Janjarasskul et al. (2020) developed edible films, based on whey protein (WPI) to package food or dry ingredients. The authors report that the films showed good permeability to water vapor and oxygen and low values for the mechanical properties of tensile strength, elongation, and modulus of elasticity and concluded that the films have the potential to be used as leave-in packaging for food dried in portions.

Zhang et al. (2020b) produced films composed of isolated whey protein (WPI) and psyllium seed gum (PSG), and the authors note that the WPI/PSG composite films had a greater contact angle with water and vapor permeability of water, demonstrating greater hydrophobicity, as well as less oxygen permeability and light transmittance compared to individual WPI or PSG films. It was also observed that the tensile strength, elongation at break, and the modulus of elasticity of the composite film WPI/PSG were higher than in the individual films.

### 4.3.2 Sodium Caseinate

Sodium caseinate is a mixture of casein monomers and small aggregates formed after elimination of colloidal calcium phosphate from casein micelles (Pankaj et al. 2014a). The technical and functional properties of caseinates make them proteins used mainly to improve the emulsifying properties of food matrices and the development of edible films (Broyard and Gaucheron 2015).

Due to the high proportion of polar groups, sodium caseinate films are good barriers to nonpolar substances, such as oxygen, carbon dioxide, and aromatic compounds. On the other hand, they present low resistance to traction and do not present a good barrier to water vapor due to their relatively high hydrophilicity (Belyamani et al. 2014; Jiang et al. 2020; Lin et al. 2020). Different types of physical, chemical, and biochemical modifications can be used to improve the structure and functionality of caseinates (Broyard and Gaucheron 2015; Picchio et al. 2018).

Brzoska et al. (2018) studied the effects of plasticizers sorbitol and glycerol and the concentration of lipids (oleic acid or mixtures of oleic acid–beeswax) in films based on sodium caseinate. According to the authors, films plasticized with sorbitol were more rigid than those with glycerol and it was also possible to observe a decline in the rate of moisture transmission with the addition of 10% bee wax.

### 4.3.3 Gelatin

Gelatin is a biodegradable and renewable material, obtained by the chemical or thermal degradation of collagen, and is widely found in skin, bones, and connective tissue (Chiellini et al. 2001; Rbini et al. 2020). In addition, skin, bone, or animal tissues constitute a significant by-product from animal processing industry, generating waste and pollution problems, and these by-products are capable of providing a valuable source of gelatin (Badii and Howell 2006).

Gelatin presents a unique sequence of amino acids with high content of glycine, proline, and hydroxyproline. The content of proline and hydroxyproline is particularly important in terms of the gelling effect (Gomez-Guillen et al. 2011). It is a material of relatively low cost and easy to obtain, being widely used by the food, cosmetic, and pharmaceutical industries due to its functional and technological properties (Badii and Howell 2006).

Its characteristics of biocompatibility, high availability, good processability, biodegradability, low cost, and absence of toxicity make gelatin a suitable material for the production of food packaging (Badhe et al. 2017; Hosseini and Gómez-Guillén 2018; Rbini et al. 2020). As one of the potential biomolecules for film development, gelatin has been extensively studied due to its good film-forming properties and its application as an encapsulating material, as well as a carrier of active ingredients (Gómez-Guillén et al. 2009; Shankar et al. 2017).

The films made with gelatin have good oxygen barrier properties. On the other hand, it has moderate mechanical properties, and relatively poor thermal stability (Gómez-Guillén et al. 2009; Wang et al. 2014) and its high-water permeability limit its industrial applications. However, gelatin functionality can be improved by mixing with other ingredients, such as plasticizers and food additives. The use of plasticizers improves the stability, resistance, and flexibility in gelatin films, which allows its use as packaging material (Park et al. 2008).

Nur Hanani et al. 2012 produced films with bovine, swine, and fish gelatin in different concentrations. The authors observed that films with a higher concentration of gelatin had their mechanical properties improved. Fish gelatin films showed the lowest water vapor permeability values, while films derived from swine gelatin showed less water solubility compared to those formed with fish and bovine skin, regardless of the concentrations used.

In the study of Dammak et al. (2017), an active gelatin-based film was developed, and microparticles coated with chitosan were prepared to encapsulate the rutin. The authors comment that results for water vapor permeability and the mechanical properties of tensile strength, elongation, and modulus of elasticity showed a decrease in values when the concentration of chitosan was greater than 0.5%. The microstructure analysis of the films revealed different micropores soaked in oil, resulting from the incorporation of the microparticles in the gelatin matrix, and they concluded that the film can be a good alternative to incorporate fat-soluble active compounds to design an active packaging.

Khedri et al. (2021) investigate the effects of bioactive casein phosphopeptides (CPPS) on gelatin-based films and reported that the addition of CPPs to the films significantly decreased solubility and water vapor permeability. The films also showed improved mechanical properties when compared to the standard and the FTIR and DSC analyses indicated adequate interactions between the functional groups of gelatins and CPPS.

#### 4.3.3.1 Fish Gelatin

Gelatin prepared from fish skin has advantages over bovine and swine, such as the deficit in outbreaks of bovine spongiform encephalopathy (BSE) and religious restrictions (Islam and Judaism) (Rattaya et al. 2009; Kwak et al. 2017). Fish-derived gelatin (FG) is known to have a relatively low amino acid content (hydroxy-proline and proline) compared to mammalian gelatin, presenting less strength in the formation of gels and, therefore, are more suitable for application in films (Karim and Bhat 2009).

Fish gelatin films exhibit good physical and mechanical properties and have an oxygen barrier superior to synthetic films (Nilsuwan et al. 2017). However, they have disadvantages such as high-water solubility and low-water barrier property, due to its high hydrophilicity (Vanin et al. 2005). One way to improve the mechanical properties and moisture susceptibility of films is to modify the polymer network by inducing intermolecular and intramolecular chemical bonds through chemical, enzymatic, or physical treatment (Kwak et al. 2017).

Syahida et al. (2020) studied fish gelatin films added with different concentrations of palm wax (PW) (0–60%). The results showed that gelatin/palm wax films with a higher concentration of PW were thicker, opaque, and more flexible than the control films without PW. The tensile strength increased with the incorporation of 15% PW and also showed a lower water vapor permeability and greater contact angle than the control film.

### 4.3.4 Fish Protein

Proteins attract attention due to their structure, which has the ability to form strong three-dimensional networks (Rhim et al. 2013; Benbettaïeb et al. 2016a). Animal proteins, such as myofibrillar proteins, are more promising due to their ability to form slightly transparent films with excellent barrier properties to ultraviolet light compared to commercial polyvinyl chloride film, for example (Kaewprachu et al. 2016, 2017).

Fish myofibrillar proteins (FMPs) have become a resource that they can be used to produce films with good transparency and resistance. These proteins are completely filamentous and elastic, and play a functional role because they have functional groups to form intra- and intermolecular bonds in the food system (Limpan et al. 2012; Blanco-Pascual et al. 2014). The use of fish myofibrillar proteins is also taken into account due to the economic aspects and the need to find adequate ways to take advantage of the remaining residues and by-products of fish production and processing (Hamaguchi et al. 2007; Pires et al. 2013).

Myofibrillar proteins extracted from the muscles of several species of fish have been used successfully for film production. In general, films have poor mechanical properties compared to synthetic polymers. Shabanpour et al. (2018) produced films of fish myofibrillar proteins adding bacterial cellulose nanofibers. The authors report that the addition of nanofibers led to a 49% increase in tensile strength, as well as a reduction in water vapor permeability, swelling, and solubility properties, and the evaluation of the thermal properties of these films suggests an improvement in the thermal stability of them.

The search for alternatives to improve the properties of materials of biological origin is necessary to make them suitable for use in food packaging without compromising the food product quality. Strategies to improve the performance of these bio-based polymers, such as the use of chemicals (López De Dicastillo et al. 2016), enzymes (Al-Hassan and Norziah 2017; Kaewprachu et al. 2017), mixed with other materials (Mujtaba et al. 2017; Romani et al. 2019), and others, have been widely studied, and physical strategies, such as cold plasma, are ecological and dry processes and its use can be explored.

Romani et al. (2019) investigated the use of cold plasma as a surface modification strategy for the treatment of films prepared from fish myofibrillar proteins (Fig. 4.4). The authors point out that films treated for 2 min with cold plasma showed an increase in elongation at break and a decrease in tensile strength. However, the plasma treatment could be a promising alternative in order to improve the properties of myofibrillar protein films for food packaging.

**Fig. 4.4** Fish myofibrillar protein film (Source: Romani et al. (2019))

### 4.3.5 Corn Protein Films

Zein is a prolamin polymer derived from corn, which is a co-product obtained during the extraction of corn starch (Ghanbarzadeh et al. 2006). The zein is insoluble in water, and this characteristic occurs due to the low content of polar amino acids and the high content of nonpolar amino acids present in the structure (Yong Cho et al. 2002).

Many studies in the pharmaceutical and food areas illustrate the potential of zein as a material for packaging production due to its biocompatibility, biodegradability, and nontoxicity (Han et al. 2014; Paliwal and Palakurthi 2014; Wang et al. 2019), in addition to its ability to incorporate bioactive compounds (Ozcalik and Tihminlioglu 2013; Yin et al. 2014; Wang et al. 2019). With the composition of 1/4 of the hydrophilic amino acid residues and 3/4 of the lipophilic amino acids, zein has inherent hydrophobic properties (Sun et al. 2017).

The development of methods to improve the functional properties of zein films, especially the mechanical properties and wettability, present challenges for their applications, such as increasing the shelf life of food products, tissue engineering, grafting, or printing of materials (Wang et al. 2009; Paliwal and Palakurthi 2014). Kadam et al. (2017) evaluated the effect of adding $TiO_2$ nanoparticles on the thermomechanical properties of zein films, where they observed that the incorporation of nanoparticles changed the properties of the films and improved some of their mechanical properties; however, it reduced the elongation at break in 50%.

The study of Gu et al. (2013) sought to improve the flexibility of zein films by adding gliadin. The authors report that there was a significant improvement in the flexibility of the films, as the amount of gliadin added to the solution increased, the water vapor permeability decreased. Indicating that the interactions of zein and gliadin established a solid network structure, gliadin conferred extensibility, and zein contributed to the resistance of the films.

Nogueira and Martins et al. (2018) developed and characterized zein/hake protein isolate (*Cynoscion guatacupa*) bilayer films (Fig. 4.5). The authors comment that the bilayer film was malleable and easy to handle, and with improved mechanical and thermal properties, they also report that the structural assessment of the bilayer revealed interaction between the different polymer layers and it was not necessary to use any type of adhesive to form a single structure.

### 4.3.6 Soy Protein Films

One of the most important by-products of the soy oil industry is the soy protein isolate (SPI), but its economic value is lower compared to the lipid components (Siracusa 2012; Koshy et al. 2015). SPI contains a large number of polar amino acids that provide hydrophilicity to SPI-based materials, and this effect is reflected in the film's fragility against moisture, resulting in a poor water vapor barrier and

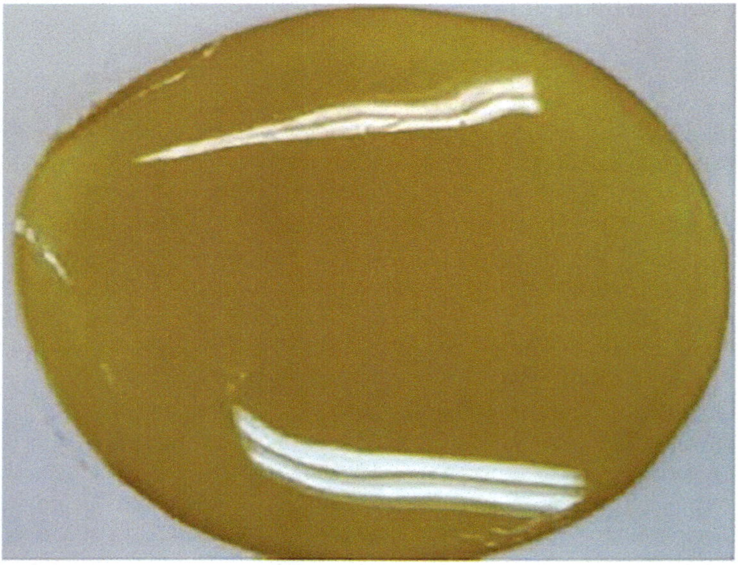

**Fig. 4.5** Bilayer film of isolated protein from hake and zein (Source: Nogueira and Martins et al. (2018))

insufficient mechanical properties when compared to some synthetics films (González et al. 2011).

Wang et al. (2016) developed antioxidant films of isolated soy protein with aqueous extract of Chinese chestnut (*Castanea molíssima*). The authors report that the films presented a good barrier to ultraviolet light and oxygen and the hydrophilic properties were improved. Advances in research involving the protein polymer have been recognized, but further studies are needed. Then, more alternatives can be presented to improve the physical and mechanical properties of the protein's films in general.

## 4.4 Biopolymers Produced by Biotechnological Processes

An alternative for obtaining biodegradable polymers is through biotechnological processes, using enzymes, fungi, and bacteria. The most well known are polylactic acid, polyhydroxyalkanoates, and polyhydroxybutyrate, which have great diffusion and industrial application in the development of commercial biodegradable packaging (Jõgi and Bhat 2020). These biopolymers and their derivatives have characteristics and properties similar to conventional petroleum-based polymers, and among the types of biopolymers presented above, polyesters have greater industrial application in replacing conventional packaging.

### 4.4.1 Polylactic Acid (PLA)

PLA aliphatic polyester is a bio-based polymer produced industrially, being applied in the development of packaging for food and medical materials due to its characteristics of thermoplasticity, transparency, biodegradability, and biocompatibility (Nofar et al. 2019). Discovered in 1845, PLA is produced from the chemical synthesis of lactic acid obtained from bacterial glucose fermentation, which can be derived from corn starch, sugar beet, sugar cane, and other renewable biomass and waste (Srivastava et al. 2018).

Initially, new polymerization techniques were studied to obtain PLA with high molecular weight (higher than 100.000 g/mol), which offer better physical and mechanical properties (Hamad et al. 2018). Currently, synthesis to obtain PLA occurs through two polymerization techniques, the direct condensation of lactic acid monomer, or the opening of the cyclic lactide dimer ring (Nofar et al. 2019), in which the last technique is commonly used industrially for generating PLA polymers with lower production costs and high molecular weight (Hamad et al. 2018; Nofar et al. 2019).

Regarding its properties, PLA has characteristics similar to conventional polymers from petroleum, such as polyethylene terephthalate (PET), polyethylene (PE), polystyrene (PS), polypropylene (PP), and polycarbonate (PC) (Hamad et al. 2018; Koh et al. 2018). PLA when applied as food packaging offers good malleability, resistance to UV light, higher oxygen limit index, lower water absorption rates, and lower processing temperature, when compared to conventional thermoplastics, such as PET (Srivastava et al. 2018; Asgher et al. 2020).

Furthermore, several studies show that the incorporation of additives (e.g., essential oils, plant extracts, and synthetic active compounds) and nanomaterials (e.g., cellulose nanofibers, nanoclay, and metallic nanoparticles) can act improving the physical and mechanical properties of packaging PLA basis. In the incorporation of rosemary, myrtle, and thyme oils in the PLA matrix, Yahyaoui et al. (2016) observed that the films obtained did not show changes in the transparency index, providing materials with less moisture transfer and better mechanical properties, when compared to PLA films without the addition of the oil.

In the study of Mohsen and Ali (2018), the addition of nanoclay provided a reduction of up to 32.38% in the elongation of PLA films; however, the tensile strength and Young's modulus increased 30.06 MPa and 6.4 GPa, respectively. On the other hand, the authors point out that the incorporation of nanoclay in the PLA matrix reduced the oxygen permeability coefficient and the water vapor transmission rate, demonstrating that this nanomaterial can offer better barrier properties when incorporated in PLA films and, consequently, greater protection against moisture and gases when applied on food, extending the shelf life of packaged products.

Bearing in mind that PLA is capable of releasing compounds in a controlled manner, in the area of food packaging, this biopolymer is commonly incorporated with antioxidant and antimicrobial compounds, which promote an increase in the shelf life of food products (Zhang et al. 2020a). Table 4.2 shows some of the studies

**Table 4.2** Incorporation of active compounds in PLA matrix

| Active compound | Effects in foods | Reference |
|---|---|---|
| Thymol, kesum and curry oils | Increased shelf life of chicken meat | Mohamad et al. (2020) |
| Beta-carotene and lycopene | Preservation and reduction of sunflower oil peroxides | Stoll et al. (2019) |
| Nanochitosan | Increased shelf life of chilled white shrimp | Fathima et al. (2018) |
| Green tea extract | Reduced lipid oxidation of smoked salmon | Martins et al. (2018) |
| α-tocopherol | Oxidative stability of soybean oil | Manzanarez-López et al. (2011) |
| Natamycin | Mold inhibition on the cheese surface | Lantano et al. (2014) |
| Fennel oil | Extended shelf life of chilled oysters | Miao et al. (2019) |

using PLA incorporated with active compounds that promote a longer shelf life of some food products.

Despite the many applications and advantages of using PLA in the area of food packaging, the reduction in production costs is one of the problems encountered by the polymeric industries, since its value is quite high when compared to oil-derived polymers (Koh et al. 2018). In addition, PLA has a low resistance to breakage compared to commercial polymers, making it more brittle (Zhong et al. 2020). For this reason, the alternatives studied to supply the mechanical and barrier deficiencies found in the PLA are the mixture with other biopolymers, resulting in low-cost biopolymeric blends with better mechanical and barrier properties (Koh et al. 2018; Zhong et al. 2020).

### 4.4.2 Polyhydroxyalkanoates and Polyhydroxybutyrate

Polyhydroxyalkanoates (PHAs) are biopolymers synthesized through microbial fermentation, being thermostable, having a melting temperature around 180 °C (Srivastava et al. 2018). On an industrial scale, *Cupriavidus necator* is the culture commonly used to obtain PHA in which, in restricted conditions of nutrients and in high concentrations of carbon source, its cells increase in weight and size, accumulating intracellular PHA (Jõgi and Bhat 2020).

PHA has great applicability in packaging development due to its good chemical properties, biocompatibility, resistance to UV light, and ability to be recyclable, being similar to petrochemical thermoplastics (Muneer et al. 2020; Zhong et al. 2020). Through the microbial synthesis of PHA, it is possible to obtain several classes of biopolymers, which are classified according to the size of its carbon chain. Among these PHA classes, the most well-known polymers are PHB and PHBV, which have a high crystallinity index (55–80%) and short chain, representing the most basic forms of commercial PHA available (Muneer et al. 2020).

Regarding its applicability, packaging produced from PHB and PHBV has a high barrier to aromatic compounds and water vapor, preserving and maintaining the quality of packaged foods for a longer time (Muneer et al. 2020; Zhong et al. 2020). In addition, PHBV can be applied in the development of blow-molded beverage packaging, resisting processing temperatures above 200 °C, making it an interesting biopolymer for the development of food packaging (Muneer et al. 2020). PHB films incorporated with eugenol oil were produced, and showed greater crystallinity, and antioxidant and antimicrobial activity, being able to be applied as food packaging (Rech et al. 2020). On the other hand, the incorporation of graphene nanocomposites in the PHB matrix promoted a reduction in the permeability to water vapor and oxygen, and an increase in the melting point, thermal stability, and tensile strength of the developed films (Manikandan et al. 2020). Therefore, the PHA matrix can be easily incorporated with additives that promote better physical, chemical, and mechanical characteristics of the packaging.

However, these polymers have low physical and mechanical resistance, making their industrial and commercial application difficult, when compared to other biodegradable polymers or petroleum derivatives (Meereboer et al. 2020). Therefore, PHB and PHBV when associated with other polymers demonstrate better strength tensile, less fragility, and greater flexibility, being suitable for the development of materials for the medical and food areas. In addition, its high production cost limits its application, having as an alternative the use of food waste as a carbon source during the fermentation process, reducing the costs of the synthesis of PHA polymers (Srivastava et al. 2018).

## 4.5 Techniques for Improving the Performance of Biodegradable Films

The main restriction in the application of natural polymers in the development of food packaging is due to the poor mechanical and barriers properties in relation to conventional petroleum-based packaging, and therefore, several techniques are developed and used to improve the performance of biodegradable packaging. One of the most used techniques involves mixing two or more natural polymers in the same film-forming solution, developing biopolymeric blends (Asgher et al. 2020). As previously mentioned, biodegradable films are commonly impregnated with nanocomposites (e.g., nanocellulose, nanofibers, and nanoclay) or bioactive compounds (e.g., essential oils, antioxidant and antimicrobial extracts) to provide better mechanical and barrier properties. On the other hand, the crosslinking technique, obtained by physical (e.g., UV lights, hot or cold plasma), enzymatic (e.g., transglutaminase), and chemical agents (e.g., calcium chloride, citric acid, and lactic acid), modifies the structures of natural polymers, improving the physical, mechanical, and barrier properties of biodegradable packaging, expanding the possibility of replacing conventional packaging (Rezaee et al. 2020).

### 4.5.1 Blends of Natural Polymers

The development of eco-friendly materials that present optimized performance and properties has been obtained over the years, such as the synthesis of blends made from different natural sources. The practice of developing blends for food packaging is a simple alternative for obtaining materials with improved characteristics and properties suitable for application as food packaging (Nogueira and Martins 2019). In general, the formation of blends consists of the mixture of two or more compatible polymers in the same filmogenic solution, obtaining dry films through the technique of casting, thermocompression, or extrusion (Zhong et al. 2020).

Blends from natural polymers are materials developed in order to reduce the impacts caused by the overuse of plastics from petroleum. In addition to the sustainable appeal, the mixture of biopolymers generates materials with improved physical, mechanical, and barrier characteristics due to the intermolecular interactions of polymeric networks, as reported in several studies (Romani et al. 2018; Nofar et al. 2019; Nogueira and Martins 2019; Filipini et al. 2020). Blends can present lower degrees of crystallinity in relation to conventional films, promoting important characteristics for food packaging, such as greater density, transparency, tensile strength, barrier to water vapor, and less flammability (Imre and Pukánszky 2013).

Another relevant point in relation to blends is the modification of the thermal degradation index, the mixture of biopolymers, promote greater stability at low temperatures of the films, enabling their application on chilled or frozen foods (Filipini et al. 2020; Wang et al. 2020). The changes in the thermal properties of blends and greater resistance to changes at high temperatures expand its industrial application, and it can be produced by blowing, extrusion, blow molding, or injection (Zhong et al. 2020). Some examples using blending technology to improve the treatment of film properties can be seen in Table 4.3.

A mixing material that gained attention was curdlan gum for having unique physicochemical properties such as insolubility in water and alcohol (Mohsin et al.

**Table 4.3** Effects of biopolymer blends compared to single polymer film

| Polymeric matrix | Effects | References |
|---|---|---|
| Cassava starch/Chitosan<br>Corn starch/Chitosan | The blend resulted in higher tensile strength | Luchese et al. (2017a) |
| Gelatin/chitosan | The blend resulted in higher elongation property | Bonilla et al. (2018) |
| Starch/gelatin | The incorporation of gelatin increased the mechanical properties of the films | Kumar et al. (2019) |
| Xanthan-Curdlan | Synergistically increased tensile strength | Mohsin et al. (2020) |
| Whey protein isolate/Chitosan | The blend resulted in the improvement of the mechanical properties | Tavares et al. (2021) |
| Isolated soy protein/Poly (lactic acid) (PLA) | The incorporation of PLA improved functional properties | González and Alvarez Igarzabal (2013) |

2019). According to Wu et al. (2018b), its dispersion in aqueous solutions should help in the formation of gels with better qualitative and mechanical characteristics. In the study of Mohsin et al. (2020) were evaluated the properties of films obtained by combining xanthan and curdlan gums in different proportions. It was observed that individually both polymers have low tensile strength; however, when mixed this property increased. It was also observed that elongation, thickness, and solubility are inversely proportional, with higher values of xanthan gum and lower for curdlan, so the combination can be performed to obtain films with specific properties by varying the concentration of each polymer.

The gelatin/chitosan blend increased the elongation property, compared to the use of each matrix separately; however, for traction and water vapor permeability, the best results were chitosan> chitosan/gelatin> gelatin (Bonilla et al. 2018). When chitosan was combined with corn starch, higher values of tensile strength and elongation were obtained, and higher values of tensile strength were also observed when chitosan was combined with cassava starch compared to the polymer separately (Luchese et al. 2017a).

## 4.5.2 Crosslinking

The crosslinking process is mainly aimed to modify certain polymer properties, such as chemical and thermal stability, rigidity, permeability, chelation efficiency, protein, and cellular immobilization capacity (Neto et al. 2005; Nair et al. 2020). Nair et al. (2020), state that the number of methods and crosslinking agents found in the literature is many, but the choice of parameters is closely linked to the purpose of the application, given that each agent, method, and type of polymer influence the results obtained from the crosslinking differently.

However, the modifications of the polymers by crosslinking process can be classified by physical or chemical methods (Fig. 4.6).

### 4.5.2.1 Chemical Crosslinking

The chemical crosslinking method occurs due to the numerous hydroxyl groups (−OH) present in the polymers, and they are susceptible to chemical modifications, such as sulfonation, carboxymethylation, phosphorylation, or hydroxyethylation (Ren et al. 2017a; Dimassi et al. 2018; Musso et al. 2019). From specific agents, all chemical groups along the polymer's backbone can be crosslinked to form "chemical" hydrogels, or "physical" hydrogels from interactions with each other due to ionic and hydrophobic interactions, and molecular entanglements (Pellá et al. 2018). A representation of a crosslinking method is shown in Fig. 4.7.

Several chemical agents are used for the purpose of crosslinking polymers (Xu et al. 2015). The study made by Nair et al. (2020) addresses the main crosslinking agents applied to collagen, as well as their advantages and disadvantages. Among the chemical agents mentioned by the authors are GTA (glutaraldehyde), EDC-NHS ((1-ethyl-3-(3-dimethylaminopropyl) carbodiimide)-

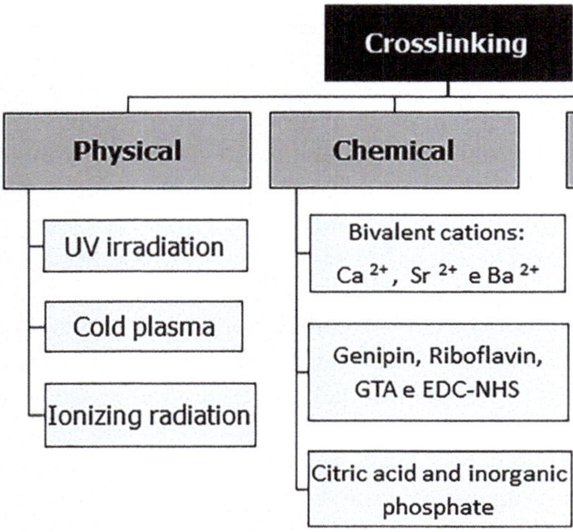


**Fig. 4.6** Classification of crosslinking modification methods

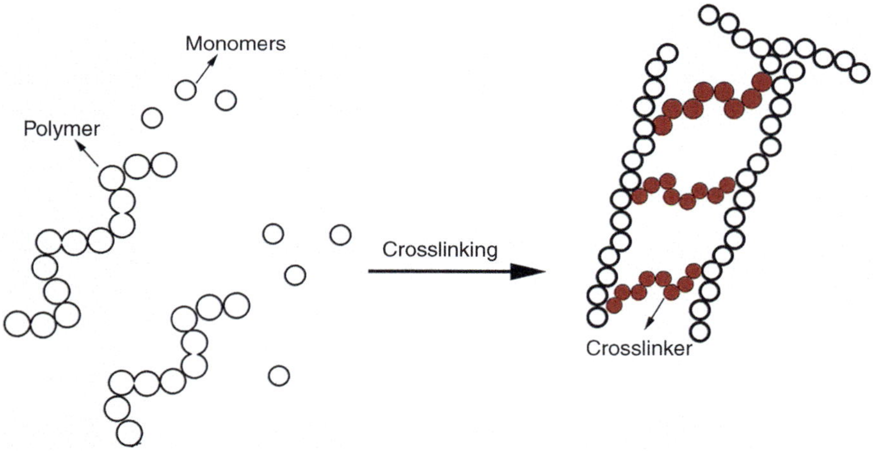


**Fig. 4.7** Representation of crosslinking method

NHS (N-hydroxysuccinimide)), genipin, transglutaminase 2, and physical agent, ultraviolet radiation (UV), and DHT (dehydrothermal treatment).

As mentioned throughout this chapter, some natural polymers have poor mechanical and barrier properties, so the crosslinking process is applied to modify and improve these properties. For example, alginate is a hydrophilic polymeric matrix; therefore, its films present low barrier properties. According to Benavides et al. (2012) and Rhim and Wang (2014), making a crosslink with polyvalent cations is an alternative to improve its water barrier properties, mechanical resistance, cohesiveness, and rigidity.

The addition of certain cations to the alginate solution leads to the formation of a gel through ion exchange (Lu et al. 2006). Among the most used cations, calcium ($Ca^{2+}$), strontium ($Sr^{2+}$), and barium ($Ba^{2+}$) (Lee and Rogers 2012) stand out. Calcium ions are more used to crosslink alginate, especially calcium chloride (SenturkParreidt et al. 2018). $Ca^{2+}$ sources influence the rate of gel formation and its concentration. Thus, it is possible to obtain films with less permeability to water vapor and greater resistance (Cazón et al. 2017).

Another agent used for crosslink films is citric acid, which has the advantage of being nontoxic, low cost, and capable of improving the functional properties of the film (Amanatidou et al. 2000). Kumar et al. (2019) showed that films made with citric acid crosslinked starch showed good tensile and barrier properties and also improved solubility in water; however, the films showed low elasticity. Some crosslinking agents are not used for food packaging, as copper, zinc, lead, and cobalt (SenturkParreidt et al. 2018).

According to González et al. (2011), the polymer chain of chitosan can be crosslinked by different agents, such as metallic ions ($Ag^{+}$, $Hg^{2+}$, and $Cu^{2+}$), and the authors also state that studies demonstrate greater resistance to dissolution in acid medium, less hydrophilicity, and reduced chemical reactivity of the material prepared from crosslinked chitosan. The characteristics of chitosan and the reaction conditions will affect the crosslinking process (Berger et al. 2004).

According to Xu et al. (2015), chemical modifications in the structure of polysaccharides and proteins (starch, cellulose, gelatin, gums, etc.) can be carried out by various crosslinking agents, and in general, the mechanism is used for the same purpose, to improve the properties of polymers to enhance their use in biodegradable packaging.

#### 4.5.2.2 Enzymatic Crosslinking

The application of enzymatic crosslinking occurs mainly in protein matrix (e.g., collagen, gelatin, and casein) due to the damage of structures caused during the dissociation and/or extraction process of the biopolymer (Duan et al. 2020). The use of enzymes to promote the crosslinking process has been showing positive effects in the formation of films, generating materials resistant to water and with better mechanical, barrier, thermal, and morphological properties (Liu et al. 2017).

According to Wu et al. (2018a, b), the application of enzymatic crosslinking has a nontoxic characteristic, high efficiency, and viability, since the enzymes are GRAS-certified and are usually used in the food industry (e.g., transglutaminase, laccase, tyrosinase, and peroxidase). In this context, enzymatic crosslinking comprises the approximation and formation of inter- or intramolecular covalent bonds of protein molecules, without the formation of toxic and harmful compounds to human health, providing safe control to the process when compared to the chemical crosslinking (Isaschar-Ovdat and Fishman 2018).

Due to its wide availability, ease, and low cost to obtain in relation to the others, the commercially available enzyme transglutaminase (MTg) is widely used in the food industries to promote changes in the texture of protein products, inducing gelation cold and/or joining protein molecules (Romeih and Walker 2017; Isaschar-Ovdat and Fishman 2018). MTg shows optimum activity at pH between 5 and 7 and in the temperature range of 40–50 °C, being inactivated by exposure to temperatures above 70 °C (Romeih and Walker 2017).

Gelatin films crosslinked with transglutaminase showed a solubility reduction of up to 80% (Liu et al. 2017). Films composed of coconut protein and guar gum showed better mechanical, barrier, and thermal properties when treated with MTg. The crosslinking promoted by the enzyme increased the interaction between the two biopolymers. On the other hand, when inserting the enzyme oxidase tyrosinase in casein films, Juvonen et al. (2011) obtained materials with lower values of solubility and smaller contact angle, enabling the application of this biomaterial as packaging for foods with higher moisture content.

#### 4.5.2.3 Plasma Technology

Plasma is considered the fourth state of matter in the world. With the increase in the energy level, the state of matter is changed from solid to liquid, then to gas, and finally to plasma (Pankaj et al. 2014c). The term "plasma" refers to an almost neutral ionized gas, which can be explained by the existence of an equal number of positive and negative charges carried by different species, composed mainly of ions, electrons, photons, and atoms in their fundamental states or excited (Pankaj et al. 2014b; Sharma and Singh 2020).

Plasma is generally divided into two categories, thermal and nonthermal (cold plasma) that depend on the condition in which they are generated (Pankaj and Thomas 2016). The thermal process occurs when operated at high pressure and temperature and requires great power to produce it, and this occurs when electrons and other types of gases are in thermodynamic equilibrium. Nonthermal process is when the operation occurs at low pressure, requiring less operating energy (Hati et al. 2018). Plasma can be generated by radio frequency (RF), direct current (CD), or microwave (MW) applied to the precursor gas used in the application (Sharma and Singh 2020).

There are several options to choose the carrier gas. Gases such as air, nitrogen, oxygen, helium, or argon can be used individually or combined at plasma technology, and the basic advantage of using air as a transport gas is its availability and low cost (Niemira 2012). Misra et al. (2014) reported that plasma chemistry depends on several factors, such as the composition of the feed gas, humidity, energy, applied voltage, and surrounding stage.

The interaction of the compounds generated by the plasma gas discharge with the different groups of the polymer molecules will form polar hydrophilic groups improving the adhesion and wettability properties, allowing the retention of active molecules and other compounds on the polymer surface, as representing in Fig. 4.8. These changes in polymer surfaces through the use of cold plasma have the advantage of not using hazardous solvents, in addition the treatment is uniform and does not cause thermal damage when in contact with the materials, which allows it to be used in temperature-sensitive biomaterials (Morent et al. 2011).

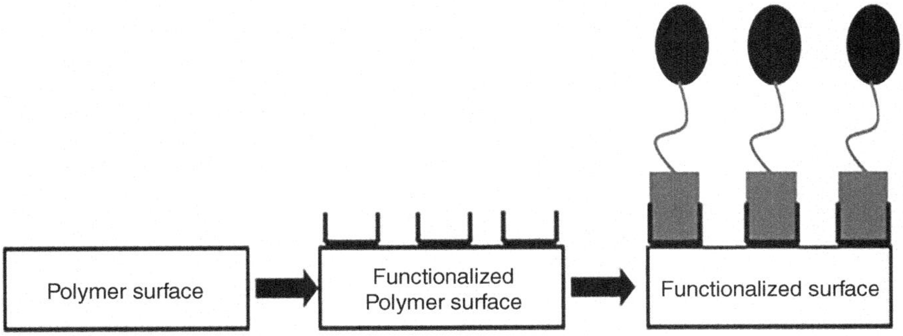


**Fig. 4.8** Representation of the functionalization of surfaces by plasma technology (Source: Adapted from Goddard and Hotchkiss (2007))

The use of cold plasma to improve the properties of biopolymers has attracted attention in research such as the use of chitosan (Demina et al. 2012; Chang and Chian 2013), polylactic acid (Pankaj et al. 2014b; Benetto et al. 2015), starch (Albuquerque et al. 2014; Benetto et al. 2015; Pankaj et al. 2015a), zeina (Pankaj et al. 2014c; Gezer et al. 2015), gelatin (Demina et al. 2012; Pankaj et al. 2015b), fish protein (Romani et al. 2019), and defatted soy flour (Oh et al. 2016) are some examples where treatment with cold plasma was performed.

Cold plasma technology has applications in several fields, such as textile processes, sterilization, biological sciences, packaging, and drug distribution systems (Samanta et al. 2010; Luna et al. 2011; Hagiwara et al. 2013). It presents several promising applications for the food industries such as food decontamination, toxin degradation, food process causing modification of the functionality of food components, equipment sterilization, air filtration, wastewater remediation, improved germination performance of seeds, improvement in the physicochemical properties of grains, degradation of agrochemical residues, and packaging surface treatments (Pankaj and Thomas 2016).

According to Ren et al. (2017b), cold plasma treatment is an environmentally friendly process, since it is free of chemicals and does not produce waste. It is an example of a dry process suitable for materials sensitive to heat that do not pollute the environment (Chu et al. 2002; Pankaj et al. 2014a; Mahmoud 2016).

There are several rapidly expanding technologies that have been used to generate cold plasma. Since they can operate at atmospheric pressure or in a vacuum, ionized gas can be as simple as air or nitrogen, or it can be a more complex mixture using noble gases, as helium, argon, or neon (Niemira 2012).

Among the plasma systems most used to modify the surfaces of the films, the low-pressure luminescent discharge (DC) stands out, which is formed when a sufficiently high potential difference is applied between two electrodes placed in a gas, the latter will divide into positive ions and electrons, giving rise to a discharge of it (Pankaj et al. 2014a). The atmospheric pressure plasma jet method (APPJ) is

**Table 4.4** Source of plasma and polymers used

| Plasma source | Polymers | Reference |
|---|---|---|
| Low pressure luminescent discharge (DC) | Fish myofibrillar proteins | Romani et al. (2019) |
| | Whey protein and gluten | Moosavi et al. (2020) |
| | Fish proteins | Romani et al. (2019) |
| Dielectric barrier discharges (DBD) | Polylactic acid (PLA) | Pankaj et al. (2014b) |
| | Bovine gelatin | Pankaj et al. (2015b) |
| | PLA/microfibrillated cellulose (MFC) | Meriçer et al. (2016) |
| | Zein/Chitosan | Chen et al. (2019) |
| | Casein | Wu et al. (2020) |
| | PLA coated zein | Chen et al. (2020) |
| | Sodium caseinate | Jahromi et al. (2020) |
| | Isolated pea protein | MahdavianMehr and Koocheki (2020) |
| Atmospheric pressure cold plasma (ACP) | Isolated whey protein | Segat et al. (2015) |
| | Defatted soybean meal (DSM) | Oh et al. (2016) |
| | Zein | Dong et al. (2018) |

generated when rare gases, such as helium or argon, are used to generate cold plasma even under atmospheric pressure (Ojha et al. 2021).

In dielectric barrier discharge (DBD), plasma is generated between two metal electrodes, at least one of which is covered by a dielectric layer that has the role of limiting the discharge current by preventing the transition of the arc and randomly distributing its coils on the electrode surface for homogeneous treatment (Tendero et al. 2006; Pankaj et al. 2014c). Another method used for cold plasma generation is cold plasma at atmospheric pressure (ACP), which can be obtained by exposing a gas or mixture of gases to an electric field, and it accelerates the charged particles, resulting in collisions with heavy species (as ions and neutrals) (Attri et al. 2013; Sadhu et al. 2017).

Studies show that treatment with cold plasma can induce various physical and chemical changes in the plasma surfaces. Some sources of plasma are more used in the modification of film surfaces, and Table 4.4 presents some polymers and the source of plasma used to obtain improvements in the characteristics of the films.

Romani et al. (2019) functionalized fish protein films, using cold plasma (luminescent discharge) and coated with carnauba wax. The application of luminescent discharge plasma followed by carnauba wax coating resulted in an increase of 175% in the tensile strength and 65% less water vapor permeability in fish protein films compared to control films.

Moosavi et al. (2020) studied the influence of different cold plasma treatments to modify the properties of whey and gluten protein film. The authors report that the films treated with low-pressure vacuum luminescent discharge plasma showed an increase in the tensile strength 10 min of processing, from 6.9 to 10.7 MPa and from 1.8 to 2.5 MPa for whey and gluten protein films, respectively.

The exposed above indicate that the plasma source, working gas, type of protein, sample volume, and treatment medium play an important role in plasma application, inducing denaturation/modification of proteins. The targeted functionalization can offer another innovative approach for specific modification of the functional properties of films.

#### 4.5.2.4 UV Irradiation

The application of irradiation by UV light can promote the crosslinking of biopolymers, causing the modification and improvement of the characteristics of biodegradable materials. UV irradiation is a physical crosslinking technique that has several advantages over other types of crosslinking, highlighting its simplicity, environmental friendly, and low costs, eliminating the use of chemical agents in the process (Rezaee et al. 2020). In addition, UV irradiation has no negative effects on human health, being applied as a technique to reduce and inactivate a wide range of microorganisms (Bigi et al. 2020). UV irradiation is classified into UV-A (315–400 nm), UV-B (280–315 nm), and UV-C (200–280 nm) according to the wavelength region, and this technique can cause modifications in the polymer chains from exposure of the film-forming solution or pre-formed film (Fathi et al. 2018) as shown in Fig. 4.9.

During the exposure of films to UV irradiation, the molecules become excited, causing chain splitting, crosslinking, and oxidation (Sionkowska et al. 2010). In addition, the increase in chemical reactions from photo-oxidative degradation modifies the surfaces of polymers, generating reactive sites in which they can immobilize various active compounds (e.g., metallic nanoparticles, bioactive compounds, and antimicrobial agents), which can improve the characteristics of the packaging (Alonso et al. 2009). However, the time of exposure to UV irradiation must be accurately determined for each type of material, since the permanence to UV light for long periods can damage and impoverish the characteristics of biopolymers due to photodegradation that, in the first instance, destroys interactions of

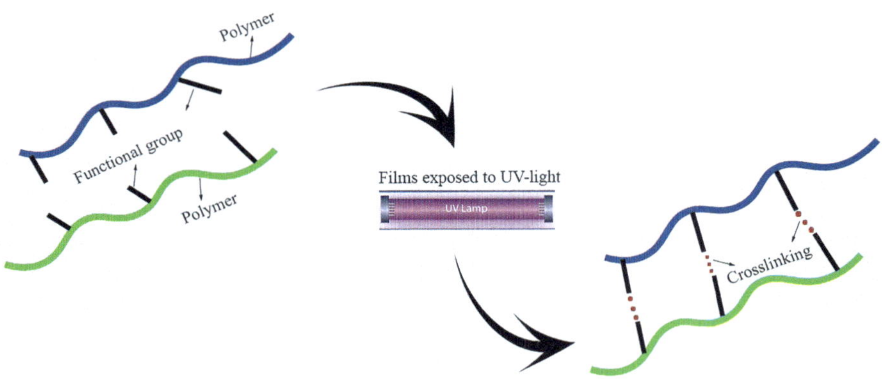


**Fig. 4.9** Representation of the crosslinking system in polymers using UV-light

**Table 4.5** Effects of UV irradiation on biodegradable films

| Biopolymer | Effects on materials | Reference |
|---|---|---|
| Isolated sesame protein | Increased crystallinity index<br>Compact structure formation<br>Reduced moisture content, solubility and water vapor permeation<br>Increased hydrophobic density of films<br>Greater tensile strength and Young modulus | Fathi et al. (2018) |
| Whey protein isolate | Modification of the tertiary structures of proteins<br>Reduction of solubility<br>Increased surface hydrophobicity | Kristo et al. (2012) |
| Whey protein concentrate | Structural modifications of proteins<br>Different color of the films<br>Increased tensile strength of films | Díaz et al. (2016) |
| Blend of collagen/ chitosan | Reduction in tensile strength and elongation of films<br>Increased of Young modulus | Sionkowska et al. (2006) |
| Blend of chitosan/ keratin | Reduction of traction, elongation and Young modulus<br>Increased surface polarity of films | Sionkowska et al. (2010) |

noncovalent polymer chains and then breaks the covalent bonds of the matrix, resulting in reductions in physical, mechanical, and barrier properties (Sionkowska et al. 2015).

In relation to protein-based films, the nondestructive and moderate effects of UV irradiation under suitable conditions cause desirable changes in the physical and mechanical properties of protein materials (Rezaee et al. 2020). Several studies demonstrate the efficiency of using UV irradiation in collagen and gelatin films. This efficiency occurs because proteins are one of the main targets of UV light, inducing several physical–chemical processes to the biopolymeric network, promoting expansion and chemical stability (Stylianou et al. 2014).

Table 4.5 shows some effects of the UV light on the surface of the biopolymer-based materials.

## 4.6 Conclusion

One of the main functionalities of the packaging is the preservation and extension of the shelf life of food products, reducing waste and promoting greater safety in the consumption of food products. Biodegradable packaging becomes an alternative to conventional packaging, since it can reduce the impacts caused by the excessive use of synthetic materials. However, biodegradable materials have poor mechanical and barrier properties in relation to synthetic packaging, hindering their diffusion in the industrial area. Despite this, studies show that the application of some techniques can expand the applicability of biodegradable materials. However, it is increasingly necessary to develop more studies in relation to the effects of applying these techniques to improve the performance of bio-based films, in order to improve their properties, making them more competitive in the packaging market.

## References


Abdul Khalil HPS, Saurabh CK, Tye YY et al (2017) Seaweed based sustainable films and composites for food and pharmaceutical applications: a review. Renew Sustain Energy Rev 77:353–362. https://doi.org/10.1016/j.rser.2017.04.025

Acosta S, Jiménez A, Cháfer M et al (2015) Physical properties and stability of starch-gelatin based films as affected by the addition of esters of fatty acids. Food Hydrocoll 49:135–143. https://doi.org/10.1016/j.foodhyd.2015.03.015

Albuquerque MDF, Bastos DC, Simão RA (2014) Surface modification of starch films by plasma. Macromol Symp 343:96–101. https://doi.org/10.1002/masy.201300199

Al-Hassan AA, Norziah MH (2017) Effect of transglutaminase induced crosslinking on the properties of starch/gelatin films. Food Packag Shelf Life 13:15–19. https://doi.org/10.1016/j.fpsl.2017.04.006

Alonso D, Gimeno M, Olayo R et al (2009) Cross-linking chitosan into UV-irradiated cellulose fibers for the preparation of antimicrobial-finished textiles. Carbohydr Polym 77:536–543. https://doi.org/10.1016/j.carbpol.2009.01.027

Amanatidou A, Slump RA, Gorris LGM, Smid EJ (2000) High oxygen and high carbon dioxide modified atmospheres for shelf-life extension of minimally processed carrots. J Food Sci 65:61–66. https://doi.org/10.1111/j.1365-2621.2000.tb15956.x

Arham R, Mulyati MT, Metusalach M, Salengke S (2016) Physical and mechanical properties of agar based edible film with glycerol plasticizer. Int Food Res J 23:1669–1675. https://doi.org/10.31227/osf.io/tq2pf

Asgher M, Qamar SA, Bilal M, Iqbal HMN (2020) Bio-based active food packaging materials: sustainable alternative to conventional petrochemical-based packaging materials. Food Res Int 137:109625. https://doi.org/10.1016/j.foodres.2020.109625

Attri P, Arora B, Choi EH (2013) Retracted article: utility of plasma: a new road from physics to chemistry. RSC Adv 3:12540–12567. https://doi.org/10.1039/C3RA41277F

Avérous L (2008) Polylactic acid: synthesis, properties and applications. In: Monomers, polymers and composites from renewable resources. Elsevier, pp 433–450

Aziz MSA, Salama HE, Sabaa MW (2018) Biobased alginate/castor oil edible films for active food packaging. LWT—Food Sci Technol 96:455–460. https://doi.org/10.1016/j.lwt.2018.05.049

Badhe RV, Bijukumar D, Chejara DR et al (2017) A composite chitosan-gelatin bi-layered, biomimetic macroporous scaffold for blood vessel tissue engineering. Carbohydr Polym 157:1215–1225. https://doi.org/10.1016/j.carbpol.2016.09.095

Badii F, Howell NK (2006) Fish gelatin: structure, gelling properties and interaction with egg albumen proteins. Food Hydrocoll 20:630–640. https://doi.org/10.1016/j.foodhyd.2005.06.006

Barak S, Mudgil D (2014) Locust bean gum: processing, properties and food applications—A review. Int J Biol Macromol 66:74–80. https://doi.org/10.1016/j.ijbiomac.2014.02.017

Bátori V, Åkeson D, Zamani A, Taherzadeh MJ (2017) Pectin-based composites. Handb Compos from Renew Mater 1–8:487–517. https://doi.org/10.1002/9781119441632.ch100

Belyamani I, Prochazka F, Assezat G, Debeaufort F (2014) Mechanical and barrier properties of extruded film made from sodium and calcium caseinates. Food Packag Shelf Life 2:65–72. https://doi.org/10.1016/j.fpsl.2014.07.003

Benavides S, Villalobos-Carvajal R, Reyes JE (2012) Physical, mechanical and antibacterial properties of alginate film: effect of the crosslinking degree and oregano essential oil concentration. J Food Eng 110:232–239. https://doi.org/10.1016/j.jfoodeng.2011.05.023

Benbettaïeb N, Chambin O, Assifaoui A et al (2016a) Release of coumarin incorporated into chitosan-gelatin irradiated films. Food Hydrocoll 56:266–276. https://doi.org/10.1016/j.foodhyd.2015.12.026
Benbettaïeb N, Gay J-P, Karbowiak T, Debeaufort F (2016b) Tuning the functional properties of polysaccharide-protein bio-based edible films by chemical, enzymatic, and physical cross-linking. Compr Rev Food Sci Food Saf 15:739–752. https://doi.org/10.1111/1541-4337.12210
Benbettaïeb N, Karbowiak T, Brachais CH, Debeaufort F (2016c) Impact of electron beam irradiation on fish gelatin film properties. Food Chem 195:11–18. https://doi.org/10.1016/j.foodchem.2015.03.034
Benetto E, Jury C, Igos E et al (2015) Using atmospheric plasma to design multilayer film from polylactic acid and thermoplastic starch: a screening life cycle assessment. J Clean Prod 87:953–960. https://doi.org/10.1016/j.jclepro.2014.10.056
Berger J, Reist M, Mayer JM et al (2004) Structure and interactions in covalently and ionically crosslinked chitosan hydrogels for biomedical applications. Eur J Pharm Biopharm 57:19–34. https://doi.org/10.1016/S0939-6411(03)00161-9
Bigi F, Haghighi H, Leo R et al (2020) Multivariate exploratory data analysis by PCA of the combined effect of film-forming composition, drying conditions, and UV-C irradiation on the functional properties of films based on chitosan and pectin. LWT—Food Sci Technol 137: 110432. https://doi.org/10.1016/j.lwt.2020.110432
Blanco-Pascual N, Fernández-Martín F, Montero P (2014) Jumbo squid (*Dosidicus gigas*) myofibrillar protein concentrate for edible packaging films and storage stability. LWT—Food Sci Technol 55:543–550. https://doi.org/10.1016/j.lwt.2013.10.025
Bonilla J, Poloni T, Lourenço RV, Sobral PJA (2018) Antioxidant potential of eugenol and ginger essential oils with gelatin/chitosan films. Food Biosci 23:107–114. https://doi.org/10.1016/j.fbio.2018.03.007
Bouissil S, Pierre G, El Alaoui-Talibi Z et al (2019) Applications of algal polysaccharides and derivatives in therapeutic and agricultural fields. Curr Pharm Des 25:1187–1199. https://doi.org/10.2174/1381612825666190425162729
Bourtoom T (2008) Plasticizer effect on the properties of biodegradable blend from rice starch-chitosan. Songklanakarin J Sci Technol 30:149–155
Bourtoom T (2009) Edible protein films: properties enhancement. Int Food Res J 16:1–9
Brasselet C, Pierre G, Dubessay P et al (2019) Modification of chitosan for the generation of functional derivatives. Appl Sci 9:1321. https://doi.org/10.3390/app9071321
Broyard C, Gaucheron F (2015) Modifications of structures and functions of caseins: a scientific and technological challenge. Dairy Sci Technol 95:831–862. https://doi.org/10.1007/s13594-015-0220-y
Brzoska N, Müller M, Nasui L, Schmid M (2018) Effects of film constituents on packaging-relevant properties of sodium caseinate-based emulsion films. Prog Org Coatings 114:250–258. https://doi.org/10.1016/j.porgcoat.2017.10.016
Cano A, Jiménez A, Cháfer M et al (2014) Effect of amylose:amylopectin ratio and rice bran addition on starch films properties. Carbohydr Polym 111:543–555. https://doi.org/10.1016/j.carbpol.2014.04.075
Cazón P, Velazquez G, Ramírez JA, Vázquez M (2017) Polysaccharide-based films and coatings for food packaging: a review. Food Hydrocoll 68:136–148. https://doi.org/10.1016/j.foodhyd.2016.09.009
Chang SH, Chian CH (2013) Plasma surface modification effects on biodegradability and protein adsorption properties of chitosan films. Appl Surf Sci 282:735–740. https://doi.org/10.1016/j.apsusc.2013.06.044
Chen G, Dong S, Zhao S et al (2019) Improving functional properties of zein film via compositing with chitosan and cold plasma treatment. Ind Crops Prod 129:318–326. https://doi.org/10.1016/j.indcrop.2018.11.072
Chen G, Chen Y, Jin N et al (2020) Zein films with porous polylactic acid coatings via cold plasma pre-treatment. Ind Crops Prod 150:112382. https://doi.org/10.1016/j.indcrop.2020.112382

Chiellini E, Cinelli P, Corti A, Kenawy ER (2001) Composite films based on waste gelatin: thermal-mechanical properties and biodegradation testing. Polym Degrad Stab 73:549–555. https://doi.org/10.1016/S0141-3910(01)00132-X
Chopin N, Guillory X, Weiss P et al (2014) Design polysaccharides of marine origin: chemical modifications to reach advanced versatile compounds. Curr Org Chem 18:867–895. https://doi.org/10.2174/138527281807140515152334
Chu PK, Chen JY, Wang LP, Huang N (2002) Plasma-surface modification of biomaterials. Mater Sci Eng R Rep 36:143–206. https://doi.org/10.1016/S0927-796X(02)00004-9
Coman V, Teleky BE, Mitrea L et al (2020) Bioactive potential of fruit and vegetable wastes. Adv Food Nutr Res 91:157–225. https://doi.org/10.1016/bs.afnr.2019.07.001
Contardi M, Heredia-Guerrero JA, Guzman-Puyol S et al (2019) Combining dietary phenolic antioxidants with polyvinylpyrrolidone: transparent biopolymer films based on: P-coumaric acid for controlled release. J Mater Chem B 7:1384–1396. https://doi.org/10.1039/c8tb03017k
Cruz MV, Jacobowski AC, Macedo MLR et al (2019) Immobilization of antimicrobial trypsin inhibitors onto cashew gum polysaccharide/PVA films. Int J Biol Macromol 127:433–439. https://doi.org/10.1016/j.ijbiomac.2019.01.010
Dai L, Zhang J, Cheng F (2019) Effects of starches from different botanical sources and modification methods on physicochemical properties of starch-based edible films. Int J Biol Macromol 132:897–905. https://doi.org/10.1016/j.ijbiomac.2019.03.197
Dammak I, Bittante AMQB, Lourenço RV et al (2017) Properties of gelatin-based films incorporated with chitosan-coated microparticles charged with rutin. Int J Biol Macromol 101:643–652. https://doi.org/10.1016/j.ijbiomac.2017.03.163
Delattre C, Laroche C, Michaud P (2007) Production of bacterial and fungal polysaccharides. In: Advances in fermentation technology. Asiatech Publisher, New Delhi
Demina T, Zaytseva-Zotova D, Yablokov M et al (2012) DC discharge plasma modification of chitosan/gelatin/PLA films: surface properties, chemical structure and cell affinity. Surf Coatings Technol 207:508–516. https://doi.org/10.1016/j.surfcoat.2012.07.059
Díaz O, Candia D, Cobos Á (2016) Effects of ultraviolet radiation on properties of films from whey protein concentrate treated before or after film formation. Food Hydrocoll 55:189–199. https://doi.org/10.1016/j.foodhyd.2015.11.019
Dick M, Costa TMH, Gomaa A et al (2015) Edible film production from chia seed mucilage: effect of glycerol concentration on its physicochemical and mechanical properties. Carbohydr Polym 130:198–105. https://doi.org/10.1016/j.carbpol.2015.05.040
Dimassi S, Tabary N, Chai F et al (2018) Sulfonated and sulfated chitosan derivatives for biomedical applications: a review. Carbohydr Polym 202:382–396. https://doi.org/10.1016/j.carbpol.2018.09.011
Dong S, Guo P, Chen Y et al (2018) Surface modification via atmospheric cold plasma (ACP): improved functional properties and characterization of zein film. Ind Crops Prod 115:124–133. https://doi.org/10.1016/j.indcrop.2018.01.080
Drevinskas T, Naujokaitytė G, Maruška A et al (2017) Effect of molecular weight of chitosan on the shelf life and other quality parameters of three different cultivars of *Actinidia kolomikta* (kiwifruit). Carbohydr Polym 173:269–275. https://doi.org/10.1016/j.carbpol.2017.06.002
Duan S, Wang W, Li S et al (2020) Moderate laccase-crosslinking improves the mechanical and thermal properties of acid-swollen collagen-based films modified by gallotannins. Food Hydrocoll 106:105917. https://doi.org/10.1016/j.foodhyd.2020.105917
Espitia PJP, Du WX, de Jesús Avena-Bustillos R et al (2014) Edible films from pectin: physical-mechanical and antimicrobial properties—A review. Food Hydrocoll 35:287–296. https://doi.org/10.1016/j.foodhyd.2013.06.005
Evangelho JA, Silva Dannenberg G, Biduski B et al (2019) Antibacterial activity, optical, mechanical, and barrier properties of corn starch films containing orange essential oil. Carbohydr Polym 222:114981. https://doi.org/10.1016/j.carbpol.2019.114981
Fathi N, Almasi H, Pirouzifard MK (2018) Effect of ultraviolet radiation on morphological and physicochemical properties of sesame protein isolate based edible films. Food Hydrocoll 85: 136–143. https://doi.org/10.1016/j.foodhyd.2018.07.018

Fathima PE, Panda SK, Ashraf PM et al (2018) Polylactic acid/chitosan films for packaging of Indian white prawn (*Fenneropenaeus indicus*). Int J Biol Macromol 117:1002–1010. https://doi.org/10.1016/j.ijbiomac.2018.05.214

Fernandes SS, Romani VP, Filipini GS, Martins VG (2020) Chia seeds to develop new biodegradable polymers for food packaging: properties and biodegradability. Polym Eng Sci 60:2214–2223. https://doi.org/10.1002/pen.25464

Fernandez-Saiz P (2011) Chitosan polysaccharide in food packaging applications. In: Multifunctional and nanoreinforced polymers for food packaging, pp 571–593. https://doi.org/10.1533/9780857092786

Filipini GS, Romani VP, Guimarães Martins V (2020) Blending collagen, methylcellulose, and whey protein in films as a greener alternative for food packaging: physicochemical and biodegradable properties. Packag Technol Sci. https://doi.org/10.1002/pts.2541

Garrido T, Etxabide A, Guerrero P, De La Caba K (2016) Characterization of agar/soy protein biocomposite films: effect of agar on the extruded pellets and compression moulded films. Carbohydr Polym 151:408–416. https://doi.org/10.1016/j.carbpol.2016.05.089

Gezer PG, Brodsky S, Hsiao A et al (2015) Modification of the hydrophilic/hydrophobic characteristic of zein film surfaces by contact with oxygen plasma treated PDMS and oleic acid content. Colloids Surfaces B Biointerfaces 135:433–440. https://doi.org/10.1016/j.colsurfb.2015.07.006

Ghanbarzadeh B, Almasi H (2013) Biodegradable polymers. In: Chamy R (eds) Biodegradation—life of science. InTech

Ghanbarzadeh B, Oromiehie AR, Musavi M et al (2006) Effect of plasticizing sugars on rheological and thermal properties of zein resins and mechanical properties of zein films. Food Res Int 39: 882–890. https://doi.org/10.1016/j.foodres.2006.05.011

Goddard JM, Hotchkiss JH (2007) Polymer surface modification for the attachment of bioactive compounds. Prog Polym Sci 32:698–725. https://doi.org/10.1016/j.progpolymsci.2007.04.002

GökkayaErdem B, Dıblan S, Kaya S (2019) Development and structural assessment of whey protein isolate/sunflower seed oil biocomposite film. Food Bioprod Process 118:270–280. https://doi.org/10.1016/j.fbp.2019.09.015

Gómez-Guillén MC, Pérez-Mateos M, Gómez-Estaca J et al (2009) Fish gelatin: a renewable material for developing active biodegradable films. Trends Food Sci Technol 20:3–16. https://doi.org/10.1016/j.tifs.2008.10.002

Gomez-Guillen MC, Gimenez B, Lopez-Caballero ME, Montero MP (2011) Functional and bioactive properties of collagen and gelatin from alternative sources: a review. Food Hydrocoll 25:1813–1827. https://doi.org/10.1016/j.foodhyd.2011.02.007

Gonsalves AAA, Araujo CRM, Soares NA et al (2011) Different strategies for crosslinking of chitosan. Quim Nova 34:1215–1223. https://doi.org/10.1590/S0100-40422011000700021

González A, Alvarez Igarzabal CI (2013) Soy protein—poly (lactic acid) bilayer films as biodegradable material for active food packaging. Food Hydrocoll 33:289–296. https://doi.org/10.1016/j.foodhyd.2013.03.010

González A, Strumia MC, Alvarez Igarzabal CI (2011) Cross-linked soy protein as material for biodegradable films: synthesis, characterization and biodegradation. J Food Eng 106:331–338. https://doi.org/10.1016/j.jfoodeng.2011.05.030

Gouveia TIA, Biernacki K, Castro MCR et al (2019) A new approach to develop biodegradable films based on thermoplastic pectin. Food Hydrocoll. 97:105175. https://doi.org/10.1016/j.foodhyd.2019.105175

Gu L, Wang M, Zhou J (2013) Effects of protein interactions on properties and microstructure of zein-gliadin composite films. J Food Eng 119:288–298. https://doi.org/10.1016/j.jfoodeng.2013.05.022

Guerrero P, Retegi A, Gabilondo N, De La Caba K (2010) Mechanical and thermal properties of soy protein films processed by casting and compression. J Food Eng 100:145–151. https://doi.org/10.1016/j.jfoodeng.2010.03.039

Gupta B, Tummalapalli M, Deopura BL, Alam MS (2014) Preparation and characterization of in-situ crosslinked pectin-gelatin hydrogels. Carbohydr Polym 106:312–318. https://doi.org/10.1016/j.carbpol.2014.02.019

Hafemann E, Battisti R, Marangoni C, Machado RAF (2019) Valorization of royal palm tree agroindustrial waste by isolating cellulose nanocrystals. Carbohydr Polym 218:188–198. https://doi.org/10.1016/j.carbpol.2019.04.086
Hagiwara K, Hasebe T, Hotta A (2013) Effects of plasma treatments on the controlled drug release from poly (ethylene-co-vinyl acetate). Surf Coatings Technol 216:318–323. https://doi.org/10.1016/j.surfcoat.2012.11.064
Hamad K, Kaseem M, Ayyoob M et al (2018) Polylactic acid blends: the future of green, light and tough. Prog Polym Sci 85:83–127. https://doi.org/10.1016/j.progpolymsci.2018.07.001
Hamaguchi PY, WuYin W, Tanaka M (2007) Effect of pH on the formation of edible films made from the muscle proteins of Blue marlin (*Makaira mazara*). Food Chem 100:914–920. https://doi.org/10.1016/j.foodchem.2005.10.045
Han JH (2014) Edible and coatings: a review. In: Innovations in food packaging, 2nd edn. Academic, San Diego
Han YL, Xu Q, Lu ZQ, Wang JY (2014) Preparation of transparent zein films for cell culture applications. Colloids Surfaces B Biointerfaces 120:55–62. https://doi.org/10.1016/j.colsurfb.2014.04.019
Hao L, Ding J, Yuan N et al (2018) Visual and flexible temperature sensor based on a pectin-xanthan gum blend film. Org Electron 59:243–246. https://doi.org/10.1016/j.orgel.2018.05.019
Hati S, Patel M, Yadav D (2018) Food bioprocessing by non-thermal plasma technology. Curr Opin Food Sci 19:85–91. https://doi.org/10.1016/j.cofs.2018.03.011
Hay ID, Rehman ZU, Moradali MF et al (2013) Microbial alginate production, modification and its applications. Microb Biotechnol 6:637–650. https://doi.org/10.1111/1751-7915.12076
Hosseini SF, Gómez-Guillén MC (2018) A state-of-the-art review on the elaboration of fish gelatin as bioactive packaging: Special emphasis on nanotechnology-based approaches. Trends Food Sci Technol 79:125–135. https://doi.org/10.1016/j.tifs.2018.07.022
Hou X, Xue Z, Liu J et al (2019) Characterization and property investigation of novel eco-friendly agar/carrageenan/TiO2 nanocomposite films. J Appl Polym Sci 136:47113. https://doi.org/10.1002/app.47113
Huang D, Zhang Z, Zheng Y et al (2020) Synergistic effect of chitosan and halloysite nanotubes on improving agar film properties. Food Hydrocoll 101:105471. https://doi.org/10.1016/j.foodhyd.2019.105471
Huntrakul K, Yoksan R, Sane A, Harnkarnsujarit N (2020) Effects of pea protein on properties of cassava starch edible films produced by blown-film extrusion for oil packaging. Food Packag Shelf Life 24:100480. https://doi.org/10.1016/j.fpsl.2020.100480
Hussain A, Zia KM, Tabasum S et al (2017) Blends and composites of exopolysaccharides; properties and applications: a review. Int J Biol Macromol 94:10–27. https://doi.org/10.1016/j.ijbiomac.2016.09.104
Imre B, Pukánszky B (2013) Compatibilization in bio-based and biodegradable polymer blends. Eur Polym J 49:1215–1233. https://doi.org/10.1016/j.eurpolymj.2013.01.019
Isaschar-Ovdat S, Fishman A (2018) Crosslinking of food proteins mediated by oxidative enzymes—A review. Trends Food Sci Technol 72:134–143. https://doi.org/10.1016/j.tifs.2017.12.011
Jahromi M, Niakousari M, Golmakani MT, Mohammadifar MA (2020) Physicochemical and structural characterization of sodium caseinate based film-forming solutions and edible films as affected by high methoxyl pectin. Int J Biol Macromol 165:1949–1959. https://doi.org/10.1016/j.ijbiomac.2020.10.057
Janjarasskul T, Krochta JM (2010) Edible packaging materials. Annu Rev Food Sci Technol 1:415–448. https://doi.org/10.1146/annurev.food.080708.100836
Janjarasskul T, Tananuwong K, Phupoksakul T, Thaiphanit S (2020) Fast dissolving, hermetically sealable, edible whey protein isolate-based films for instant food and/or dry ingredient pouches. LWT—Food Sci Technol 134:110102. https://doi.org/10.1016/j.lwt.2020.110102
Jayakumar R, Menon D, Manzoor K, Nair SV, Tamura (2010) Biomedical applications of chitin and chitosan derivatives. Carbohydr Polym 82:227–232. https://doi.org/10.1201/ebk1439816035-p5

Jiang YH, Cheng JH, Sun DW (2020) Effects of plasma chemistry on the interfacial performance of protein and polysaccharide in emulsion. Trends Food Sci Technol 98:129–139. https://doi.org/10.1016/j.tifs.2020.02.009

Jõgi K, Bhat R (2020) Valorization of food processing wastes and by-products for bioplastic production. Sustain Chem Pharm 18:100326. https://doi.org/10.1016/j.scp.2020.100326

Joshi M, Aldred P, McKnight S et al (2013) Physicochemical and functional characteristics of lentil starch. Carbohydr Polym 92:1484–1496. https://doi.org/10.1016/j.carbpol.2012.10.035

Juvonen H, Smolander M, Boer H et al (2011) Film formation and surface properties of enzymatically crosslinked casein films. J Appl Polym Sci 119:2205–2213. https://doi.org/10.1002/app.32943

Kadam DM, Thunga M, Srinivasan G et al (2017) Effect of TiO2 nanoparticles on thermo-mechanical properties of cast zein protein films. Food Packag Shelf Life 13:35–43. https://doi.org/10.1016/j.fpsl.2017.06.001

Kaewprachu P, Osako K, Benjakul S, Rawdkuen S (2016) Effect of protein concentrations on the properties of fish myofibrillar protein-based film compared with PVC film. J Food Sci Technol 53:2083–2091. https://doi.org/10.1007/s13197-016-2170-7

Kaewprachu P, Rungraeng N, Osako K, Rawdkuen S (2017) Properties of fish myofibrillar protein film incorporated with catechin-Kradon extract. Food Packag Shelf Life 13:56–65. https://doi.org/10.1016/j.fpsl.2017.07.003

Karim AA, Bhat R (2009) Fish gelatin: properties, challenges, and prospects as an alternative to mammalian gelatins. Food Hydrocoll 23:563–576. https://doi.org/10.1016/j.foodhyd.2008.07.002

Khazaei N, Esmaiili M, Djomeh ZE et al (2014) Characterization of new biodegradable edible film made from basil seed (*Ocimum basilicum* L.) gum. Carbohydr Polym 102:199–206. https://doi.org/10.1016/j.carbpol.2013.10.062

Khedri S, Sadeghi E, Rouhi M et al (2021) Bioactive edible films: development and characterization of gelatin edible films incorporated with casein phosphopeptides. LWT—Food Sci Technol 138:110649. https://doi.org/10.1016/j.lwt.2020.110649

Kim SRB, Choi YG, Kim JY, Lim ST (2015) Improvement of water solubility and humidity stability of tapioca starch film by incorporating various gums. LWT—Food Sci Technol 64:475–482. https://doi.org/10.1016/j.lwt.2015.05.009

Koh JJ, Zhang X, He C (2018) Fully biodegradable poly (lactic acid)/starch blends: a review of toughening strategies. Int J Biol Macromol 109:99–113. https://doi.org/10.1016/j.ijbiomac.2017.12.048

Koshy RR, Mary SK, Thomas S, Pothan LA (2015) Environment friendly green composites based on soy protein isolate—a review. Food Hydrocoll 50:174–192. https://doi.org/10.1016/j.foodhyd.2015.04.023

Kristo E, Hazizaj A, Corredig M (2012) Structural changes imposed on whey proteins by UV irradiation in a continuous UV light reactor. J Agric Food Chem 60:6204–6209. https://doi.org/10.1021/jf300278k

Kumar R, Ghoshal G, Goyal M (2019) Synthesis and functional properties of gelatin/CA–starch composite film: excellent food packaging material. J Food Sci Technol 56:1954–1965. https://doi.org/10.1007/s13197-019-03662-4

Kurt A, Kahyaoglu T (2014) Characterization of a new biodegradable edible film made from salep glucomannan. Carbohydr Polym 104:50–58. https://doi.org/10.1016/j.carbpol.2014.01.003

Kurt A, Toker OS, Tornuk F (2017) Effect of xanthan and locust bean gum synergistic interaction on characteristics of biodegradable edible film. Int J Biol Macromol 102:1035–1044. https://doi.org/10.1016/j.ijbiomac.2017.04.081

Kwak HW, Woo H, Kim IC, Lee KH (2017) Fish gelatin nanofibers prevent drug crystallization and enable ultrafast delivery. RSC Adv 7:40411–40417. https://doi.org/10.1039/c7ra06433k

Lantano C, Alfieri I, Cavazza A et al (2014) Natamycin based sol-gel antimicrobial coatings on polylactic acid films for food packaging. Food Chem 165:342–347. https://doi.org/10.1016/j.foodchem.2014.05.066

Laroche C, Delattre C, Mati-Baouche N, Salah R, Ursu AV, Moulti-Mati F, Michaud P, Pierre G (2018) Bioactivity of chitosan and its derivatives. Curr Org Chem 22:641–667. https://doi.org/10.2174/1385272821666170811114145
Lee P, Rogers MA (2012) Effect of calcium source and exposure-time on basic caviar spherification using sodium alginate. Int J Gastron Food Sci 1:96–100
Li L, Yuan TZ, Setia R et al (2019) Characteristics of pea, lentil and faba bean starches isolated from air-classified flours in comparison with commercial starches. Food Chem 276:599–607. https://doi.org/10.1016/j.foodchem.2018.10.064
Limpan N, Prodpran T, Benjakul S, Prasarpran S (2012) Influences of degree of hydrolysis and molecular weight of Poly(vinyl alcohol) (PVA) on properties of fish myofibrillar protein/PVA blend films. Food Hydrocoll 29:226–233. https://doi.org/10.1016/j.foodhyd.2012.03.007
Lin HC, Wang BJ, Weng YM (2020) Development and characterization of sodium caseinate edible films cross-linked with genipin. LWT—Food Sci Technol 118:108813. https://doi.org/10.1016/j.lwt.2019.108813
Liu F, Chiou BS, Avena-Bustillos RJ et al (2017) Study of combined effects of glycerol and transglutaminase on properties of gelatin films. Food Hydrocoll 65:1–9. https://doi.org/10.1016/j.foodhyd.2016.10.004
Liu C, An F, He H et al (2018) Pickering emulsions stabilized by compound modified areca taro (*Colocasia esculenta* (L.) Schott) starch with ball-milling and OSA. Colloids Surfaces A Physicochem Eng Asp 556:185–194. https://doi.org/10.1016/j.colsurfa.2018.08.032
Liu F, Chang W, Chen M et al (2020) Film-forming properties of guar gum, tara gum and locust bean gum. Food Hydrocoll 98:105007. https://doi.org/10.1016/j.foodhyd.2019.03.028
López De Dicastillo C, Rodríguez F, Guarda A, Galotto MJ (2016) Antioxidant films based on cross-linked methyl cellulose and native Chilean berry for food packaging applications. Carbohydr Polym 136:1052–1060. https://doi.org/10.1016/j.carbpol.2015.10.013
Lu JW, Zhu YL, Guo ZX et al (2006) Electrospinning of sodium alginate with poly (ethylene oxide). Polymer (Guildf) 47:8026–8031. https://doi.org/10.1016/j.polymer.2006.09.027
Luchese CL, Spada JC, Tessaro IC (2017a) Starch content affects physicochemical properties of corn and cassava starch-based films. Ind Crops Prod 109:619–626. https://doi.org/10.1016/j.indcrop.2017.09.020
Luchese CL, Sperotto N, Spada JC, Tessaro IC (2017b) Effect of blueberry agro-industrial waste addition to corn starch-based films for the production of a pH-indicator film. Int J Biol Macromol 104:11–18. https://doi.org/10.1016/j.ijbiomac.2017.05.149
Luchese CL, Benelli P, Spada JC, Tessaro IC (2018) Impact of the starch source on the physicochemical properties and biodegradability of different starch-based films. J Appl Polym Sci 135:46564. https://doi.org/10.1002/app.46564
Luna SM, Silva SS, Gomes ME et al (2011) Cell adhesion and proliferation onto chitosan-based membranes treated by plasma surface modification. J Biomater Appl 26:101–116. https://doi.org/10.1177/0885328210362924
MahdavianMehr H, Koocheki A (2020) Effect of atmospheric cold plasma on structure, interfacial and emulsifying properties of Grass pea (*Lathyrus sativus* L.) protein isolate. Food Hydrocoll 106:105899. https://doi.org/10.1016/j.foodhyd.2020.105899
Mahmoud KH (2016) Optical properties of hydroxyethyl cellulose film treated with nitrogen plasma. Spectrochim Acta Part A Mol Biomol Spectrosc 157:153–157. https://doi.org/10.1016/j.saa.2015.12.029
Majid I, Ahmad Nayik G, Mohammad Dar S, Nanda V (2018) Novel food packaging technologies: innovations and future prospective. J Saudi Soc Agric Sci 17:454–462. https://doi.org/10.1016/j.jssas.2016.11.003
Makaremi M, Yousefi H, Cavallaro G et al (2019) Safely dissolvable and healable active packaging films based on alginate and pectin. Polymers (Basel) 11:1–18. https://doi.org/10.3390/polym11101594
Manikandan NA, Pakshirajan K, Pugazhenthi G (2020) Preparation and characterization of environmentally safe and highly biodegradable microbial polyhydroxybutyrate (PHB) based

graphene nanocomposites for potential food packaging applications. Int J Biol Macromol 154: 866–877. https://doi.org/10.1016/j.ijbiomac.2020.03.084

Manzanarez-López F, Soto-Valdez H, Auras R, Peralta E (2011) Release of α-Tocopherol from poly(lactic acid) films, and its effect on the oxidative stability of soybean oil. J Food Eng 104: 508–517. https://doi.org/10.1016/j.jfoodeng.2010.12.029

Martău GA, Mihai M, Vodnar DC (2019) The use of chitosan, alginate, and pectin in the biomedical and food sector—biocompatibility, bioadhesiveness, and biodegradability. Polymers (Basel) 11: 1837. https://doi.org/10.3390/polym11111837

Martins C, Vilarinho F, Sanches Silva A et al (2018) Active polylactic acid film incorporated with green tea extract: development, characterization and effectiveness. Ind Crops Prod 123:100–110. https://doi.org/10.1016/j.indcrop.2018.06.056

Meereboer KW, Misra M, Mohanty AK (2020) Review of recent advances in the biodegradability of polyhydroxyalkanoate (PHA) bioplastics and their composites. Green Chem 22:5519–5558. https://doi.org/10.1039/d0gc01647k

Menzel C, González-Martínez C, Chiralt A, Vilaplana F (2019) Antioxidant starch films containing sunflower hull extracts. Carbohydr Polym 214:142–151. https://doi.org/10.1016/j.carbpol.2019.03.022

Meriçer Ç, Minelli M, Angelis MGD et al (2016) Atmospheric plasma assisted PLA/microfibrillated cellulose (MFC) multilayer biocomposite for sustainable barrier application. Ind Crops Prod 93:235–243. https://doi.org/10.1016/j.indcrop.2016.03.020

Miao L, Walton WC, Wang L et al (2019) Characterization of polylactic acids-polyhydroxybutyrate based packaging film with fennel oil, and its application on oysters. Food Packag Shelf Life 22: 100388. https://doi.org/10.1016/j.fpsl.2019.100388

Misra NN, Patil S, Moiseev T et al (2014) In-package atmospheric pressure cold plasma treatment of strawberries. J Food Eng 125:131–138. https://doi.org/10.1016/j.jfoodeng.2013.10.023

Mochizuki M (2009) Synthesis, properties and structure of polylactic acid fibres. In: Handbook of textile fibre structure. Elsevier, pp 257–275

Mohamad N, Mazlan MM, Tawakkal ISMA et al (2020) Development of active agents filled polylactic acid films for food packaging application. Int J Biol Macromol 163:1451–1457. https://doi.org/10.1016/j.ijbiomac.2020.07.209

Mohsen AH, Ali NA (2018) Mechanical, color and barrier, properties of biodegradable nanocomposites polylactic acid/nanoclay. J Bioremediat Biodegrad 9. https://doi.org/10.4172/2155-6199.1000455

Mohsin A, Zhang K, Hu J et al (2018) Optimized biosynthesis of xanthan via effective valorization of orange peels using response surface methodology: a kinetic model approach. Carbohydr Polym 181:793–800. https://doi.org/10.1016/j.carbpol.2017.11.076

Mohsin A, Sun J, Khan IM et al (2019) Sustainable biosynthesis of curdlan from orange waste by using Alcaligenes faecalis: a systematically modeled approach. Carbohydr Polym 205:626–635. https://doi.org/10.1016/j.carbpol.2018.10.047

Mohsin A, Zaman WQ, Guo M et al (2020) Xanthan-Curdlan nexus for synthesizing edible food packaging films. Int J Biol Macromol 162:43–49. https://doi.org/10.1016/j.ijbiomac.2020.06.008

Moon RJ, Martini A, Nairn J et al (2011) Cellulose nanomaterials review: structure, properties and nanocomposites. Chem Soc Rev 40:3941–3994. https://doi.org/10.1039/c0cs00108b

Moosavi MH, Khani MR, Shokri B et al (2020) Modifications of protein-based films using cold plasma. Int J Biol Macromol 142:769–777. https://doi.org/10.1016/j.ijbiomac.2019.10.017

Morent R, Geyter N, Desmet T et al (2011) Plasma surface modification of biodegradable polymers: a review. Plasma Process Polym 8:171–190. https://doi.org/10.1002/ppap.201000153

Mostafavi FS, Zaeim D (2020) Agar-based edible films for food packaging applications—A review. Int J Biol Macromol 159:1165–1176. https://doi.org/10.1016/j.ijbiomac.2020.05.123

Mu R, Hong X, Ni Y et al (2019) Recent trends and applications of cellulose nanocrystals in food industry. Trends Food Sci Technol 93:136–144. https://doi.org/10.1016/j.tifs.2019.09.013

Mujtaba M, Salaberria AM, Andres MA et al (2017) Utilization of flax (*Linumu sitatissimum*) cellulose nanocrystals as reinforcing material for chitosan films. Int J Biol Macromol 104:944–952. https://doi.org/10.1016/j.ijbiomac.2017.06.127
Muneer F, Rasul I, Azeem F et al (2020) Microbial polyhydroxyalkanoates (PHAs): efficient replacement of synthetic polymers. J Polym Environ 28:2301–2323. https://doi.org/10.1007/s10924-020-01772-1
Muñoz LA, Aguilera JM, Rodriguez-Turienzo L et al (2012) Characterization and microstructure of films made from mucilage of Salvia hispanica and whey protein concentrate. J Food Eng 111:511–518. https://doi.org/10.1016/j.jfoodeng.2012.02.031
Murariu M, Dubois P (2016) PLA composites: from production to properties. Adv Drug Deliv Rev 107:17–46. https://doi.org/10.1016/j.addr.2016.04.003
Murmu SB, Mishra HN (2017) Optimization of the Arabic gum based edible coating formulations with sodium caseinate and tulsi extract for guava. LWT—Food Sci Technol 80:271–279. https://doi.org/10.1016/j.lwt.2017.02.018
Murmu SB, Mishra HN (2018) The effect of edible coating based on Arabic gum, sodium caseinate and essential oil of cinnamon and lemon grass on guava. Food Chem 245:820–828. https://doi.org/10.1016/j.foodchem.2017.11.104
Musso YS, Salgado PR, Mauri AN (2019) Smart gelatin films prepared using red cabbage (*Brassica oleracea* L.) extracts as solvent. Food Hydrocoll 89:674–681. https://doi.org/10.1016/j.foodhyd.2018.11.036
Nair M, Best SM, Cameron RE (2020) Crosslinking collagen constructs: achieving cellular selectivity through modifications of physical and chemical properties. Appl Sci 10:1–14. https://doi.org/10.3390/app10196911
Narasimman P, Sethuraman P (2016) An overview on the fundamentals of pectin. Int J Adv Res 4:1855–1860. https://doi.org/10.21474/ijar01/2593
Nayak K, Ansari MT, Sami F et al Natural polysaccharides in drug delivery and biomedical applications. Academic, London
Nechita P, Iana-Roman MR (2020) Review on polysaccharides used in coatings for food packaging papers. Coatings 10:1–24. https://doi.org/10.3390/COATINGS10060566
Nesic AR, Seslija SI (2017) The influence of nanofillers on physical–chemical properties of polysaccharide-based film intended for food packaging. In: Food Packaging, pp 637–697. https://doi.org/10.1016/b978-0-12-804302-8.00019-4
Nešić A, Cabrera-Barjas G, Dimitrijević-Branković S et al (2020) Prospect of polysaccharide-based materials as advanced food packaging. Molecules 25:135. https://doi.org/10.3390/molecules25010135
Neto CGT, Dantas TNC, Fonseca JLC, Pereira MR (2005) Permeability studies in chitosan membranes. Effects of crosslinking and poly (ethylene oxide) addition. Carbohydr Res 340:2630–2636. https://doi.org/10.1016/j.carres.2005.09.011
Niemira BA (2012) Cold plasma decontamination of foods. Annu Rev Food Sci Technol 3:125–142. https://doi.org/10.1146/annurev-food-022811-101132
Nilsuwan K, Benjakul S, Prodpran T (2017) Properties, microstructure and heat seal ability of bilayer films based on fish gelatin and emulsified gelatin films. Food Biophys 12:234–243. https://doi.org/10.1007/s11483-017-9479-2
Nofar M, Sacligil D, Carreau PJ et al (2019) Poly (lactic acid) blends: processing, properties and applications. Int J Biol Macromol 125:307–360. https://doi.org/10.1016/j.ijbiomac.2018.12.002
Nogueira D, Martins VG (2018) Biodegradable bilayer films prepared from individual films of different proteins. J Appl Polym Sci 135:1–13. https://doi.org/10.1002/app.46721
Nogueira D, Martins VG (2019) Use of different proteins to produce biodegradable films and blends. J Polym Environ 27:2027–2039. https://doi.org/10.1007/s10924-019-01494-z
Norcino LB, Mendes JF, Natarelli CVL et al (2020) Pectin films loaded with copaiba oil nanoemulsions for potential use as bio-based active packaging. Food Hydrocoll 106:105862. https://doi.org/10.1016/j.foodhyd.2020.105862

Nur Hanani ZA, Roos YH, Kerry JP (2012) Use of beef, pork and fish gelatin sources in the manufacture of films and assessment of their composition and mechanical properties. Food Hydrocoll 29:144–151. https://doi.org/10.1016/j.foodhyd.2012.01.015

Nur Hanani ZA, Roos YH, Kerry JP (2014) Use and application of gelatin as potential biodegradable packaging materials for food products. Int J Biol Macromol 71:94–102. https://doi.org/10.1016/j.ijbiomac.2014.04.027

Oh YA, Roh SH, Min SC (2016) Cold plasma treatments for improvement of the applicability of defatted soybean meal-based edible film in food packaging. Food Hydrocoll 58:150–159. https://doi.org/10.1016/j.foodhyd.2016.02.022

Ojha S, Fröhling A, Durek J et al (2021) Principles and application of cold plasma in food processing. Innov Food Process Technol 1857:519–540. https://doi.org/10.1016/b978-0-08-100596-5.23033-3

Oliveira MA, Furtado RF, Bastos MSR et al (2018) Performance evaluation of cashew gum and gelatin blend for food packaging. Food Packag Shelf Life 17:57–64. https://doi.org/10.1016/j.fpsl.2018.05.003

Oymaci P, Altinkaya SA (2016) Improvement of barrier and mechanical properties of whey protein isolate based food packaging films by incorporation of zein nanoparticles as a novel bionanocomposite. Food Hydrocoll 54:1–9. https://doi.org/10.1016/j.foodhyd.2015.08.030

Ozcalik O, Tihminlioglu F (2013) Barrier properties of corn zein nanocomposite coated polypropylene films for food packaging applications. J Food Eng 114:505–513. https://doi.org/10.1016/j.jfoodeng.2012.09.005

Paixão LC, Lopes IA, Barros Filho AKD, Santana AA (2019) Alginate biofilms plasticized with hydrophilic and hydrophobic plasticizers for application in food packaging. J Appl Polym Sci 136:1–11. https://doi.org/10.1002/app.48263

Paliwal R, Palakurthi S (2014) Zein in controlled drug delivery and tissue engineering. J Control Release 189:108–122. https://doi.org/10.1016/j.jconrel.2014.06.036

Pankaj SK, Thomas S (2016) Cold plasma applications in food packaging. Elsevier Inc.

Pankaj SK, Bueno-Ferrer C, Misra NN et al (2014a) Applications of cold plasma technology in food packaging. Trends Food Sci Technol 35:5–17. https://doi.org/10.1016/j.tifs.2013.10.009

Pankaj SK, Bueno-Ferrer C, Misra NN et al (2014b) Characterization of polylactic acid films for food packaging as affected by dielectric barrier discharge atmospheric plasma. Innov Food Sci Emerg Technol 21:107–113. https://doi.org/10.1016/j.ifset.2013.10.007

Pankaj SK, Bueno-Ferrer C, Misra NN et al (2014c) Zein film: effects of dielectric barrier discharge atmospheric cold plasma. J Appl Polym Sci 131:9541–9546. https://doi.org/10.1002/app.40803

Pankaj SK, Bueno-Ferrer C, Misra NN et al (2015a) Dielectric barrier discharge atmospheric air plasma treatment of high amylose corn starch films. LWT—Food Sci Technol 63:1076–1082. https://doi.org/10.1016/j.lwt.2015.04.027

Pankaj SK, Bueno-Ferrer C, Misra NN et al (2015b) Characterization of dielectric barrier discharge atmospheric air cold plasma treated gelatin films. Food Packag Shelf Life 6:61–67. https://doi.org/10.1016/j.fpsl.2015.09.002

Park JW, Scott Whiteside W, Cho SY (2008) Mechanical and water vapor barrier properties of extruded and heat-pressed gelatin films. LWT—Food Sci Technol 41:692–700. https://doi.org/10.1016/j.lwt.2007.04.015

Pawar SN, Edgar KJ (2011) Chemical modification of alginates in organic solvent systems. Biomacromolecules 12:4095–4103. https://doi.org/10.1021/bm201152a

Pellá MCG, Lima-Tenório MK, Tenório-Neto ET et al (2018) Chitosan-based hydrogels: from preparation to biomedical applications. Carbohydr Polym 196:233–245. https://doi.org/10.1016/j.carbpol.2018.05.033

Picchio ML, Linck YG, Monti GA et al (2018) Casein films crosslinked by tannic acid for food packaging applications. Food Hydrocoll 84:424–434. https://doi.org/10.1016/j.foodhyd.2018.06.028

Pires C, Ramos C, Teixeira B et al (2013) Hake proteins edible films incorporated with essential oils: physical, mechanical, antioxidant and antibacterial properties. Food Hydrocoll 30:224–231. https://doi.org/10.1016/j.foodhyd.2012.05.019

Priyadarshi R, Sauraj KB, Negi YS (2018) Chitosan film incorporated with citric acid and glycerol as an active packaging material for extension of green chilli shelf life. Carbohydr Polym 195: 329–338. https://doi.org/10.1016/j.carbpol.2018.04.089

Quiroga ALB (2015) Dossiê gomas. Food ingredients Bras 1–22

Rattaya S, Benjakul S, Prodpran T (2009) Properties of fish skin gelatin film incorporated with seaweed extract. J Food Eng 95:151–157. https://doi.org/10.1016/j.jfoodeng.2009.04.022

Rbini K, Boanini E, Menichetti A et al (2020) Quercetin loaded gelatin films with modulated release and tailored anti-oxidant, mechanical and swelling properties. Food Hydrocoll 109:106089. https://doi.org/10.1016/j.foodhyd.2020.106089

Rech CR, Brabes KCS, Silva BEB et al (2020) Biodegradation of polyhydroxybutyrate films incorporated with eugenol in different soil types. Case Stud Chem Environ Eng 2:100014. https://doi.org/10.1016/j.cscee.2020.100014

Reddy N, Yang Y (2013) Thermoplastic films from plant proteins. J Appl Polym Sci 130:729–738. https://doi.org/10.1002/app.39481

Ren L, Yan X, Zhou J et al (2017a) Influence of chitosan concentration on mechanical and barrier properties of corn starch/chitosan films. Int J Biol Macromol 105:1636–1643. https://doi.org/10.1016/j.ijbiomac.2017.02.008

Ren Y, Ding Z, Wang C et al (2017b) Influence of DBD plasma pretreatment on the deposition of chitosan onto UHMWPE fiber surfaces for improvement of adhesion and dyeing properties. Appl Surf Sci 396:1571–1579. https://doi.org/10.1016/j.apsusc.2016.11.215

Rezaee M, Askari G, EmamDjomeh Z, Salami M (2020) UV-irradiated gelatin-chitosan bio-based composite film, physiochemical features and release properties for packaging applications. Int J Biol Macromol 147:990–996. https://doi.org/10.1016/j.ijbiomac.2019.10.066

Rhim JW, Wang LF (2014) Preparation and characterization of carrageenan-based nanocomposite films reinforced with clay mineral and silver nanoparticles. Appl Clay Sci 97–98:174–181. https://doi.org/10.1016/j.clay.2014.05.025

Rhim JW, Park HM, Ha CS (2013) Bio-nanocomposites for food packaging applications. Prog Polym Sci 38:1629–1652. https://doi.org/10.1016/j.progpolymsci.2013.05.008

Rodrigues AAM, Santos LF, Costa RR, et al. (2020) Characterization of starch from different non-traditional sources and its application as coating in 'Palmer' mango fruit. Ciência e Agrotecnologia 44. https://doi.org/10.1590/1413-7054202044011220

Rodriguez-Garcia I, Cruz-Valenzuela MR, Silva-Espinoza BA et al (2016) Oregano (*Lippia graveolens*) essential oil added within pectin edible coatings prevents fungal decay and increases the antioxidant capacity of treated tomatoes. J Sci Food Agric 96:3772–3778. https://doi.org/10.1002/jsfa.7568

Rogovina SZ, Prut EV, Berlin AA (2019) Composite materials based on synthetic polymers reinforced with natural fibers. Polym Sci Ser A 61:417–438. https://doi.org/10.1134/S0965545X19040084

Romani VP, Hernández CP, Martins VG (2018) Pink pepper phenolic compounds incorporation in starch/protein blends and its potential to inhibit apple browning. Food Packag Shelf Life 15: 151–158. https://doi.org/10.1016/j.fpsl.2018.01.003

Romani VP, Olsen B, Pinto Collares M et al (2019) Improvement of fish protein films properties for food packaging through glow discharge plasma application. Food Hydrocoll 87:970–976. https://doi.org/10.1016/j.foodhyd.2018.09.022

Romeih E, Walker G (2017) Recent advances on microbial transglutaminase and dairy application. Trends Food Sci Technol 62:133–140. https://doi.org/10.1016/j.tifs.2017.02.015

Ruan C, Zhang Y, Wang J, Sun Y, Gao X, Xiong G, Liang J (2019) Preparation and antioxidant activity of sodium alginate and carboxymethyl cellulose edible films with epigallocatechin gallate. Int J Biol Macromol 134:1038–1044. https://doi.org/10.1016/j.ijbiomac.2019.05.143

Ruiz-Navajas Y, Viuda-Martos M, Barber X et al (2015) Effect of chitosan edible films added with Thymus moroderi and Thymus piperella essential oil on shelf-life of cooked cured ham. J Food Sci Technol 52:6493–6501. https://doi.org/10.1007/s13197-015-1733-3

Sadhu S, Thirumdas R, Deshmukh RR, Annapure US (2017) Influence of cold plasma on the enzymatic activity in germinating mung beans (Vigna radiata). LWT—Food Sci Technol 78: 97–104. https://doi.org/10.1016/j.lwt.2016.12.026

Salehi F, Kashaninejad M (2015) Static rheological study of *Ocimum basilicum* seed gum. Int J Food Eng 11:97–103. https://doi.org/10.1515/ijfe-2014-0189

Samanta KK, Jassal M, Agrawal AK (2010) Antistatic effect of atmospheric pressure glow discharge cold plasma treatment on textile substrates. Fibers Polym 11:431–437. https://doi.org/10.1007/s12221-010-0431-z

Sapper M, Wilcaso P, Santamarina MP et al (2018) Antifungal and functional properties of starch-gellan films containing thyme (*Thymus zygis*) essential oil. Food Control 92:505–515. https://doi.org/10.1016/j.foodcont.2018.05.004

Saurabh CK, Gupta S, Bahadur J et al (2015) Mechanical and barrier properties of guar gum-based nano-composite films. Carbohydr Polym 124:77–84. https://doi.org/10.1016/j.carbpol.2015.02.004

Scudeler CGS, Costa TL, Cortez-Vega WR et al (2020) Development and characterization of Nile tilapia (*Oreochromis niloticus*) protein isolate-based biopolymer films incorporated with essential oils and nanoclay. Food Packag Shelf Life 25:100542. https://doi.org/10.1016/j.fpsl.2020.100542

Sedayu BB, Cran MJ, Bigger SW (2019) A review of property enhancement techniques for carrageenan-based films and coatings. Carbohydr Polym 216:287–302. https://doi.org/10.1016/j.carbpol.2019.04.021

Segat A, Misra NN, Cullen PJ, Innocente N (2015) Atmospheric pressure cold plasma (ACP) treatment of whey protein isolate model solution. Innov Food Sci Emerg Technol 29:247–254. https://doi.org/10.1016/j.ifset.2015.03.014

SenturkParreidt T, Müller K, Schmid M (2018) Alginate-based edible films and coatings for food packaging applications. Foods 7:170. https://doi.org/10.3390/foods7100170

Seyedi S, Koocheki A, Mohebbi M, Zahedi Y (2014) Lepidium perfoliatum seed gum: a new source of carbohydrate to make a biodegradable film. Carbohydr Polym 101:349–358. https://doi.org/10.1016/j.carbpol.2013.09.072

Shabanpour B, Kazemi M, Ojagh SM, Pourashouri P (2018) Bacterial cellulose nanofibers as reinforce in edible fish myofibrillar protein nanocomposite films. Int J Biol Macromol 117:742–751. https://doi.org/10.1016/j.ijbiomac.2018.05.038

Shah U, Naqash F, Gani A, Masoodi FA (2016) Art and science behind modified starch edible film and coatings: a review. Compr Rev Food Sci Food Saf 15:568–580. https://doi.org/10.1111/1541-4337.12197

Shankar S, Wang LF, Rhim JW (2017) Preparation and properties of carbohydrate-based composite films incorporated with CuO nanoparticles. Carbohydr Polym 169:264–271. https://doi.org/10.1016/j.carbpol.2017.04.025

Sharma S, Singh R k. (2020) Cold plasma treatment of dairy proteins in relation to functionality enhancement. Trends Food Sci Technol 102:30–36. https://doi.org/10.1016/j.tifs.2020.05.013

Siah WM, Aminah A, Ishak A (2015) Edible films from seaweed (*Kappaphycus alvarezii*). Int Food Res J 22:2230–2236

Sionkowska A, Wisniewski M, Skopinska J et al (2006) Thermal and mechanical properties of UV irradiated collagen/chitosan thin films. Polym Degrad Stab 91:3026–3032. https://doi.org/10.1016/j.polymdegradstab.2006.08.009

Sionkowska A, Skopinska-Wisniewska J, Planecka A, Kozlowska J (2010) The influence of UV irradiation on the properties of chitosan films containing keratin. Polym Degrad Stab 95:2486–2491. https://doi.org/10.1016/j.polymdegradstab.2010.08.002

Sionkowska A, Kaczmarek B, Gnatowska M, Kowalonek J (2015) The influence of UV-irradiation on chitosan modified by the tannic acid addition. J Photochem Photobiol B Biol 148:333–339. https://doi.org/10.1016/j.jphotobiol.2015.03.028

Siracusa V (2012) Food packaging permeability behaviour: a report. Int J Polym. https://doi.org/10.1155/2012/302029

Sivakanthan S, Rajendran S, Gamage A et al (2020) Antioxidant and antimicrobial applications of biopolymers: a review. Food Res Int 136:109327. https://doi.org/10.1016/j.foodres.2020.109327
Smith AM, Moxon S, Morris GA (2016) Biopolymers as wound healing materials. Wound Heal Biomater 2:261–287. https://doi.org/10.1016/B978-1-78242-456-7.00013-1
Smithers GW (2008) Whey and whey proteins-from "gutter-to-gold". Int Dairy J 18:695–704. https://doi.org/10.1016/j.idairyj.2008.03.008
Song F, Tang DL, Wang XL, Wang YZ (2011) Biodegradable soy protein isolate-based materials: a review. Biomacromolecules 12:3369–3380. https://doi.org/10.1021/bm200904x
Soni A, Kandeepan G, Mendiratta SK et al (2016) Development and characterization of essential oils incorporated carrageenan based edible film for packaging of chicken patties. Nutr Food Sci 46:82–95. https://doi.org/10.1108/NFS-05-2015-0065
Sousa AMM, Gonçalves MP (2015) Strategies to improve the mechanical strength and water resistance of agar films for food packaging applications. Carbohydr Polym 132:196–204. https://doi.org/10.1016/j.carbpol.2015.06.022
Souza VGL, Fernando AL, Pires JRA et al (2017) Physical properties of chitosan films incorporated with natural antioxidants. Ind Crops Prod 107:565–572. https://doi.org/10.1016/j.indcrop.2017.04.056
Souza VGL, Pires JRA, Vieira ÉT et al (2019) Activity of chitosan-montmorillonite bionanocomposites incorporated with rosemary essential oil: from in vitro assays to application in fresh poultry meat. Food Hydrocoll 89:241–252. https://doi.org/10.1016/j.foodhyd.2018.10.049
Spotti ML, Cecchini JP, Spotti MJ, Carrara CR (2016) Brea Gum (from *Circadian praecox*) as a structural support for emulsion-based edible films. LWT—Food Sci Technol 68:127–134. https://doi.org/10.1016/j.lwt.2015.12.018
Srivastava R, Bano K, Zaheer M, Kuddus M (2018) Bio-based materials for food packaging. Springer, Singapore
Stoll L, Rech R, Flôres SH et al (2019) Poly(acid lactic) films with carotenoids extracts: Release study and effect on sunflower oil preservation. Food Chem 281:213–221. https://doi.org/10.1016/j.foodchem.2018.12.100
Stylianou A, Yova D, Alexandratou E (2014) Investigation of the influence of UV irradiation on collagen thin films by AFM imaging. Mater Sci Eng C 45:455–468. https://doi.org/10.1016/j.msec.2014.09.006
Sun C, Dai L, Gao Y (2017) Interaction and formation mechanism of binary complex between zein and propylene glycol alginate. Carbohydr Polym 157:1638–1649. https://doi.org/10.1016/j.carbpol.2016.11.046
Swaroop C, Shukla M (2018) Mechanical, optical and antibacterial properties of polylactic acid/polyethylene glycol films reinforced with MgO nanoparticles. Mater Today Proc 5:20711–20718
Syahida N, Fitry I, Zuriyati A, Hanani N (2020) Effects of palm wax on the physical, mechanical and water barrier properties of fish gelatin films for food packaging application. Food Packag Shelf Life 23:100437. https://doi.org/10.1016/j.fpsl.2019.100437
Tavares L, Souza HKS, Gonçalves MP, Rocha CMR (2021) Physicochemical and microstructural properties of composite edible film obtained by complex coacervation between chitosan and whey protein isolate. Food Hydrocoll 113:106471. https://doi.org/10.1016/j.foodhyd.2020.106471
Tavassoli-Kafrani E, Shekarchizadeh H, Masoudpour-Behabadi M (2016) Development of edible films and coatings from alginates and carrageenans. Carbohydr Polym 137:360–374. https://doi.org/10.1016/j.carbpol.2015.10.074
Tendero C, Tixier C, Tristant P et al (2006) Atmospheric pressure plasmas: a review. Spectrochim Acta Part B At Spectrosc 61:2–30. https://doi.org/10.1016/j.sab.2005.10.003
Thakur VK, Thakur M (2016) Handbook of sustainable polymers. Jenny Stanford Publishing

Thakur S, Chaudhary J, Sharma B et al (2018) Sustainability of bioplastics: opportunities and challenges. Curr Opin Green Sustain Chem 13:68–75

Tsai MJ, Weng YM (2019) Novel edible composite films fabricated with whey protein isolate and zein: preparation and physicochemical property evaluation. LWT—Food Sci Technol 101:567–574. https://doi.org/10.1016/j.lwt.2018.11.068

Vanin FM, Sobral PJA, Menegalli FC et al (2005) Effects of plasticizers and their concentrations on thermal and functional properties of gelatin-based films. Food Hydrocoll 19:899–907. https://doi.org/10.1016/j.foodhyd.2004.12.003

Vestena M, Gross IP, Müller CMO, Pires ATN (2016) Nanocomposite of poly(lactic acid)/cellulose nanocrystals: effect of CNC content on the polymer crystallization kinetics. J Braz Chem Soc 27:905–911. https://doi.org/10.5935/0103-5053.20150343

Vieira MGA, Silva MA, Santos LO, Beppu MM (2011) Natural-based plasticizers and biopolymer films: a review. Eur Polym J 47:254–263. https://doi.org/10.1016/j.eurpolymj.2010.12.011

Wang HJ, Fu JX, Wang JY (2009) Effect of water vapor on the surface characteristics and cell compatibility of zein films. Colloids Surfaces B Biointerfaces 69:109–115. https://doi.org/10.1016/j.colsurfb.2008.11.015

Wang Z, Zhou J, Wang X et al (2014) The effects of ultrasonic/microwave assisted treatment on the water vapor barrier properties of soybean protein isolate-based oleic acid/stearic acid blend edible films. Food Hydrocoll 35:51–58. https://doi.org/10.1016/j.foodhyd.2013.07.006

Wang H, Hu D, Ma Q, Wang L (2016) Physical and antioxidant properties of flexible soy protein isolate films by incorporating chestnut (*Castanea mollissima*) bur extracts. LWT—Food Sci Technol 71:33–39. https://doi.org/10.1016/j.lwt.2016.03.025

Wang L, Xue J, Zhang Y (2019) Preparation and characterization of curcumin loaded caseinate/zein nanocomposite film using pH-driven method. Ind Crops Prod 130:71–80. https://doi.org/10.1016/j.indcrop.2018.12.072

Wang X, Tang Y, Zhu X et al (2020) Preparation and characterization of polylactic acid/polyaniline/nanocrystalline cellulose nanocomposite films. Int J Biol Macromol 146:1069–1075. https://doi.org/10.1016/j.ijbiomac.2019.09.233

Wihodo M, Moraru CI (2013) Physical and chemical methods used to enhance the structure and mechanical properties of protein films: a review. J Food Eng 114:292–302. https://doi.org/10.1016/j.jfoodeng.2012.08.021

Wu X, Liu A, Wang W, Ye R (2018a) Improved mechanical properties and thermal-stability of collagen fiber-based film by crosslinking with casein, keratin or SPI: effect of crosslinking process and concentrations of proteins. Int J Biol Macromol 109:1319–1328. https://doi.org/10.1016/j.ijbiomac.2017.11.144

Wu Y, Ding W, He Q (2018b) The gelation properties of tara gum blended with κ-carrageenan or xanthan. Food Hydrocoll 77:764–771. https://doi.org/10.1016/j.foodhyd.2017.11.018

Wu C, Sun J, Zheng P et al (2019) Preparation of an intelligent film based on chitosan/oxidized chitin nanocrystals incorporating black rice bran anthocyanins for seafood spoilage monitoring. Carbohydr Polym 222:115006. https://doi.org/10.1016/j.carbpol.2019.115006

Wu X, Liu Q, Luo Y et al (2020) Improved packing performance and structure-stability of casein edible films by dielectric barrier discharges (DBD) cold plasma. Food Packag Shelf Life 24:100471. https://doi.org/10.1016/j.fpsl.2020.100471

Xu H, Canisag H, Mu B, Yang Y (2015) Robust and flexible films from 100% starch cross-linked by biobased disaccharide derivative. ACS Sustain Chem Eng 3:2631–2639. https://doi.org/10.1021/acssuschemeng.5b00353

Xu T, Gao CC, Feng X et al (2019) Structure, physical and antioxidant properties of chitosan-gum arabic edible films incorporated with cinnamon essential oil. Int J Biol Macromol 134:230–236. https://doi.org/10.1016/j.ijbiomac.2019.04.189

Yahyaoui M, Gordobil O, Herrera Díaz R et al (2016) Development of novel antimicrobial films based on poly(lactic acid) and essential oils. React Funct Polym 109:1–8. https://doi.org/10.1016/j.reactfunctpolym.2016.09.001

Yayli D, Turhan S, Saricaoglu FT (2017) Edible packaging film derived from mechanically deboned chicken meat proteins: Effect of transglutaminase on physicochemical properties. Korean J Food Sci Anim Resour 37:635–645. https://doi.org/10.5851/kosfa.2017.37.5.635

Yin YC, Yin SW, Yang XQ et al (2014) Surface modification of sodium caseinate films by zein coatings. Food Hydrocoll 36:1–8. https://doi.org/10.1016/j.foodhyd.2013.08.027

Yong Cho S, Park JW, Rhee C (2002) Properties of laminated films from whey powder and sodium caseinate mixtures and zein layers. LWT—Food Sci Technol 35:135–139. https://doi.org/10.1006/fstl.2001.0826

Younes I, Rinaudo M (2015) Chitin and chitosan preparation from marine sources. Structure, properties and applications. Mar Drugs 13:1133–1174. https://doi.org/10.3390/md13031133

Zafar R, Zia KM, Tabasum S et al (2016) Polysaccharide based bionanocomposites, properties and applications: a review. Int J Biol Macromol 92:1–13. https://doi.org/10.1016/j.ijbiomac.2016.07.102

Zhang L, Huang C, Xu Y et al (2020a) Synthesis and characterization of antibacterial polylactic acid film incorporated with cinnamaldehyde inclusions for fruit packaging. Int J Biol Macromol 164:4547–4555. https://doi.org/10.1016/j.ijbiomac.2020.09.065

Zhang X, Zhao Y, Li Y et al (2020b) Physicochemical, mechanical and structural properties of composite edible films based on whey protein isolate/psyllium seed gum. Int J Biol Macromol 153:892–901. https://doi.org/10.1016/j.ijbiomac.2020.03.018

Zhong Y, Godwin P, Jin Y, Xiao H (2020) Biodegradable polymers and green-based antimicrobial packaging materials: a mini-review. Adv Ind Eng Polym Res 3:27–35. https://doi.org/10.1016/j.aiepr.2019.11.002

# Smart Freshness Indicator for Animal-Based Product Packaging: Current Status

# 5

Pimonpan Kaewprachu, Samart Sai-Ut, and Saroat Rawdkuen

**Abstract**

Freshness is one of the most significant attributes for monitoring the quality of animal-based products. During product supply chain, freshness of animal-based products decreases, along with the product quality and safety. In order to meet the consumer preferences for the fresh and safe products, smart freshness indicator has been developed. Smart freshness indicator is a device that can indicate degree of freshness and provide directly information about the quality of product based on microbial growth or metabolites in packaged products through direct color visual change. The application of smart freshness indicator has become increasingly interesting because it can provide convenience, enhance safety, and provide information to consumer. According to its benefits, much research is published every year focusing on evaluation of animal-based product freshness using freshness indicator. Therefore, smart freshness indicator could be helpful in assuring the quality and safety of packaged products.

**Keywords**

Animal-based products · Freshness indicator · Quality · Smart packaging

P. Kaewprachu
College of Maritime Studies and Management, Chiang Mai University, Samut Sakhon, Thailand

S. Sai-Ut
Department of Food Science, Faculty of Science, Burapha University, Chonburi, Thailand

S. Rawdkuen (✉)
Unit of Innovative Food Packaging and Biomaterials, School of Agro-Industry, Mae Fah Luang University, Chiang Rai, Thailand
e-mail: saroat@mfu.ac.th





A. K. Shukla (ed.), *Food Packaging: The Smarter Way*,
https://doi.org/10.1007/978-981-16-7196-8_5

## 5.1 Introduction

Animal-based products (meat, poultry, fish, and seafood) are classified as highly perishable foods. They are rich in nitrogenous substances, carbohydrates, and lipids with a pH of $\geq 6$ and water activity ($a_w$) greater than 0.97 (Erkmen and Bozoglu 2016). These factors are favorable for microorganism growth (bacteria, yeasts, and molds) and biochemical reactions. Microorganisms can cause unacceptable changes with the formation of various metabolites in animal-based products. Microbial-generated metabolites (e.g., biogenic amines (e.g., cadaverine, histamine, putrescine, tyramine), carbon dioxide, ethanol, organic acids, sulfuric compounds (hydrogen sulfide), and volatile nitrogen compounds (ammonia, dimethylamine, and trimethylamine described as total volatile basic nitrogen; TVB-N)) are mostly found in animal-based products (Rukchon et al. 2014; Smolander 2003). As is known, the spoilage process always generates during storage, transport, and retailing, and in consumers' homes, which is correlated with the organoleptic changes. However, it is hard to believe decision by evaluating the organoleptic analysis of products and traditional quality analysis method has complicated and time-consuming procedures. Thus, the development of new technologies with accurate, convenient, cost-effective, rapid, and nondestructive method to evaluate quality and safety of animal-based products is a great needed.

Intelligent or smart packaging is defined as packaging that contains an internal or external indicator to provide information about aspects of the history of the package and/or the product quality directly (Robertson 2012). Smart packaging has been currently focused because it has a function for freshness indication. A smart freshness indicator can be made from biopolymer material (solid support) incorporated with dye indicator (chemical and natural dye indicators). The consumers are currently concerned about the use of chemical dye indicators in food packaging system because of their possible toxicity and potential risk to human health. So, its use is restricted as dye indicator. Many natural dye indicators have been effectively incorporated into biopolymer films. Some include anthocyanin (Liu et al. 2019; Rawdkuen et al. 2020; Zeng et al. 2019), betalains (Kanatt 2020; Qin et al. 2020; Yao et al. 2020), and curcumin (Chen et al. 2020; Liu et al. 2018). Smart freshness indicators are able to indicate the product quality directly by changes in color in the real time, resulting from the reaction between the presence of generated metabolites and the smart freshness indicators during loss of freshness. Thus, smart freshness indicators are a promising tool to inform product quality and safety to consumers and suppliers via real-time and rapid qualitative analysis (through visible color changes by naked eye) at any given time. Recently, many research studies have successfully used natural dye indicators and then applied in a real food system. For example, smart freshness indicators can be used as an effective tool for the real-time monitoring of meat, poultry, fish, and seafood products during storage (Chen et al. 2019; Dudnyk et al. 2018; Mohammadalinejhad et al. 2020; Moradi et al. 2019).

In this chapter, the compounds indicating the quality of packaged animal-based products are presented. Subsequently, the concept of smart freshness indicators and the potential sources of dye indicators are reviewed. The current researches on the applications of smart packaging with a function of freshness indication on various animal-based products are also discussed.

## 5.2 Compounds Indicating the Quality of Packaged Animal-Based Products

Freshness is a critical variable that affects the organoleptic properties and food product acceptability (Cardello and Schutz 2003). The freshness of animal-based products is reduced in time as a result of microbial growth or chemical changes. Loss of freshness indicates that animal-based products have started to spoil. Difference in formation of metabolites in animal-based products is affected by the nature of product, associated spoilage flora, packaging system, and storage conditions (Kerry 2012; Smolander 2003). During storage, transport, and retailing, various metabolites are generated in animal-based products such as biogenic amines (e.g., cadaverine, histamine, putrescine, tyramine), carbon dioxide ($CO_2$), ethanol, organic acids, sulfuric compounds [hydrogen sulfide ($H_2S$)], and volatile nitrogen compounds [ammonia ($NH_3$), dimethylamine (DMA; $(CH_3)_2NH$), and trimethylamine (TMA; $(CH_3)_3N$) described as total volatile basic nitrogen (TVB-N)] due to microbial growth and metabolism (Rukchon et al. 2014; Smolander 2003). Animal-based products are composed of free amino acids; proteins in animal-based products are broken down into amino acids by hydrolysis. As a result, they can be partially or totally degraded into simple compounds such as $CO_2$, $H_2O$, $NH_3$, and $H_2S$ (Rukchon et al. 2014). Currently, most research studies have developed metabolite indicators associated with animal-based products such as $CO_2$, $H_2S$, and TVB-N produced during storage by microbial degradation of protein-rich foods. Some of the metabolites or compounds indicating the quality of packaged animal-based products representing potential target marker metabolites for the smart freshness indicators are discussed in detail.

### 5.2.1 Biogenic Amines

Biogenic amines are known as a toxic substance. They have been employed as indicators of hygienic quality, freshness quality index, and control the processing of muscle-based products (Ruiz-Capillas and Herrero 2019; Smolander 2003). The most important biogenic amines substance found in food are cadaverine, histamine, putrescine, and tyramine. Fish and fishery products are abundant in histamine. The hazard potential of biogenic amines should be considered due to their physiological and toxicological effects, and risk to human health. In the USA, Food and Drug Administration (FDA) has set histamine limits in fish and fishery products in general at 50 mg/kg (Ruiz-Capillas and Herrero 2019). Kung et al. (2017) studied the effect of polyethylene packaging (PEP) (in air) and vacuum packaging (VP) on histamine changes in milkfish sticks during storage. They found that milkfish sticks packaged with VP had lower levels of aerobic plate count (APC), TVB-N, and histamine than milkfish sticks packaged with PEP.

### 5.2.2 Carbon Dioxide

Carbon dioxide ($CO_2$) is mainly generated during microbial growth (Smolander 2003). It can be used to indicate the quality deterioration. Increased $CO_2$ concentrations are frequently accompanied by increasing the storage time (Lee et al. 2019). However, the indication of microbial growth by $CO_2$ may be difficult in some packaging technology system like modified atmosphere packaging (MAP) because MAP already containing high concentration of $CO_2$ (typically 20–80%) (Kerry 2012). Lee et al. (2019) developed a freshness indicator based on poly(ether-block-amide) incorporated with bromocresol green for the real-time monitoring of quality changes in chicken breasts during storage at 4 and 10 °C for 10 days. They found that changes in color (from yellow to green) of freshness indicator were correlated with $CO_2$, TVB-N, and bacterial growth in the chicken breast.

### 5.2.3 Organic Acids

Organic acids, such as lactic acid and acetic acid, are the main compounds having a role in glucose fermentation by lactic acid bacteria (Smolander 2003). Zhao et al. (2019) found that the concentration of lactic acid, acetic acid, and succinic acid in tilapia fillet increased with increasing storage time. They also suggested that organic acids were related to fish spoilage.

### 5.2.4 Ethanol

Ethanol is another major end product of fermentative metabolism of lactic acid bacteria. Zhao et al. (2019) found an increase in the amount of ethanol concentration in tilapia fillet during refrigeration storage. The formation of ethanol can contribute to unfavorable fishy odor.

### 5.2.5 Sulfuric Compounds

Hydrogen sulfide ($H_2S$) is produced from cysteine. It forms a green pigment, sulfmyoglobin, when it is bound to myoglobin (Smolander 2003). It has a remarkable effect on the sensory quality of animal-based products due to their off-flavor and low odor threshold. $H_2S$ and other sulfuric compounds have been found to be produced during the spoilage of meat, poultry, and seafood by a number of bacterial species. *Shewanella* are $H_2S$-producing and responsible for seafood spoilage markers when stored at low temperature (Zhu et al. 2015).

### 5.2.6 Volatile Nitrogen Compounds

Volatile nitrogen compounds are ammonia, dimethylamine, and trimethylamine, which are responsible for quality and spoilage of fish (Kung et al. 2017). These compounds (ammonia, dimethylamine, and trimethylamine) are described as TVB-N. TVB-N is formed during storage due to endogenous enzymes and microbial activities, resulting in the formation of compounds. TVB-N increases as the fish freshness decreases. Currently, the use of anthocyanin as natural dye indicator for the determination of volatile nitrogen compounds has been suggested in many studies (Chen et al. 2020; Dudnyk et al. 2018; Kang et al. 2020).

## 5.3 Smart Freshness Indicators

Indicators are tools or devices that provide information about the absence or presence of substance or the degree of reaction between two or more substances by changing characteristic like color (Hogan and Kerry 2008). Most often, they provide qualitative information through directly visible color change (different product's status exhibits differences in color intensities). Indicators can be divided into three categories of time–temperature indicators, gas indicators, and freshness indicators. According to Robertson (2012), all of the indicators are group of product quality and value-improving systems, which are frequently used for intelligent food packaging applications.

Smart freshness indicators are produced in the form of labels or tags, and then attached inside the packaging materials. They have been developed with the aim of the real-time monitoring of the quality and safety of packaged food products through visible color changes, which can be directly observed by the naked eye. Changes in color of indicators should be noticeable even at the minimal spoilage level of the packaged food products. The concept of smart freshness indicators is based on changes in color of the indicators due to the presence of microbial-generated metabolites during loss of freshness. The smart freshness indicators accurately track the increase in concentration of metabolites in the package headspace. They may share information about the packaged animal-based product quality and safety directly to the consumers and suppliers at any given time. They can also alert the consumers and suppliers about the degree of freshness. Thus, the consumers and suppliers can decide that the packaged product should be distributed or retailed or consumed or discarded (Fig. 5.1). The major advantages of smart freshness indicator are sensitive, can be detected by naked eye, and are easy to be integrated into packaging system. However, the disadvantage of smart freshness indicators is false-negative results. This problem should be resolved before widespread commercial uptake to prevent producers from adopting indicators in the use of real situation. Furthermore, an exact correlation between target metabolites, product type, and organoleptic quality and safety is necessary to avoid false negatives (Smolander 2003). Currently, smart freshness indicators have been proposed, for example, for $CO_2$ and TVB-N.

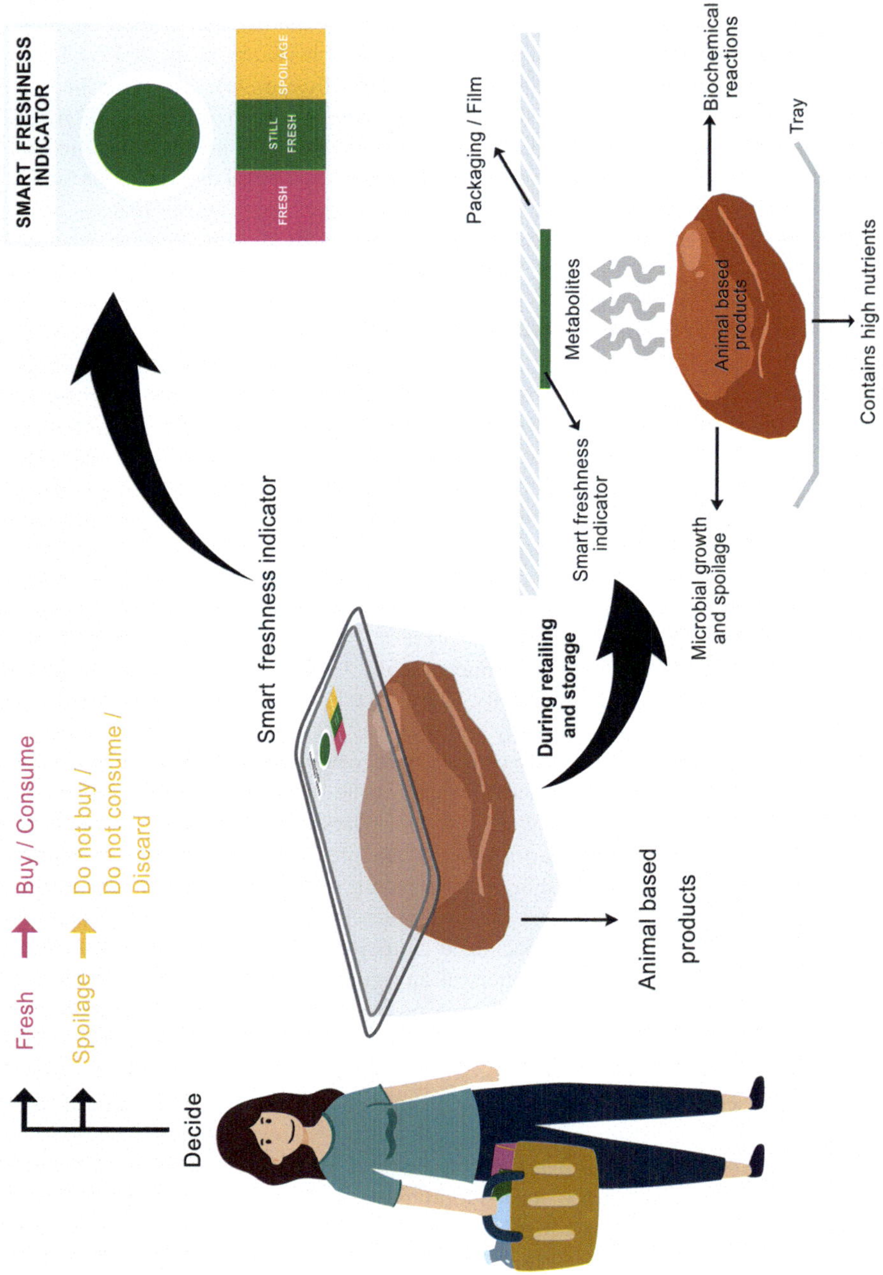


**Fig. 5.1** Concept of smart freshness indicator

## 5.4 Sources of Dye Indicators

There are two major sources of dye indicators: chemical dye indicators and natural dye indicators. Currently, the consumer preferences have shifted the attention of dye indicators from chemical to natural because natural dye indicators are safe, renewable, and have no harmful effects of the human health. Dye indicators are generally added to packaging material to produce intelligent packaging. They can alert the quality and safety of packaged products to consumer during retailing and storage through visible color changes.

### 5.4.1 Chemical Dye Indicators

Chemical dye indicators are also commonly known as acid–base indicators or pH indicators, belonging to two different families: Sulfonephthalein and Azo (Pastore et al. 2021). Most of these indicators are organic and synthetic origin. They are dye substances, which can change color as a result of exposure to different conditions (e.g., variation of pH and concentration of gases) (Pastore et al. 2021). Most of them are soluble in organic solvent but insoluble in water. They have been widely used in the industry such as food packaging industry (pH sensor device for monitoring food spoilage), cosmetics, detergent industry, and textile industry. However, the current trend of chemical dye indicators is not acceptable due to their toxicity (e.g., acute toxicity, carcinogenicity, mutagenicity, and genotoxicity), which may affect consumer's health (Sabnis 2007). The consumer preferences have shifted the attention of manufacturers from chemical to natural dye indicators. They are also limited in the pH range because they showed very narrow color, which are difficult to observe by the naked eye. For example, bromocresol green has only three color transitions: yellow (pH 3.8), green (pH range 3.8–5.4), and blue (pH 5.4). Thus, the combination of two or more chemical dye indicators is needed for effectively enhancing their pH indication. Example and color changes in chemical dye indicators frequently used in the intelligent packaging are summarized in Table 5.1.

### 5.4.2 Natural Dye Indicators

Recently, there is becoming increasingly important in smart freshness indicator made up of natural dye indicator and natural biopolymer materials, owing to their nontoxic material, safe, eco-friendly, renewable, and biodegradable that might replace chemical one. Various compounds that found in natural source have interesting potential for use as dye indicators. Currently, many research studies focus on natural dye pH sensing indicators such as anthocyanin, betalains, and curcumin due to their nontoxic, renewable, and safe. All of these compounds could be a good source of dye indicator material. Many studies have demonstrated a promising of natural dye indicators for potential development of smart packaging systems (Table 5.2).

**Table 5.1** Examples and color changes in chemical dye indicators

| Chemical dye indicators | pH range | Indication (color changes) |
|---|---|---|
| Alizarin red | 5.5–6.8<br>10.1–12.1 | From yellow to red<br>From red to purple |
| Bromocresol green | 3.8–5.4 | From yellow to blue |
| Bromocresol purple | 5.2–6.8 | From yellow to purple |
| Bromophenol blue | 3.0–4.6 | From yellow to blue-purple |
| Bromophenol red | 5.2–6.8 | From yellow to red |
| Bromothymol blue | 6.0–7.6 | From yellow to blue |
| *m*-Cresol purple | 1.2–2.8<br>7.4–9.0 | From red to yellow<br>From yellow to purple |
| *o*-Cresol red | 0.2–1.8<br>7.0–8.8 | From red to yellow<br>From yellow to reddish purple |
| Methyl orange | 3.0–4.4 | From red to yellow |
| Methyl red | 4.4–6.2 | From red to yellow |
| Methyl yellow | 2.9–4.0 | From red to yellow |
| Phenolphthalein | 8.0–10.0 | From colorless to pink |
| Phenol red | 6.8–8.4 | From yellow to red |
| Resazurin | 3.8–6.5 | From orange to purple-violet |
| Thymol blue | 1.2–2.8<br>8.0–9.6 | From red to yellow<br>From yellow to blue |

Source: Sabnis (2007)

Anthocyanin is water-soluble phenolic pigments, belonging to the family flavonoids. It is responsible for red, pink, purple, and blue color of plants. It can be extracted from barberry (Alizadeh-Sani et al. 2021), butterfly pea (Rawdkuen et al. 2020), mulberry (Zeng et al. 2019; Zhang et al. 2020), purple sweet potato (Choi et al. 2017; Jiang et al. 2020), red cabbage (Dudnyk et al. 2018; Liu et al. 2021; Vo et al. 2019), and roselle (Zhang et al. 2019). Color changes in anthocyanin are mostly due to modification of chemical structure of phenolic substances. The structure of anthocyanin changes at different pH values (Miguel 2011). Anthocyanin can change its color with wide ranges when the pH changes. Butterfly pea anthocyanin extract solution had red color, pink color, violet or purple color, blue color, and turned to green color when the pH of butterfly pea anthocyanin extract solution was 2, 3, 4–6, 7, and 7–12, respectively (Rawdkuen et al. 2020). Red cabbage, sweet potatoes, butterfly pea, mangosteen, and mapring seed have been reported to be suitable natural sources of anthocyanin extraction and can be added into biopolymer films for developing smart indicator (Rawdkuen et al. 2020). Mohammadalinejhad et al. (2020) suggested that the film based on bacterial cellulose with *Echium amoenum* anthocyanin is highly sensitive to detect TVB-N by visually distinguishable color changes (fresh: violet, use soon: gray, spoiled: yellow) of shrimp and the incorporation of *Echium amoenum* anthocyanin can be beneficial for monitoring protein-rich food spoilage.

Betalains are water-soluble pigments, which are composed of a nitrogenous core structure of betalamic acid [4-(2-oxoethylidene)-1,2,3,4-tetrahydropyridine-2,6-

**Table 5.2** Examples and color changes in natural dye indicators

| Sources | pH range | Indication (color changes) | References |
|---|---|---|---|
| *Anthocyanin* | | | |
| Mulberry | 2.0–11.0 | From bright red to dark green | Zeng et al. (2019) |
| Butterfly pea | 2.0–12.0 | From red to green | Rawdkuen et al. (2020) |
| Purple sweet potato | 2.0–12.0 | From red to green and yellow | Jiang et al. (2020) |
| Jambolan (*Syzygium cumini*) fruit | 1.0–13.0 | From red to yellow | Merz et al. (2020) |
| *Echium amoenum* | 2.0–12.0 | From red to yellow | Mohammadalinejhad et al. (2020) |
| Black carrot | 2.0–11.0 | From red to gray | Moradi et al. (2019) |
| *Lycium ruthenicum* Murr. | 2.0–10.0 | From pink to yellow | Liu et al. (2019) |
| *Betalains* | | | |
| Red pitaya (*Hylocereus polyrhizus*) peel | 3.0–12.0 | From red to yellow | Qin et al. (2020) |
| *Amaranthus* leaf | 1.0–9.0 | From pink to yellow | Kanatt (2020) |
| Cactus pears (*Opuntia ficus-indica*) | 3.0–12.0 | From purple/red to orange/yellow | Yao et al. (2020) |
| *Curcumin* | | | |
| Turmeric | 3.0–10.0 | From yellow to red and then reddish brown | Liu et al. (2018) |
| Turmeric | 5.0–11.0 | From yellow to light yellow and then reddish brown | Chen et al. (2020) |

dicarboxylic acid] (Rahimi et al. 2019). Betalamic acid can either condense with imino compounds (cyclo-DOPA) to form red/violet betacyanins, or with amines and their derivatives to form yellow/orange betaxanthins (Prieto-Santiago et al. 2020). They are obtained from plants such as amaranth, red pitaya, red beetroot, cactus pears, and prickly pear (Kanatt 2020; Qin et al. 2020; Rahimi et al. 2019; Yao et al. 2020). In alkaline conditions, they have variations in color from red to yellow and their structures are largely stable in the pH range of 3–7 (Qin et al. 2020). Betalains have broader pH stability than anthocyanins (Kanatt 2020). They also possess various biological properties such as antioxidant, antimicrobial, anti-inflammatory, anticancer, and anti-lipidemic (Kanatt 2020; Prieto-Santiago et al. 2020). Betalains have a potential for being used as dye indicator in intelligent packaging. Qin et al. (2020) developed intelligent film from red pitaya (*Hylocereus polyrhizus*) peel extract incorporated in starch/polyvinyl alcohol film. They reported that developed intelligent film could be effectively presented as visual color changes and could be used as intelligent film for monitoring food freshness. Kanatt (2020) suggested that the incorporation of betalains extracted from *Amaranthus* leaf into polyvinyl alcohol/gelatin film could help to alert the consumer to know the food begins to spoil due to visible color changes (from red to yellow).

Curcumin is a natural polyphenol compound found in the rhizome of *Curcuma longa* (turmeric) and in others *Curcuma* spp. (Hewlings and Kalman 2017). It represents the main curcuminoid of turmeric and is responsible for the intense yellow color. It is widely used as a spice, medicinal herb, and colorant. It has several pharmacological activities such as antimicrobial, antioxidant, anti-inflammatory, and anticancer properties (Hewlings and Kalman 2017; Liu et al. 2018). Liu et al. (2018) studied the color changes in curcumin solutions in a wide pH range (3–10) and found that the color of curcumin solutions changed from bright yellow to reddish brown with increasing pH values. However, the color changes in curcumin solutions are difficult to distinguish in pH range from 3 to 7 and may affect to judge freshness of product accurately. Chen et al. (2020) developed freshness indicator based on starch/polyvinyl alcohol containing curcumin, anthocyanin, and mixture of curcumin and anthocyanin. The films containing curcumin had greater color stability than the films incorporated with anthocyanin when stored at 25 °C for 180 days. They also reported that the incorporation of a mixture of curcumin and anthocyanin in a ratio of 2:8 (v/v) into films exhibited good color indication and significant change in color for fish freshness during storage at 4 °C for 10 days.

## 5.5 Applications of Smart Freshness Indicator

The quality and safety of packaged animal-based products continuously decrease with increasing storage time. Freshness is basic characteristic for judging the quality and safety of packaged animal-based products. The loss of packaged product freshness means packaged product has begun to spoil. Animal-based products are classified as highly perishable foods due to their richness in nutrients, resulting in susceptibility to microbial contamination and biochemical reactions. The metabolites generated by spoilage microorganisms affect quality and safety of packaged products. Recently, the consumers are more concerned about animal-based product quality, safety, and their health. Consumers generally know the freshness of food products through checking "best before" or "expiration date" that label on the package or sensory method. However, it is hard to believe decision by evaluating the overall appearance or texture of animal-based products. In recent years, so-called smart freshness indicator has emerged as an alternative method for monitoring animal-based product freshness. Thus, smart freshness indicator for the real-time evaluation of highly perishable food quality like meat, poultry, fish, and seafood is needed.

### 5.5.1 Meat Products

Meat products are extremely highly perishable food. They have a very short shelf life, usually 3–4 days at refrigerated storage, owing to microbial contamination and biochemical reactions. They contain adequate amounts of essential nutrients, which support the growth of microorganisms. Various metabolites are produced by microorganism associated with food spoilage, leading to off-flavor, off-odor, discoloration, and nutritional loss. These effects result in pH variation in the packaged meat products. Generally, it is difficult to detect spoilage of packaged meat through the naked eye. Thus, meat product industry requires rapid, inexpensive, simple, and nondestructive tool or device for naked eye and real-time monitoring of the degree of freshness of packaged meat products. Smart freshness indicator is an effective tool for evaluating packaged meat quality and safety. Recently, various natural dye indicators are increasingly interested in developing smart freshness indicator such as purple sweet potato (Choi et al. 2017), barberry (Alizadeh-Sani et al. 2021), roselle (Zhang et al. 2019), and red cabbage (Dudnyk et al. 2018; Liu et al. 2021; Vo et al. 2019). These natural dye indicators can effectively sense packaged meat product spoilage, exhibiting through visual color changes. Thus, the consumer can know information about the packaged meat quality and safety by the naked eye through the visible color changes in smart freshness indicator on the package. Several smart freshness indicators have been developed with the aim of monitoring meat product freshness (Table 5.3).

### 5.5.2 Poultry Products

Poultry products are classified as highly perishable food. They usually deteriorate rapidly (within 1 week of slaughter) because they are susceptible to physicochemical and biological changes. They are rich in proteins (made up of 18%) and are composed of approximately 75.5% water (Cutter et al. 2012). The presence of water in poultry tissue affords microorganism with another component to support microbial growth. Spoilage in packaged poultry products is mainly due to the growth of microorganisms such as *Escherichia coli*, *Staphylococcus aureus*, *Campylobacter* sp., *Salmonella* sp., *Pseudomonas* sp., and *Listeria monocytogenes*. Furthermore, the growth of microorganisms can produce a variety of metabolites such as $CO_2$, $NH_3$, and $H_2S$ during storage and negatively affect organoleptic characteristics and nutritional value of packaged poultry products. Thus, smart freshness indicator for the real-time evaluating of poultry product quality and safety is needed. Many studies have demonstrated the development of smart freshness indicator for monitoring poultry product quality and safety (Table 5.4).

### 5.5.3 Fish and Seafood Products

Fish and seafood products are highly susceptible to microbial and enzymatic deterioration, owing to their richness in protein nitrogenous substances, non-protein nitrogenous substances, polyunsaturated fatty acids, and other growth factors. A major characteristic in monitoring packaged fish and seafood products is freshness, which can be investigated by releasing TVB-N. During fish and seafood product spoilage, volatile amines, such as TMA, $NH_3$, and DMA, are generated by microorganisms and are responsible for the fishy odor and flavor. In general,

**Table 5.3** Smart freshness indicators for monitoring meat product quality and safety

| Meat products | Packaging/film materials | Dye indicator | Marker | Indication (color changes) | References |
|---|---|---|---|---|---|
| Lamb meat | Methylcellulose/ chitosan nanofiber | Barberry anthocyanin | TVB-N | From pink to pale green and turned to yellow | Alizadeh-Sani et al. (2021) |
| Pork | Cellulose | Naphthoquinone dyes extracted from *Arnebia euchroma* | TVB-N | From magenta to purple | Dong et al. (2020) |
| Pork | Methylcellulose | Mixture of bromothymol blue and methyl red | TVB-N | From red to goldenrod and turned to green | Chen et al. (2019) |
| Pork | Starch/polyvinyl alcohol | Roselle anthocyanin | TVB-N | From red to green and turned to yellow | Zhang et al. (2019) |
| Pork | Polyvinyl alcohol/sodium carboxymethyl cellulose | Red cabbage anthocyanin | TVB-N | From red to blue-green | Liu et al. (2021) |
| Pork | Cellulose fibers | Bromothymol blue | TVB-N | From yellow to green | Cao et al. (2019a, b) |
| Pork | Agar/potato starch | Purple sweet potato anthocyanin | TVB-N | From red to green | Choi et al. (2017) |
| Pork | Chitosan/ polyvinyl alcohol | Red cabbage anthocyanin | TVB-N | From translucently sea green to pink and turned to yellowish with pale green | Vo et al. (2019) |
| Beef | Cellulose acetate | Methyl red | TVB-N | From red to yellow | Lee and Shin (2019) |
| Beef | Filter paper | Methyl red<br>Bromocresol purple | TVB-N | From red to yellow<br>From yellow to purple | Kuswandi and Nurfawaidi (2017) |
| Beef | Pectin | Red cabbage extract | TVB-N | From purple to yellow | Dudnyk et al. (2018) |
| Minced beef | Cellulose–chitosan | Alizarin | TVB-N | From yellow to brown and turned to purple | Ezati et al. (2019a, b) |

*TVB-N* total volatile basic nitrogen

increased levels of TVB-N are accompanied by increased pH in the packaged fish and seafood products as a result of bacterial metabolism. Moradi et al. (2019) developed smart freshness indicator from bacterial cellulose nanofibers incorporated with black carrot anthocyanin for monitoring the freshness of rainbow trout and

**Table 5.4** Smart freshness indicators for monitoring poultry product quality and safety

| Poultry products | Packaging/film materials | Dye indicator | Marker | Indication (color changes) | References |
|---|---|---|---|---|---|
| Chicken breast | Sugarcane bagasse nanocellulose | Bromothymol blue/ methyl red | $CO_2$ | From green to red | Lu et al. (2020) |
| Chicken breast | Poly(ether-block-amide) | Alizarin Red S<br>Bromocresol green<br>Bromophenol blue<br>Bromothymol blue<br>m-Cresol purple<br>Cresol red<br>Curcumin<br>Thymol blue | TVB-N and $CO_2$ | From red to yellow<br>From yellow to blue<br>From yellow to blue<br>No color change<br>From red to yellow<br>From red to yellow<br>No color change<br>From red to yellow | Lee et al. (2019) |
| Chicken breast | Chitosan/ polyethylene oxide | Curcumin | TVB-N | From bright yellow to red | Yildiz et al. (2021) |
| Chicken patties | Filter paper strips | Jamun fruit (*Syzygium cumini*) skin extract | TVB-N | From violet to yellow | Talukder et al. (2020) |
| Chicken | Cellulose fibers | Bromothymol blue | TVB-N | From yellow to green | Cao et al. (2019a, b) |
| Chicken | Agarose | Bromocresol purple | TVB-N | From light yellow to purple | Soni et al. (2018) |
| Chicken fillet | Pectin | Red cabbage extract | TVB-N | From purple to yellow | Dudnyk et al. (2018) |
| Chicken | Polyvinyl alcohol/gelatin | *Amaranthus* leaf extract | TVB-N | From red to yellow | Kanatt (2020) |

*$CO_2$* carbon dioxide, *TVB-N* total volatile basic nitrogen

common carp fillet during storage at 4 °C for 15 days. They reported that freshness indicator showed accurately sensed fish fillet spoilage by distinguishing color changes at different stages. During different freshness stages (fresh, best to eat, and spoiled), the indicator had deep carmine color, charm pink color, and jelly bean blue and khaki colors, respectively. Many studies have demonstrated the development of smart freshness indicator for monitoring fish and seafood product quality and safety (Table 5.5).

**Table 5.5** Smart freshness indicators for monitoring fish and seafood product quality and safety

| Fish and seafood products | Packaging/film materials | Dye indicator | Marker | Indication (color changes) | References |
|---|---|---|---|---|---|
| Fish (mud carp) | Gelatin/polyvinyl alcohol | Mulberry anthocyanin extracts | TVB-N | From purple to gray-purple and turned to dark green | Zeng et al. (2019) |
| Fish (grass crap) | Carboxymethyl cellulose/starch | Purple sweet potato anthocyanin | TVB-N | From pink to purple and turned to blue | Jiang et al. (2020) |
| Fish (Wuchang bream) | Agar | *Arnebia euchroma* root extracts | TVB-N | From pink to purple | Huang et al. (2019) |
| Fish (*Cyprinus carpio var. specularis*) | Sodium carboxymethyl starch/κ-carrageenan | Mulberry anthocyanin | TVB-N | From red to dark blue | Zhang et al. (2020) |
| Fish | Starch/polyvinyl alcohol | Mixture of curcumin and anthocyanin | TVB-N | From yellow to light yellow and turned to reddish brown | Chen et al. (2020) |
| Rainbow trout fillet | Starch/cellulose | Alizarin | TVB-N | From orange to reddish brown | Ezati et al. (2019a, b) |
| Rainbow trout and common carp fillet | Bacterial cellulose nanofibers | Black carrot anthocyanin | TVB-N | From deep carmine to charm pink and turned to jelly bean blue | Moradi et al. (2019) |
| Shrimp | Chitosan/polyvinyl alcohol | Anthocyanin from jambolan (*Syzygium cumini*) fruit | TVB-N | From red to blue | Merz et al. (2020) |
| Shrimp | Bacterial cellulose | *Echium amoenum* anthocyanin | TVB-N | From violet to gray and turned to yellow | Mohammadalinejhad et al. (2020) |
| Shrimp | Starch/polyvinyl alcohol | Red pitaya (*Hylocereus polyrhizus*) peel | TVB-N | From purple to yellow | Qin et al. (2020) |
| Shrimp | κ-Carrageenan | Curcumin | TVB-N | From yellow to red | Liu et al. (2018) |
| Shrimp | κ-Carrageenan | *Lycium ruthenicum* Murr. | TVB-N | From light gray to bluish green and turned to yellow | Liu et al. (2019) |
| Shrimp | Chitosan/polyvinyl alcohol | Cactus pears (*Opuntia ficusindica*) extract | TVB-N | From purple to orange and turned to yellow | Yao et al. (2020) |

| | | | | | |
|---|---|---|---|---|---|
| Shrimp | Polyvinyl alcohol/okra mucilage polysaccharide | Rose anthocyanin | TVB-N | From purple to blue to dark green and turned to yellow | Kang et al. (2020) |
| Shrimp | Cellulose fibers | Bromothymol blue | TVB-N | From light yellow to blue-green | Cao et al. (2019a) |

*TVB-N* total volatile basic nitrogen

## 5.6 Conclusions

Because animal-based products are important source of nutrition for humans, they also play an important role in the economic growth. As a result, the new packaging technology has been continuously developed to meet consumer needs in the animal-based product area. Recently, the effective potential of smart freshness indicators based on natural biopolymer materials incorporated with natural dye indicators for the real-time monitoring of animal-based product quality and safety has been demonstrated by many studies. The developed smart freshness indicators have exhibited great sensitivity toward pH and target metabolites, showing color variations and thus exhibiting prospective use as visible smart freshness indicators. The use of smart freshness indicators could improve product quality and safety, and real-time monitoring and reduce food loss across the supply chain. However, smart freshness indicators also have many limitations, such as some dye indicator had broad-spectrum color changes and can be provided false-negative results, which lead to the producer rejection and consumer reliability. Therefore, the further development of extremely sensitive freshness indicators, more exact color, and cost effectiveness is of great significance for commercialization opportunity and increasing consumer acceptance.

## References

Alizadeh-Sani M, Tavassoli M, Mohammadian E, Ehsani A, Khaniki GJ, Priyadarshi R, Rhim JW (2021) pH-responsive color indicator films based on methylcellulose/chitosan nanofiber and barberry anthocyanins for real-time monitoring of meat freshness. Int J Biol Macromol 166: 741–750

Cao L, Ma Q, Liang T, Sun G, Chi W, Zhang C, Li J, Wang L (2019a) A semen cassia gum-based film with visual-olfactory function for indicating the freshness change of animal protein-rich food. Int J Biol Macromol 133:243–252

Cao L, Sun G, Zhang C, Liu W, Li J, Wang L (2019b) An intelligent film based on cassia gum containing bromothymol blue-anchored cellulose fibers for real-time detection of meat freshness. J Agric Food Chem 67:2066–2074

Cardello AV, Schutz HG (2003) The concept of food freshness: uncovering its meaning and importance to consumers. In: Cadwallader KR, Weenen H (eds) Freshness and shelf life of foods. American Chemical Society, Washington, pp 22–41

Chen HZ, Zhang M, Bhandari B, Yang CH (2019) Development of a novel colorimetric food package label for monitoring lean pork freshness. LWT 99:43–49

Chen HZ, Zhang M, Bhandari B, Yang CH (2020) Novel pH-sensitive films containing curcumin and anthocyanins to monitor fish freshness. Food Hydrocoll 100:105438

Choi I, Lee JY, Lacroix M, Han J (2017) Intelligent pH indicator film composed of agar/potato starch and anthocyanin extracts from purple sweet potato. Food Chem 218:122–128

Cutter CN, Senevirathne RN, Chang VP, Cutaia RB, Fabrizio KA, Geiger AM, Valadez AM, Yoder SF (2012) Major microbiological hazards associated with packaged fresh and processed meat and poultry. In: Kerry JP (ed) Advances in meat, poultry and seafood packaging. Woodhead Publishing, Cambridge, pp 3–58

Dong H, Ling Z, Zhang X, Zhang X, Ramaswamy S, Xu F (2020) Smart colorimetric sensing films with high mechanical strength and hydrophobic properties for visual monitoring of shrimp and pork freshness. Sens Actuators B Chem 309:127752

Dudnyk I, Janeček ER, Vaucher-Joset J, Stellacci F (2018) Edible sensors for meat and seafood freshness. Sensors Actuators B Chem 259:1108–1112
Erkmen O, Bozoglu TF (2016) Food microbiology: principles into practice. Wiley, New York
Ezati P, Tajik H, Moradi M (2019a) Fabrication and characterization of alizarin colorimetric indicator based on cellulose-chitosan to monitor the freshness of minced beef. Sensors Actuators B Chem 285:519–528
Ezati P, Tajik H, Moradi M, Molaei R (2019b) Intelligent pH-sensitive indicator based on starch-cellulose and alizarin dye to track freshness of rainbow trout fillet. Int J Biol Macromol 132: 157–165
Hewlings SJ, Kalman DS (2017) Curcumin: a review of its effects on human health. Foods 6:92
Hogan SA, Kerry JP (2008) Smart packaging of meat and poultry products. In: Kerry J, Butler P (eds) Smart packaging technologies for fast moving consumer goods. Wiley, West Sussex, pp 33–54
Huang S, Xiong Y, Zou Y, Dong Q, Ding F, Liu X, Li H (2019) A novel colorimetric indicator based on agar incorporated with *Arnebia euchroma* root extracts for monitoring fish freshness. Food Hydrocoll 90:198–205
Jiang G, Hou X, Zeng X, Zhang C, Wu H, Shen G, Li S, Luo Q, Li M, Liu X, Chen A, Wang Z, Zhang Z (2020) Preparation and characterization of indicator films from carboxymethyl-cellulose/starch and purple sweet potato (*Ipomoea batatas* (L.) lam) anthocyanins for monitoring fish freshness. Int J Biol Macromol 143:359–372
Kanatt SR (2020) Development of active/intelligent food packaging film containing *Amaranthus* leaf extract for shelf life extension of chicken/fish during chilled storage. Food Packag Shelf Life 24:100506
Kang S, Wang H, Xia L, Chen M, Li L, Cheng J, Li X, Jiang S (2020) Colorimetric film based on polyvinyl alcohol/okra mucilage polysaccharide incorporated with rose anthocyanins for shrimp freshness monitoring. Carbohydr Polym 229:115402
Kerry JP (2012) Application of smart packaging systems for conventionally packaged muscle-based food products. In Kerry JP (ed) Advances in meat, poultry and seafood packaging. Woodhead Publishing, Cambridge, pp 522–564
Kung HF, Lee YC, Lin CW, Huang YR, Cheng CA, Lin CM, Tsai YH (2017) The effect of vacuum packaging on histamine changes of milkfish sticks at various storage temperatures. J Food Drug Anal 25:812–818
Kuswandi B, Nurfawaidi A (2017) On-package dual sensors label based on pH indicators for real-time monitoring of beef freshness. Food Control 82:91–100
Lee EJ, Shin HS (2019) Development of a freshness indicator for monitoring the quality of beef during storage. Food Sci Biotechnol 28:1899–1906
Lee K, Park H, Baek S, Han S, Kim D, Chung S, Yoon JY, Seo J (2019) Colorimetric array freshness indicator and digital color processing for monitoring the freshness of packaged chicken breast. Food Packag Shelf Life 22:100408
Liu J, Wang H, Wang P, Guo M, Jiang S, Li X, Jiang S (2018) Films based on κ-carrageenan incorporated with curcumin for freshness monitoring. Food Hydrocoll 83:134–142
Liu J, Wang H, Guo M, Li L, Chen M, Jiang S, Li X, Jiang S (2019) Extract from *Lycium ruthenicum* Murr. incorporating κ-carrageenan colorimetric film with a wide pH-sensing range for food freshness monitoring. Food Hydrocoll 94:1–10
Liu D, Cui Z, Shang M, Zhong Y (2021) A colorimetric film based on polyvinyl alcohol/sodium carboxymethyl cellulose incorporated with red cabbage anthocyanin for monitoring pork freshness. Food Packag Shelf Life 28:100641
Lu P, Yang Y, Liu R, Liu X, Ma J, Wu M, Wang S (2020) Preparation of sugarcane bagasse nanocellulose hydrogel as a colourimetric freshness indicator for intelligent food packaging. Carbohydr Polym 249:116831
Merz B, Capello C, Leandro GC, Moritz DE, Monteiro AR, Valencia GA (2020) A novel colorimetric indicator film based on chitosan, polyvinyl alcohol and anthocyanins from

jambolan (*Syzygium cumini*) fruit for monitoring shrimp freshness. Int J Biol Macromol 153: 625–632
Miguel MG (2011) Anthocyanins: antioxidant and/or anti-inflammatory activities. J Appl Pharm Sci 1:7–15
Mohammadalinejhad S, Almasi H, Moradi M (2020) Immobilization of Echium amoenum anthocyanins into bacterial cellulose film: a novel colorimetric pH indicator for freshness/spoilage monitoring of shrimp. Food Control 113:107169
Moradi M, Tajik H, Almasi H, Forough M, Ezati P (2019) A novel pH-sensing indicator based on bacterial cellulose nanofibers and black carrot anthocyanins for monitoring fish freshness. Carbohydr Polym 222:115030
Pastore A, Badocco D, Bogialli S, Cappellin L, Pastore P (2021) Behavior of sulfonephthalein and azo dyes as effective pH sensors in hybrid materials. Microchem J 160:105605
Prieto-Santiago V, Cavia MM, Alonso-Torre SR, Carrillo C (2020) Relationship between color and betalain content in different thermally treated beetroot products. J Food Sci Technol 57:3305–3313
Qin Y, Liu Y, Zhang X, Liu J (2020) Development of active and intelligent packaging by incorporating betalains from red pitaya (*Hylocereus polyrhizus*) peel into starch/polyvinyl alcohol films. Food Hydrocoll 100:105410
Rahimi P, Abedimanesh S, Mesbah-Namin SA, Ostadrahimi A (2019) Betalains, the nature-inspired pigments, in health and diseases. Crit Rev Food Sci Nutr 59:2949–2978
Rawdkuen S, Faseha A, Benjakul S, Kaewprachu P (2020) Application of anthocyanin as a color indicator in gelatin films. Food Biosci 36:100603
Robertson GL (2012) Food packaging: principles and practice, 3rd edn. CRC Press, New York
Ruiz-Capillas C, Herrero AM (2019) Impact of biogenic amines on food quality and safety. Foods 8:62
Rukchon C, Nopwinyuwong A, Trevanich S, Jinkarn T, Suppakul P (2014) Development of a food spoilage indicator for monitoring freshness of skinless chicken breast. Talanta 130:547–554
Sabnis RW (2007) Handbook of acid-base indicators. CRC Press, New York
Smolander M (2003) The use of freshness indicators in packaging. In: Ahvenainen R (ed) Novel food packaging techniques. Woodhead Publishing, Cambridge, pp 128–143
Soni A, Mendiratta SK, Talukder S, Kumar RR, Chand S, Jaiswal RK (2018) Development of colorimetric on-package indicator for monitoring of chicken meat freshness during refrigerated storage (4±1° C). J Anim Res 8:847
Talukder S, Mendiratta SK, Kumar RR, Agrawal RK, Soni A, Luke A, Chand S (2020) Jamun fruit (*Syzygium cumini*) skin extract based indicator for monitoring chicken patties quality during storage. J Food Sci Technol 57:537–548
Vo TV, Dang TH, Chen BH (2019) Synthesis of intelligent pH indicative films from chitosan/poly (vinyl alcohol)/anthocyanin extracted from red cabbage. Polymers 11:1088
Yao X, Hu H, Qin Y, Liu J (2020) Development of antioxidant, antimicrobial and ammonia-sensitive films based on quaternary ammonium chitosan, polyvinyl alcohol and betalains-rich cactus pears (*Opuntia ficus-indica*) extract. Food Hydrocoll 106:105896
Yildiz E, Sumnu G, Kahyaoglu LN (2021) Monitoring freshness of chicken breast by using natural halochromic curcumin loaded chitosan/PEO nanofibers as an intelligent package. Int J Biol Macromol 170:437–446
Zeng P, Chen X, Qin YR, Zhang YH, Wang XP, Wang JY, Ning ZX, Ruan QJ, Zhang YS (2019) Preparation and characterization of a novel colorimetric indicator film based on gelatin/polyvinyl alcohol incorporating mulberry anthocyanin extracts for monitoring fish freshness. Food Res Int 126:108604

Zhang J, Zou X, Zhai X, Huang X, Jiang C, Holmes M (2019) Preparation of an intelligent pH film based on biodegradable polymers and roselle anthocyanins for monitoring pork freshness. Food Chem 272:306–312

Zhang C, Sun G, Cao L, Wang L (2020) Accurately intelligent film made from sodium carboxymethyl starch/κ-carrageenan reinforced by mulberry anthocyanins as an indicator. Food Hydrocoll 108:106012

Zhao X, Wu JE, Chen L, Yang H (2019) Effect of vacuum impregnated fish gelatin and grape seed extract on metabolite profiles of tilapia (*Oreochromis niloticus*) fillets during storage. Food Chem 293:418–428

Zhu J, Huang X, Zhang F, Feng L, Li J (2015) Inhibition of quorum sensing, biofilm, and spoilage potential in *Shewanella* baltica by green tea polyphenols. J Microbiol 53:829–836

# Packaging of Dairy Products: Emerging Strategies

# 6

Marta Biegańska

**Abstract**

Consumer demand for dairy products has changed significantly in recent years. Globalization, urbanization, and changing lifestyles with increasing number of consumers with higher incomes in developing countries keep pushing demand for innovative dairy products. The current trend observed on the dairy market regarding healthy products forces producers to increase the variety of premium and specialty products. Consumers are more aware nowadays of health benefits of dairy products and require controlled quality, convenience (e.g., on-the-go dairy), extended shelf life, and eco-friendly products. To meet these growing requirements, manufacturers are introducing product innovations to the market, as well as modern packaging. Beside the typical packaging functions like protection, containment, communication, marketing, and ergonomics novel packaging should also have an active role. As dairy products are highly perishable, they require proper packaging in order to protect and prolong shelf life of the product inside. Among different types of packaging, active and intelligent packaging is a powerful tool, alongside with novel packaging materials (e.g., paper-based, bioplastics). Moreover, novel packaging can be a tool for detecting food fraud, but also facilitates traceability. This review will cover potential use of smart packaging with indicators (e.g., gas/integrity, freshness, time-temperature), antimicrobial coatings, data carriers, sensors in dairy packaging.

M. Biegańska (✉)
Department of Industrial Products and Packaging Quality, Poznań University of Economics and Business, Poznań, Poland
e-mail: marta.bieganska@ue.poznan.pl





A. K. Shukla (ed.), *Food Packaging: The Smarter Way*,
https://doi.org/10.1007/978-981-16-7196-8_6

**Keywords**

Dairy products · Shelf life · Smart packaging · Active packaging · Intelligent packaging · Data carriers · Sensors

## 6.1 Introduction

Packaging has been used from the moment when primitive man began to stockpile food. Over the years and with the development of civilization, it has become more and more sophisticated and complex. Packaging in the past and now is designed to fulfill specific functions, such as protection, containment, communication, marketing, and ergonomics. The primary function of any packaging is to protect the product from the surrounding environment. When it comes to food stuffs, it should be a barrier to temperature abuse, light, moisture changes, gases, pressure, odors, microorganism, pests, dirt, and dust (Ščetar et al. 2019; Alizadeh-Sani et al. 2019; Mirza Alizadeh et al. 2020; Elsherif et al. 2020). The second important function of any packaging is its communication. In recent years, packaging has become a major brand communication vehicle. From a consumers' point of view, it is the first point of contact with a particular brand—a so called "silent seller". In other words, packaging has become an important feature of brand differentiation and factor in consumers' purchase decisions (Omeni and Daniel 2020). However, this basic and also passive role of a packaging in modern, urbanized world is not enough. In this view, it mostly allows for protecting and preserving food helping to reduce food waste. Moreover, consumers are more aware of safety hazards related to food and demand safe, high quality and convenient products with extended shelf life (Sajdakowska et al. 2020; Chávez-Martínez et al. 2020). Food safety issues pose a significant challenge worldwide as foodborne illnesses result in productivity loss and increased treatment cost of $15 billion annually (e.g., hospitalization costs) (Sindi et al. 2020). As dairy products provide suitable conditions for the development of a number of microorganism, including pathogens, and because contamination of those products can occur at various production stages ensuring safety and prolonging shelf life are essential (Delorme et al. 2020).

Due to busy and more complex lifestyle, an on-the-go eating and drinking trend is visible even on the dairy market. Nowadays dairy packaging is very important as it is a vehicle for product differentiation, enables communication of the content, and is a powerful branding tool (Mania et al. 2018a; Velasco and Spence 2018; Šerešová and Kočí 2020).

In order to adapt to these changing trends and consumer expectations, the dairy market is forced to significantly increase the variety of products. This leads to increase of snack-sized, resealable, portion-controlled packaging. However, single-use packaging contributes to the increase in the amount of packaging waste (Šerešová and Kočí 2020). On the other hand, as dairy products are highly perishable, food waste and food loss in the food chain are also a problem. According to FAO as much as 1.3 billion tons of food are wasted globally each year, a 1/3 of the world production (Garnier et al. 2017; Bilska and Kołozyn-Krajewska 2019; Lipińska et al. 2019). This results in up to $1 trillion of economic losses annually. There is a difference between food loss and food waste. The first being loss occurring during production, postharvest treatment, and storage, whereas the latter is the food disposed of by retailers and consumers. To avoid or at least reduce food waste, the package should be convenient for its users. It ought to be easy to handle and refill without any mechanical damage to the product. Moreover, it should be easy to open, resealable, and allow for maximum emptying (Mania et al. 2018a). Packaging in its protection function can aid reduce food waste, but with novel packaging solutions it can also help to lower food losses along production and supply chains which are in line with the United Nations Sustainable Development goal to halve per capita global food waste (Wikström et al. 2019).

In recent years, consumers have shown rising health awareness that is anticipated to trigger some market trend, e.g., products with high protein content, lactose-free, sugar-free, nondairy alternatives or fortified and functional dairy. Those trends and growing market and consumer demand, due to changes in lifestyle and market globalization, have forced manufacturers to seek new packaging solutions like smart packaging (SP) (Mirza Alizadeh et al. 2020). The correlation between SP and main packaging functions, as well as different areas active and intelligent packaging cover is shown in Fig. 6.1.

*Smart packaging* combines the functions of both active and intelligent packaging. Both systems can work separately or synergistically as smart packaging technologies, as shown in Fig. 6.2.

*Active packaging* systems are developed with active components in the packaging (e.g., antimicrobial coatings, gas emitters or scavengers, moisture absorbers) in order to maintain or extend the product quality and shelf life, whereas *intelligent packaging* aims at monitoring the state and quality of the product and shares information on

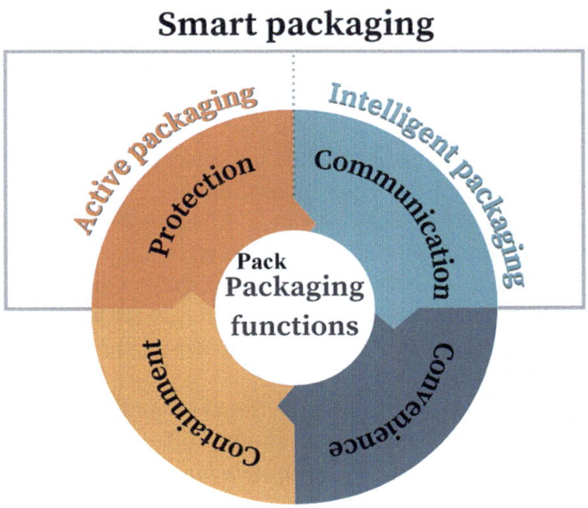


**Fig. 6.1** Main packaging functions (Source: Own work)

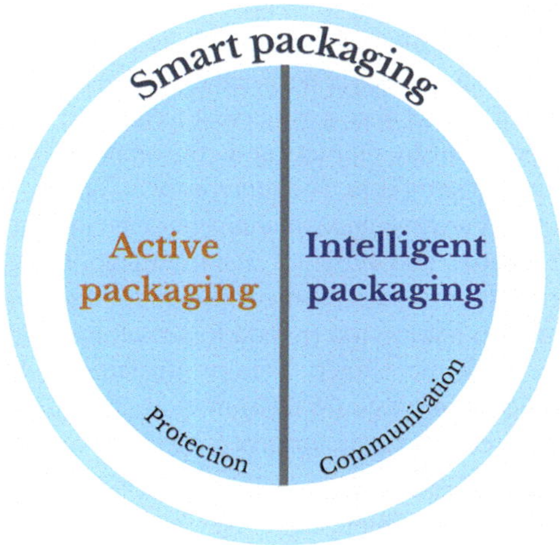


**Fig. 6.2** Smart packaging (Source: Own work)

the storing and transportation conditions throughout the distribution chain (e.g., indicators, sensors, and data carriers) (Drago et al. 2020). They will be discussed in more detail further in the chapter.

Different packaging materials are being used nowadays depending on product type, processing and storing conditions, handling requirements, and way of use/consumption. Most often used are: glass and plastic bottles, multilayer materials (e.g., Tetra Pak®, SIG Combibloc), pouches, plastic trays, cans, tubs, and buckets. They have to meet some requirements like to be printable, safe for food contact, inert, not to interact with the packed product (except for active packaging), nontoxic, be compatible with the product, show appropriate impact resistance, have barrier properties toward light, gases, and odor, have the right shape and size, be cost-effective, and have marketing appeal (Ščetar et al. 2019).

## 6.2 Dairy Market Overview

Healthy eating habits include consumption of a wide variety of products of high nutritional density. High-quality protein, minerals, and group B vitamins can be found in milk and dairy products. These are essential in the diet of children, pregnant women, and the elderly (Owusu-Kwarteng et al. 2020; Delorme et al. 2020; Smirnova et al. 2020; Tricarico et al. 2020; Chávez-Martínez et al. 2020).

Global dairy market is estimated at over $400 billion worth, which is circa 14% of worldwide agricultural trade, where milk is the third agri-food commodity (Tan and Ngan 2020).

As dairy products are made from raw milk (cow, sheep, goat, buffalo, camel, and other milking animals), it shows differences in composition due to seasonal fluctuations, origin or rearing (Faccia et al. 2020). Commercially available milk comes in different types such as ultra-high-temperature (UHT) milk, pasteurized

**Fig. 6.3** An overview of dairy products (Source: Based on (Najib et al. 2020; Byberg and Lemming 2020)

milk, semiskimmed milk, skimmed milk, fortified milk, standardized milk, and reconstituted milk (Byberg and Lemming 2020; Faccia 2020a). An overview of dairy products is shown in the Fig. 6.3.

Milk contains important micronutrients and for this reason is regarded as an important part of a balanced diet. However, in its raw state, it is highly susceptible to microbial and enzymatic spoilage (Odueke et al. 2016; Smirnova et al. 2020).

One of dairy products is kefir; it can be made from different types of milk, like cow, goat, sheep, buffalo or camel during microbial fermentation. Kefir has an acidic taste, contains many bacterial species known for their probiotic properties, and is creamy. During storage, its sensory and microbial properties can still change improving its shelf life. It has many health benefits and exhibits antimicrobial activity, due to presence of lactic acid bacteria (LAB) capable of producing bacteriocins, against for example *S. aureus, E. coli, Salmonella enteritidis, B. cereus,* and *L. monocytogenes* (Sindi et al. 2020). Most common genera of LAB are *Lactobacillus* and *Bifidobacterium*. They owe their biological functions to bioactive metabolites and among health benefits anticarcinogenic, antiobesity, antioxidant effect, and stimulation of immune response need to be mentioned (Mohamed et al. 2020). Nevertheless, it has short shelf life and high storage and packaging costs which have led to the trend of having dry kefir in powder form (through freeze or spray drying) (Farag et al. 2020).

Another most often consumed worldwide nutritious and healthy product is yoghurt (Sajdakowska et al. 2020). It is a fermented dairy product originating in the Balkans and the Middle East. Its shelf life is usually 3 weeks or less. Thanks to presence of lactic acid bacteria, it has a probiotic effect (Odueke et al. 2016). Lactic acid bacteria LAB are essential in production of fermented dairy products. The quality of fermented product depends on the bacterial strain used in the process (Mohamed et al. 2020). There are several types of yoghurts that gain in popularity like drinkable yoghurt, spoonable, Greek-style yoghurt (Newbold and Koppel 2018).

Cheese is one of the most important dairy products that come in many varieties (e.g., flavor, texture, size, and origin). It undergoes many changes (e.g., microbial, biochemical) during production and ripening. The manufacturing of cheese is one of the oldest technologies for prolonged milk nutrients consumption. It is most often classified as hard, semihard, soft, and fresh. Although, fresh cheeses are considered to be soft cheeses due to their high moisture content and fairly short shelf life (Youssef et al. 2017). Cheese comes in three main types: natural cheese, powdered cheese, and processed (melted analog) cheese. Natural cheese is prepared from pasteurized milk and its aging continues "on the shelf". Taking that into account, it requires packaging with high barrier properties. Powdered cheese used mainly for snack foods flavoring is produced by dehydration of natural cheese. Finally, processed cheese is manufactured by melting natural cheese with additives (e.g., water, salt, whey, emulsifiers) to prolong its shelf life (Mania et al. 2018a).

## 6.3 Dairy Product Portfolios and Changing Market Demands

Milk and dairy supply chains play a significant role in world food demand. The growing world population and the increase in wealth of people in developing countries have a significant impact on this demand. On the other hand, markets in developed countries seek nutritionally balanced and environmentally sustainable products forcing manufacturers to develop product innovations such as functional products (Faccia 2020b). Dairy supply chains are complex and difficult to model due to high variability of raw materials supply, large variety of products, their perishability, and changing demand. A simplified logistic chain is shown in Fig. 6.4. As dairy products, especially milk and its by-products, demonstrate high nutritional value; the use of fortifying additives to increase that value was only a matter of time (Guarnaschelli et al. 2020).

Traditional milk preservation methods include among others pasteurization, ultra-high-temperature, and sterilization. Table 6.1 presents heat treatment methods in the dairy industry. Traditional dairy preservation techniques are applied to inhibit microbial growth, increase shelf life, and maintain quality and safety. However, they can alter organoleptic properties and decrease levels of nutrients. To avoid these adverse effects, emerging technologies such as high hydrostatic pressure (HHP), pulsed electric fields (PEFs), ultrasound (US), irradiation, and cold plasma (CP) are gaining more attention. Irradiation as a nonthermal technology for food preservation

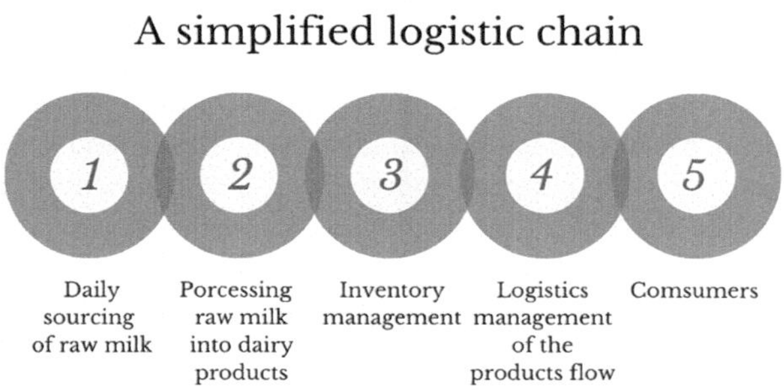


**Fig. 6.4** A simplified dairy logistic chain (Source: Based on (Guarnaschelli et al. 2020))

**Table 6.1** Heat treatment methods in the dairy industry

| Heat treatment | Temperature [°C] | Time | Product |
|---|---|---|---|
| Thermization | 57–68 | 5 s–<br>30 min | Raw milk<br>Some cheeses |
| Pasteurization | 72–80 | 15–30 s | Drinking milk<br>Cheese |
| Pasteurized with extended shelf life (ESL) | 125–140 | 1–10 s | Drinking milk with ESL at chilled conditions |
| Ultra-high-temperature (UHT) | 135–150 | 1–10 s | Drinking milk with long shelf life at ambient temperature |
| In-container sterilization | 110–120<br>125 | 10–20 min<br>5 min | Condensed milk, drinking milk with long shelf life at ambient temperature |
| | 90–95 | 5–10 min | Yoghurt |
| | 72–80 | 15–30 s | Low-heat skim milk powder (SMP) |
| | 85<br>90<br>105 | 1 min<br>30 s<br>30 s | Medium-heat SMP<br>Preheating in UHT<br>Whole milk powder |
| | 90<br>120<br>135 | 5 min<br>1 min<br>30 s | High-heat SMP |
| | >120 | >40 s | High-high-heat or high-high-heat-stable SMP |

Source: Based on (International Dairy Federation 2018)

has been utilized for many years and acknowledged by World Health Organization (WHO) as beneficial for the provision of quality, ensuring safety and extending shelf life. Irradiation at low doses and/or in frozen conditions can be effective in pathogens elimination without off-flavors in the irradiated product (Odueke et al. 2016). Another promising innovative technology with commercialization potential is ultra-violet (UV) radiation that has shown effectiveness on microbial safety and intrinsic quality properties of dairy products (Delorme et al. 2020).

UV light comprises the wavelengths range from 100 to 400 nm in which four types of rays can be distinguished: UVA (315–400 nm), UVB (280–315 nm), UVC (200–280 nm), and vacuum-UV (100–200 nm). Among those types, UVC rays are known for their microbial elimination (e.g., bacteria, viruses, and fungi). However, further studies are needed to optimize conditions of UV treatment of dairy products to overcome modifications of physicochemical and sensory features (Delorme et al. 2020).

Increasingly diverse consumers' needs regarding dairy products lead to changing production patterns in the dairy industry. They can be described as mass customization as manufacturers offer a wide variety of products meant for different consumers with different needs and expectations. Products attributes regarded as most important by consumers in relevance to ice-cream were taste, appearance, shape, texture, health benefits, and packaging material (Wedowati et al. 2020). Although those results were related to ice cream, but to a large extent they may also apply to other dairy products.

### 6.3.1 Functional Dairy Products

Fortification traditional dairy products with active ingredients lead to functional dairy products with enhanced nutritional value and even improved health benefits. Functional food market has grown significantly in the last couple of decades, mostly because of use of prebiotics and probiotics (Lai et al. 2020; Faccia 2020b). Functional dairy products include:

- Low-fat dairy
- Lactose-free dairy
- Products fortified with added functional ingredients (Lai et al. 2020)

For example, addition of Chinese sweet tea extract (*Rubus suavissimus* S. Lee leaves) to buffalo milk yoghurt improved the biological activity of the product (antioxidation and antihypertensive). Another study showed that fortification of buffalo milk yoghurt with fenugreek (*Trigonella foenum-graecum*) and *Moringa oleifera* seed flour leads to modification of mineral compounds and showed higher antibacterial activity against some undesired species, compared to unfortified yoghurt (Faccia 2020b). Lai et al. (2020) reported that fortification of yoghurt with algal oil (25% of fat content) to increase omega-3 fatty acids content did not seem to cause an off-flavor. Moreover, addition of inulin acting as a prebiotic for the increase of probiotic gut bacteria is another fortification example, just like the fortification with whey protein in sheep and goat milk dairy products. Fortification may take the form of additives constituting by-products from other food industries like vegetable, fruit, marine or cereal. A wide range of examples were presented by Iriondo-Dehond et al. (2018) among which authors mentioned fruit skin, peels, leaves, and pulp or bran cereals and fish oil.

In order to lower fat content in yoghurts, fat replacers are used. Different types of starch (e.g., corn, sweet potato, potato, chickpea, and Turkish beans) can serve as fat lowering agents. Moreover, functional additives can also be added to other dairy products such as cheese. Faccia (2020b) studied kiwi juice as cheese coagulant and although, longer coagulation and syneresis were observed, compared to calf rennet, investigated cheese exhibited higher concentrations of polyphenols and phytosterols. Crockett (2015) studied Allanblackia oil as a potential functional additive. It demonstrates good storage stability characteristics and relatively high melting point which make it a promising consistency improving agent in butter substitutes and vegetable-based dairy products.

People suffering from lactose intolerance exhibit decrease in enzyme (-β-galactosidase) production that is responsible for lactose hydrolyzation. This in turn leads to gut digestion of lactose which shows unpleasant symptoms like abdominal cramps, flatulence, and diarrhea. Around 75% of adult global population suffers from enzyme decrease. Such state leads to increased lactose-free dairy demand. However, milk from sheep and goat seems to be better tolerable by lactose-intolerant consumers, there is a trend in removing lactose from this type of milk as well like in lactose-free Ricotta cheese (Lai et al. 2020). There are also products that contain little or no lactose like Gouda, Parmesan, Cheddar, Swiss cheese or butter. Fresh cheese, however, can contain lactose at levels resulting in reaction among lactose-intolerant consumers.

Among market-available lactose-free products are potable milk, UHT milk, yoghurts, cheese, ice-cream, flavored milk, dairy powders, creams, desserts, and Dulce de Leche (Dekker et al. 2019).

#### 6.3.1.1 Nanoceuticals

Addition of nutraceuticals like carotenoids or lycopene, omega-3 fatty acids, minerals, vitamin $D_2$, phytosterols, and some probiotic bacteria species to conventional dairy products can improve their bioavailability. They are also generally recognized as safe (GRAS) by the Food and Drug Administration (FDA). They can be incorporated into the products in a form of nanoclusters, nanocages, and nanodrops. Compared to fortification described above, the main difference is in the size of the bioactive compound ranging from 1 to 100 nm in length. These nanoscale compounds show better solubility, biological activity, and bioavailability. They are also more stable during heat processing and storage to normal-size nutraceuticals (Kuswandi 2017; Poonia 2019).

Encapsulated edible nanoparticles containing drugs, vitamins, micronutrients incorporated into the packaging surface could deliver those particles through controlled release in the human body at targeted organs and improve health (Kuswandi 2016; Singh et al. 2019).

#### 6.3.1.2 Novel Fermented Dairy Products

Although, probiotic bacteria (LAB) are already known in kefir and yoghurt, the increased popularity of the probiotic concept having benefits to human health. For this reason, an emerging trend to use them in other dairy products such as cheese, dairy desserts, and ice cream can be detected. Changing dietary patterns have led to creation of a market niche for probiotic foods (Ávila et al. 2020). Addition of herb extracts, fresh spices or fruit fibers to yoghurts can increase the phytochemicals content and health benefits. Dimitrellou et al. (2020) investigated yoghurts supplemented with aronia, blueberry, and grape juices to increase the amount of phenolic compounds and antioxidant activity. Addition of investigated juices increase red color of prepared yoghurts, but had no significant effect on physico-chemical properties.

### 6.3.2 Products Beyond Dairy

Going beyond dairy is another emerging trend on the market. According to Stora Enso *Gainomaxs'* sports nutrition, UHT drinks are displayed together with other sports nutrition products outside dairy shelf (Stora Enso n.d.). Another example is Credition Dairy's (Devon, UK) brand *Arctic Coffee* that comes in several flavors and carton sizes of 330 mL and 1 L and gives consumers the experience of on-the-go coffee with milk (Elopak 2020).

One of possibilities is the use of dairy products in making functional bakery products. Graça et al. (2019) studied the addition of yoghurt and curd cheese to wheat bread to enhance its functional and nutritional value. As dairy products can help increase the amount of minerals like Ca and P, vitamin A and B12, as well as protein and essential amino acids. Research showed that yoghurt addition had a positive effect on the rheology of the dough. Both breads obtained good sensorial acceptability with yoghurt and curd cheese additions of 50 g and 30 g, respectively.

#### 6.3.2.1 Plant-Based Dairy Alternatives

Dairy alternatives are made from plant-based milk instead of animal milk. With growing consumer demand, more and more such products are available on the market. The reasons behind this trend are both health and sustainability. Global dairy alternative market is estimated to grow to 20% of the value of the dairy milk market in 2021. Among plant-based alternatives are soy-based drinks, almond-based drinks, oat-based drinks, rice-based drinks, tofu spreads (Jeske et al. 2018; Leialohilani and de Boer 2020; Paul et al. 2020). In 2020, Arla has launched a range of dairy-free plant-based drinks in Denmark under the brand name Jörd made from oats, barley, and hemp (Elopak 2020).

## 6.4 Dairy Packaging Innovations

Development of dairy packaging is a result of advances in material technologies and consumer demands. Commonly used materials include:

- Glass [inert, recyclable, impermeable to gases and water vapor, brittle, heavy in weight; bottles and jars]

- Metal [recyclable, rigid, gas barrier properties, requires coatings to prevent corrosion; cans or thin film wrappings]
- Paper/cardboard [recyclable, biodegradable, light in weight, printable, permeability for breezing, water vapor and oxygen, susceptible to tearing; secondary packaging, e.g., boxes, wrappings]
- Waxed papers [good resistance, good heat-sealing characteristics, low-cost, moisture barrier properties]
- Poly(vinylidene chloride) (PVDC) [gas, water vapor, fatty and oil products barrier properties, used in MAP]
- Cellophane [clear films, used in coatings and laminates, poor strength properties]
- Low-density polyethylene (LDPE, LD-PE, PELD, PE-LD) [recyclable, low-cost, flexible, moisture barrier properties, resistant to most solvents, low gas barrier properties; cups, trays, tubs]
- High-density polyethylene (HDPE, HD-PE, PEHD, PE-HD) [recyclable, low-cost, moderately flexible, odorless, poor clarity, low oxygen and other gases barrier properties; cups, trays, tubs]
- Poly(vinyl chloride) (PVC) [use of plasticizers, low moisture barrier properties]
- Multilayer packaging (plastic/paperboard/aluminum) [aseptic, moisture, gas, odor, light barrier properties; pouches, sachets, cartons, bottles] (Roohi et al. 2018; Ščetar et al. 2019)

Packaging manufacturers keep on developing their products to fulfill growing market and consumer demand. Ampacet (NY, USA) has launched their sustainable 3R solutions (Reuse, Reduce, and Recycle) to help meet sustainability goals. To give an example Ampacet's Safari White PET masterbatch offers high level of opacity and reduced mineral loading of less than 4%. It is suitable for ultra-high temperature (UHT) processed milk which has an unrefrigerated shelf life of 6–9 months. Safari PET preserves flavor and nutrients of fresh milk and protects it from light exposure in 400–550 nm wavelength (Anonymous 2019).

Wipak (Helsinki, Finland) offers monomaterial film BIAXOP ECO 70 XX XPP which is recyclable and has excellent transparency, with an integrated barrier layer for an even better level of product protection and tight seal. When an easy-open PEEL-solution for lidding filmsand flow packs is required, BIAXEN ECO 65 XX XFP offers a wide sealing window and early sealing initiation temperature. Polyolefin-based resealable lidding REPAK® TOP BP 70 XX and PE-based bottom NICE ECO XX 17 are also recyclable. Wipak has also launched rPET with a high recycled content in the semirigid MULTICLEAR-R 250 bottom film with great optical and mechanical properties. With high-quality recycled PET granulate, the solution allows for a high carbon footprint reduction and the material is fully approved by the European Food Safety Authority (EFSA) for use as a food contact material. In the area of renewable materials are PAPER TOP® PD BE 90 XX PEEL lidding film and PAPER BTM® Q 330 XX bottom film. Wipak's PAPER BTM® range, provides different solutions with up to 90% paper share, all enable consumers to separate film and paper before disposal (Wipak 2021).

Danish organic cooperative dairy Naturmælk has introduced its vast range of milk, yoghurt, buttermilk, and cream varieties in Pure-Pak® cartons with Natural Brown Board, and Mini Pure-Pak® cartons from Elopak (Oslo, Norway). Moreover, new Pure-Pak®Imagine carton is even more environmentally friendly than other cartons as it has 10 times less plastic compared to PET bottle and 46% less plastic than a carton with a closure. Pure-Pak®Imagine is designed with an easy-pour opening with an easy-fold feature, no plastic screw cap, and is 100% based on wood. This type of closure also provides maximum emptying capabilities of the carton minimizing food waste. It has a unique top fin-guiding consumer on how to open the carton and form an easy-pour spout which is similar to carton openings in the 70s and 80s (Elopak 2020).

Sustainability and environment awareness of consumers force packaging development. Another example is Chadwicks' (Greater Manchester, UK) paper-based lids for dairy products. Those lids are fully recyclable, offer good barrier properties (oxygen, water vapor, aromas), and high mechanical strength and stiffness. They are also printable and show excellent optical properties. Those paper-based lids consist of polyethylene-coated paper with linen or pin dot emboss. The embossing includes print, paper, and coextruded barrier polymer (Chadwicks 2021).

Among novel packaging systems, modified atmosphere packaging (MAP), edible cheese packaging or biodegradable packaging should be mentioned which will be described in more detail in this chapter.

### 6.4.1 MAP Packaging

Modified atmosphere packaging (MAP) has been previously used for fruit and vegetables, meat fish, and bakery products. This technology is based on altering the gaseous combination inside the packaging to extend product's shelf life. This is obtained by using specially composed gas mixtures of carbon dioxide, oxygen, and nitrogen. This approach allows for decreased microbial growth and reduction of lipids oxidation, thus prolonging shelf life and quality of dairy products. Lipid oxidation is responsible for quality deterioration in dairy products as oxidation of unsaturated fatty acids results in the formation of chemical compounds that give off-flavors and affect shelf life and storage stability (Clarke et al. 2020). Addition of $CO_2$ to a gas barrier packaging allowed for extending shelf life of cottage cheese of 200–400% (Newbold and Koppel 2018). This technology can be applied with success to prolong shelf life of soft cheese, semihard cheese, hard cheese, cottage cheese, whey cheese, fresh cheese, and surface mold-ripened cheese. Depending on the cheese type in order to inhibit the growth of anaerobic microorganism, $CO_2$ concentrations inside the packaging ought to vary between 20 and 60%. Usually to minimize lipids oxidation processes in MAP of dairy products, the level of $O_2$ should be as low as possible. In this case, combinations of $CO_2$ and $N_2$ should be 30/70, 40/60, 60/40, 50/50 accordingly. Although, this method allows for extending shelf life of cheeses before choosing the right gas ratio cheese type, production process and packaging material need to be considered (Garnier et al. 2017; Ščetar et al. 2019).

### 6.4.2 Edible Coatings

To meet consumer acceptability and satisfaction, cheese requires an adequate packaging to protect its quality, and safety films and coatings have been used for a long time. Covering cheese methods before ripening or during storage can prolong shelf life and minimize defects in the finished product. Consumer preferences regarding appearance and sensory features of cheese make the selection of packaging and coating an important factor in buying decisions. Moreover, environmental awareness and sustainability are also forcing manufacturers to seek recyclable or biodegradable packaging materials to minimize package waste. In recent years, there has been a growing interest in edible film and coating in the dairy sector. Edible coatings/films are thin layers of food biopolymer-based material for covering food, like cheese. Films are prepared separately and applied in the form of a wrapping material, whereas coatings are suspensions or emulsions applied directly and forming a film on the surface of cheese. Some polysaccharides, lipids, and proteins have been used as edible cheese coatings (with or without plasticizers, antimicrobial agents, and nanoparticles) to prolong shelf life (Table 6.2). Application of a film or coating can protect cheese from moisture losses, gas exchange, microbial and chemical deterioration, and physical damage. Traditionally waxes have been used as cheese coatings; however, a combination of waxes with emulsions based on polysaccharides and proteins may lead to the formation of a new generation of edible coatings (Youssef et al. 2017).

Alizadeh-Sani et al. (2019) studied storage stability of corn starch-based edible films as a wrapping of milk cake. It is a product prepared directly from milk or from a granular variety of *khoa*. During storage, it is susceptible to oxidative rancidity and absorbs odors from its surrounding environment. Traditionally milk cake is wrapped in butter paper and placed in paperboard secondary packaging. Compared to control, starch-based films demonstrated slower decrease in sensory scores under refrigerated conditions. They exhibited storage stability of 18 days without signs of spoilage. The films thickness, water activity, moisture, water solubility, and water vapor transmission rate were significantly increase, but mechanical properties and transmittance decreased with storage period.

### 6.4.3 Biodegradable Packaging

Biodegradable food packaging materials are films that after use decompose with the help of microorganisms through composting where they are degraded into $CO_2$, water, methane, and biomass. Different microorganisms use such biodegradable materials as carbon source. The biodegradation leads either to disintegration of the polymer or its fragmentation without harmful effect on the environment. They are obtained from renewable, abundant, and low-cost materials and can be alternative food packaging materials (Roohi et al. 2018; Adeyeye et al. 2019; Kuswandi and Moradi 2019). Those "natural polymers" or biopolymers are mainly of microorganisms' origin and are in nature carbohydrates. They create hydrophilic

**Table 6.2** Edible coating materials in dairy industry

| Material/additive | Compound | Properties | Product |
|---|---|---|---|
| Biopolymers | Chitosan | • Nontoxic<br>• Biodegradable<br>• Forms strong flexible films<br>• Shows antimicrobial activity | • Fior di latte cheese<br>• Ricotta cheese<br>• Ras cheese<br>• Saloio semihard cheese |
| | Galactomannans | • Controlled gas transfer<br>• Structural stability<br>• Low moisture retention<br>• Freeze-thaw stable (cold water soluble gums) | |
| | Alginates | • Sodium alginate forms a fairly strong film | • Mozzarella<br>• Fior di latte cheese |
| | Carrageenans | • Have different solubility in water (depending on the degree of sulfation: δ, ι, κ)<br>• Show similar puncture force<br>• κ gives strongest film among all three types | |
| | Carboxy methyl cellulose (CMC) | • Forms films with different strength (depending on molecular weight and degree of substitution)<br>• Film strength is lower compared to ones from alginates at similar concentrations | |
| | Starch and derivatives | • Can create clear and flexible coatings when modified<br>• Addition of plasticizers improves film properties, but reduces barrier characteristics | |
| | Proteins (e.g., whey proteins, and zein) | • Good mechanical stability<br>• Excellent gas and lipid barrier properties<br>• Poor water resistance<br>• Susceptibility to cracking | |
| Plasticizers | Glycerol, sorbitol, ethylene glycol | • Improve the elastic modulus and mechanical properties of films<br>• Increase in resistance to permeation of vapors and gases | |
| Surfactants and lipids | ND | • Improve the emulsions stability<br>• In coatings reduce surface tension and improve wettability | |

(continued)

**Table 6.2** (continued)

| Material/additive | Compound | Properties | Product |
|---|---|---|---|
| Antimicrobial agents<br>Nanoparticles<br>Bionanocomposites | Organic acids and their salts<br>– Sorbates<br>– Benzoates<br>– Propionates<br>– Lysozyme<br>Bacteriocins<br>– Nisin<br>– Natamycin<br>– Primaricin<br>Essential oils<br>Silver (silver zeolite)<br>Silver-montmorillonite<br>– Embedded into agar<br>– Used in MAP<br>– Dipped in sodium alginate and immersed in calcium chloride<br>– Chitosan/poly (vinyl alcohol)/ titanium nanoparticles<br>– Chitosan/CMC/ zinc nanoparticles | • Antimicrobial agents against Gram(+) and Gram(−) microorganisms<br>• Antimicrobial agent against Gram(−) microorganisms<br>• Antifungal agent<br>• Show wide spectrum effects on harmful microorganisms<br>• Reduces significantly the number of viable *E. coli*, *S. aureus*, and *A. niger* after 30 min of application<br>• Antimicrobial activity against three strains of Pseudomonas spp.<br>• Showed shelf life extension above 5 days<br>• Enhanced shelf life up to 10 days<br>• Good mechanical and antimicrobial properties<br>• Good mechanical properties<br>• Good antimicrobial activity against S. aureus, *P. aeruginosa*, *E. coli*, and *C. albicans* | • Gorgonzola<br>• Mozzarella<br>• Fior di latte cheese<br>• Fior di latte cheese<br>• White soft cheese |

Source: Based on (Youssef et al. 2017; Adeyeye et al. 2019)

films with good barrier properties to $O_2$, $CO_2$, and lipids. However, some biopolymers' performance is not yet comparable with traditional plastic materials. They are brittle and exhibit poor gas and vapor barrier properties. These drawbacks could be overcome through the use of nanomaterials (Adeyeye et al. 2019).

*Starch-based* packaging materials are biodegradable and with the use of plasticizers (e.g., glycerol, polyethers, and urea) can create films with appropriate for food packaging mechanical strength like tensile strength, high elongation. It can also be extruded and formed as thermoplastic materials and by using water steam it can be converted into foam, just like polystyrene (PS) packaging. Among advantages of starch in edible films are preparation simplicity, low-cost, good barrier properties to lipids and oxygen. Its drawbacks are poor water resistance and brittleness (Roohi et al. 2018; Adeyeye et al. 2019; Kuswandi and Moradi 2019).

*Chitosan* is a biopolymer with antimicrobial properties, is insoluble in water and soluble in acid solutions, and it also shows poor gas and water vapor barrier properties. Chitosan is a nontoxic, biodegradable, biocompatible packaging material (Adeyeye et al. 2019). It is mostly obtained from shrimp shells, fungi, yeast or green microalgae (Irkin and Esmer 2015).

*Curdlan* is yet another polysaccharide with excellent film-forming ability, abundant, nontoxic, and biodegradable. It is also water insoluble and demonstrates great thermal characteristics which could be incorporated in multilayer packaging materials (Adeyeye et al. 2019; Alizadeh-Sani et al. 2019).

Among biodegradable packaging materials are also ones produced *using traditional chemical synthesis* like polylactic acid (PLA), polyhydroxybutyrate (PHB), and polycaprolactone. PLA is made from renewable resources (sugarcane, corn) and has biocompatible and biodegradable properties. It requires chemical modification to improve its mechanical and physical properties. Despite that, it has the potential to be produced on a large scale and become an eco-friendly packaging material. PHB is created by microorganism and is compatible to dairy products, meat, and beverages. It is an excellent for food packaging films. Moreover, it is nontoxic, water insoluble, and shows better physical characteristics as food packaging material compared to polypropylene. However, due to its high degree of crystallinity, high melting temperature is brittle. Both of those described above polymers can be used in cheese or curdled milk packaging (Roohi et al. 2018; Adeyeye et al. 2019).

*Microbial gums* are nontoxic and have the potential to be produced on an industrial scale. As packaging materials, they are used to make gels, thickening agents, and film solutions. FDA recognized xanthan gum as food additive in 1969. It is an anionic, water-soluble polymer, and is stable at a wide pH and temperature range. Gellan gum is a suitable hydrocolloid for making edible films that are transparent and show good mechanical properties. Pullan has the ability to form colorless, transparent, odorless, and highly water permeable films. It is not used extensively in the packaging industry because of its high cost; however, polysaccharide blends with, e.g., alginate, chitosan, and starch create films easily dissolved in water. This could be utilized in development of edible food coatings with resistance to oxygen permeation (Alizadeh-Sani et al. 2019).

*Protein-based* packaging materials are also biodegradable, low-cost, and abundant. For example, whey, a cheese-making by-product can be filtered and dried to obtain pure whey protein, which can be then used to create protein-based films. Such films can be used for layered plastic films that could form packaging materials in a combination with recyclable plastics. After use, those layers can be chemically separated and plastics recycled (Kuswandi and Moradi 2019). Another protein with film-forming properties is gelatin. It is water soluble, odorless, tasteless, colorless, and transparent and is obtained through partial collagen hydrolysis in powdered or granulated form (Ramos et al. 2016). Due to its water solubility and swelling in contact with high moisture products, it needs to be modified prior to use in dairy applications.

The use of biodegradable materials in food packaging would reduce greenhouse gas emissions and increase the use of renewable sources of raw materials for their production (Sanyang and Sapuan 2015). However, at present, the production of this type of packaging on an industrial scale is insufficient due to insufficient facility of production (Roohi et al. 2018).

### 6.4.4 Biomaterials

Biochemically improved packaging materials are eco-friendly, but they do not necessarily have to be biodegradable as well. These bio-based materials use renewable resources as raw materials to decrease the use of traditional fossil fuels such as crude oil. They are in general nonrecyclable materials, but can be incinerated for energy recovery. As polyethylene terephthalate (PET) is a widely used petroleum-based food packaging material, it can be produced with the use of renewable resources giving rPET. Other examples of bio-based polyesters include:

- Polyethylene furanoate (PEF)
- Polytrimetylenefurandicarboxylate (PTF)

Both of them show improved barrier properties and higher mechanical strength (Kuswandi and Moradi 2019).

### 6.4.5 Nanomaterials

Use of nanoparticles in changing packaging materials characteristic and developing new materials was the result of market demand for packaging assisting in extending shelf life. Nano-based films can have improved barrier properties (e.g., $CO_2$, $O_2$, ethylene, moisture, lipids) which play a significant role in protecting food products from oxidation and spoilage (Poonia 2019). Nano-reinforced traditional polymers used in food packaging usually contain up to 5% w/w nanoparticles (Davarcioglu 2017; Pereda et al. 2018). Moreover, nanoparticles in the polymer matrix of common packaging materials increase physical properties of those materials such as strength, heat resistance, stiffness, shatterproof, and dimensional stability. In order to improve those properties, metal oxides, nanoparticles, nanoclays, carbon nanotubes, and metallic nanoparticles are used. Nanocomposites with nanoclays and layered silicates have improved diffusion paths (tortuous pathways) which in turn increase their barrier properties. However, many advantages of nanoparticles in plastic packaging have been shown in many studies and their main drawback is the polymer transparency reduction (Kuswandi and Moradi 2019). Amjadi et al. (2019) studied gelatin-based nanocomposite-containing chitosan nanofiber and ZnO nanoparticles as potential cheese wrappings. They showed significant decrease in pathogenic bacteria growth (*E. coli*, *S. aureus*, and *P. aeruginosa*). Prepared films did not alter the organoleptic properties of cheese until the end of storage that lasted 12 days.

## 6.5 Active Packaging for Dairy Products

The passive packaging function – protection – has been changed into an active one with the development and application of active packaging. They can take a form active component incorporated in the packaging material or labels, pads, sachets, and other forms to actively alter the conditions inside the packaging. Among active packaging solutions are gas scavengers or emitters, gas absorbers, antimicrobial agents, moisture control agents, antisticking films, and light absorbers (Jafarizadeh-Malmiri et al. 2019; Ščetar et al. 2019; Kuswandi and Moradi 2019). Carbon dioxide scavengers with iron oxide, calcium hydroxide, and activated charcoal in a form of sachets or films can be used in cheese packaging. Also flavor or odor absorbers such as cellulose triacetate, citric acid, ascorbate, clays, zeolites, activated carbon films enable smell and taste preservation, off-odors removal in dairy products (Vigneshwaran et al. 2019).

### 6.5.1 Active Coatings

Preparation of active coatings requires incorporation of active agents into polymer matrix (e.g., PE, PP, LDPE, PET, PS, and PLA) through embedding, immobilization or layer-by-layer deposition techniques. Such coatings can be either migratory or nonmigratory. In migratory active coatings, active substances are meant to migrate to the products for their specific function, e.g., antimicrobial, antioxidant, biocatalytic, nutraceutical (Irkin and Esmer 2015).

#### 6.5.1.1 Enzymatic Coatings

The use of enzymes in active packaging solutions is predominantly used to control the growth of spoilage and/or pathogenic microorganisms. These substances can be incorporated into coatings by embedding and blending. Depending on the preparation technique of such active coatings (covalent immobilization, layer-by-layer deposition, and photografting), they can form migratory or nonmigratory materials. An LDPE film with glucose oxidase and catalase showed up to 97% activity even after exposure to 325 °C and remained its oxygen scavenging properties (Bastarrachea et al. 2015).

Agrillo et al. (2019) studied the antimicrobial effect of activated PET. The activation with small synthetic peptides resulted in the growth inhibition of aerobic plate count within first 24 h and was also effective against *L. monocytogenes* biofilm formation. Also enzymes can act as antimicrobial agents and be applied to different solid materials by means of immobilization. Mirabelli et al. (2018) used hen egg white lysozyme and immobilized it in crystalline form on hydrogel composite membranes (HCMs) to develop an antimicrobial biofilm. The enzyme was crystallized by using HCMs made of poly(vinyl acetate) (PVA) and poly(ethylene glycol) diglycidyl ether (PEGDE) as cross-linker and polypropylene membranes sheets. Lysozyme immobilized on PVA-HCMs in the molecular form demonstrated antimicrobial properties against *Micrococcus lysodeikticus* within 90-min incubation.

#### 6.5.1.2 Antimicrobial Nanocoatings

Intensely studied is antimicrobial packaging using such agents as essential oils, organic acids, peptides, enzymes or biopolymers (Bastarrachea et al. 2015). The reason for this is that bacterial contamination can unfavorably affect the quality, safety, and shelf life of milk and other dairy products. The chemical composition of dairy products makes them an ideal medium for microbial growth (Hutchings et al. 2020). These antimicrobial agents such as nanoparticles of silver, magnesium oxide, copper oxide, titanium dioxide or carbon nanotubes can be coated, laminated, incorporated or immobilized onto polymer surface (Davarcioglu 2017; Kuswandi and Moradi 2019; Poonia 2019). Their performance was tested on Fior di Latte, soft cheese, soft-ripened cheese, and soft white cheese samples, as well as on milk powder or butter (Poonia 2019). Due to antimicrobial activity, zeolite-X and silver ions were investigated to create starch composite films by Elsherif et al. (2020). Starch films with zeolite-X and nanosilver were investigated as active packaging materials for pasteurized milk. Films with both zeolite and nanosilver demonstrated good antimicrobial properties and prolonged shelf life of milk up to 15 compared days to control. Silver nanoparticles are known for their antimicrobial activity. Their use in nanoscale provides relatively larger surface area compared to larger particles above 100 nm which in turn increases the antimicrobial activity at lower concentrations. Silver particles are also capable of inhibiting growth of thermoduric bacteria like *S. thermophilus* that can survive short-term heating (Basavegowda et al. 2020). Although they themselves can along with lactic acid bacteria inhibit the growth of some pathogenic bacteria through lactic acid production that is desired in fermented dairy products, but they can also lead to milk thickening (Braun et al. 2020). Singh et al. (2018) developed a PET packaging film with the addition of silver nanoparticles (AgNPs). The film was obtained by extruding virgin PET with recycled PET (waste-based in concentrations 5, 10, 15, and 20%) and then made into pouches to protect fresh white cheese at different temperatures (6, 25, and 40 °C) for up to 30 days. The antimicrobial activity of the PET/AgNPs packaging was tested against *S. aureus*, *E. coli*, and *C. tropicalis*. In this study, they also tested PET/AgNPs film with chitosan (Cs). The obtained results confirmed antimicrobial activity of active PET films. The PET/AgNPs/Cs film with the addition of 5% AgNPs demonstrated antimicrobial activity of fresh white cheese within 7 days at 40 °C.

Also bimetallic and trimetallic nanoparticles (NPs) have high antimicrobial and antioxidant properties. Intensely studied are silver and titanium dioxide ($TiO_2$) nanoparticles. Moreover, clay nanoparticles used in packaging materials have the ability to protect food from moisture, $CO_2$, and oxygen that can cause spoilage and decrease shelf life (Ligaj et al. 2020). Jin (2017) studied polylactide (PLA) films containing zero-valent iron nanoparticles antimicrobial properties. The obtained films were used as coatings on a polyolefin film and used as goat cream cheese wrappings. They were then tested against microorganisms *E. coli*, *B. subtilis*, *S, epidermis*, *R. rubra,* and *G. candidum*. The addition of 3% of zero-valent iron resulted in full microbial growth inhibition and lower concentrations showed little antimicrobial effect. The prepared wrapping protected goat cream cheese for 6 weeks (stored at chilled temperature). Moreover, the active packaging did not influence the natural microflora of the cheese. These findings are promising for active ripening and cottage cheese packaging which change their desired sensory properties during storage.

#### 6.5.1.3 Essential Oil Coatings

The use of essential oils (EO) as antimicrobial agents in active packaging can prolong shelf life of dairy products. They are also generally recognized as safe (GRAS) and show wide spectrum of pathogen inhibition. EOs are of natural plant origin and are a mixture of volatile compounds like terpens, sesquiterpens, polyphenols, lactones, esters, alcohol, and others (Munekata et al. 2020). In cheese preservation, thymol and linalool essential oils have been known. Cheddar cheese stored at 15 °C over 21-days period in films containing thymol, carvacrol, and linalool effectively reduced the population of *S. aureus* (Jin 2017). Coatings with thyme and clove essential oils in addition of 1.5% demonstrated a decrease in *E. coli* O157:H7 strain in a semihard Turkish cheese (Kashar) during 60 days of chilled storage. Similar antimicrobial properties on Kashar were reported with the use of coatings with ginger (1.5%) EOs after 30 days of storage (Munekata et al. 2020). Another research proved antimicrobial properties of cinnamaldehyde in inhibiting fungi growth on white cheese samples. No fungi were present after 26 days of storage at chilled conditions (Irkin and Esmer 2015). Studies on the addition of EOs to dairy products have shown their antimicrobial properties. These substances could also be used in active packaging since they were successfully used in the products. Treating cheese with natural herbs is known in Italy (Casoperuto, Marzolino, Romano pepato, PiacentinuEnnese), Switzerland, France, Netherlands, Egypt, Syria, Morocco, and Turkey. For example, black cumin seed oil showed antimicrobial activity against pathogenic bacteria in soft cheese, whereas cayenne and green pepper demonstrated growth inhibition toward *S. aureus* in Egyptian Kareish cheese. Moreover, extracts of garlic, lemon grass, cinnamon, sage, salvia, rosemary, thyme, basil or oregano are able to inhibit the growth of *L. monocytogenes* in cheese (e.g., feta cheese, Iranian white cheese). Essential oils can be applied to packaging materials to extend their shelf life and to overcome the problem of defining precise quantity to be added to dairy products (Ritota and Manzi 2020).

#### 6.5.1.4 Bacteriocins

Most of bacteriocins are produced by lactic acid bacteria (LAB); they are thermostable, hypoallergenic, and degrade in human gastrointestinal tract. They are low-molecular-weight peptides or proteins comprised of 30–60 amino acids (Jin 2017). One of bacteriocins is nisin, which is heat-stable, nontoxic, and exhibits a wide spectrum of antimicrobial activity, but also has a GRAS status. It is allowed for use in a range of dairy products like pasteurized cheese, mini red Babybel cheese, ricotta cheese, skimmed milk, soft cheese or BlatackeZlato cheese. Nisin can be immobilized on the surface of traditionally used polymers such as polyamide and polyethylene and in such have the ability to reduce LAB bacteria counts on, e.g., sliced cheese packed in MAP and stored at refrigerated conditions (Irkin and Esmer 2015). Nisin has shown to extend shelf life of raw and pasteurized milk stored at chilled conditions over a 4-day period and cheddar cheese over a 12-week period at 4 °C (Youssef et al. 2017).

### 6.5.2 Oxygen Scavengers

There are many dairy compounds that are susceptible to oxidative degradation such as lipids, vitamins (A, D, E, and C), carotenoids, chlorophyll, and anthocyanins. The process leads to quality deterioration and can take place at any point of the distribution chain. Usually in order to preserve dairy products from oxidation, preservatives are being used. However, due to health awareness and seeking for products with clean labels (as much natural as possible) by consumers, there is a growing necessity for development of different solutions. Antioxidant active packaging can be an alternative to traditionally used preservatives. These are packaging materials with incorporated antioxidants like oxygen scavengers (e.g., $TiO_2$, ferrous compounds metallic salts) in a form of sachets or labels (Bastarrachea et al. 2015; Kuswandi and Moradi 2019). Their action is based on release of antioxidant compound to food and subsequently scavenging undesired substances (Jafarizadeh-Malmiri et al. 2019). They can act either as migratory or nonmigratory antioxidants. Migratory antioxidant active coatings are designed to release the antioxidant in a controlled way during product's shelf life. Nonmigratory active coatings act as scavengers (e.g., oxygen, free radicals, and transition metals) and they do not require direct food contact as they can remove the undesired substances from the packaging headspace altering the conditions inside the packaging and prolonging shelf life. An advantage of nonmigratory active coating in comparison to migratory ones is that the active agents can be applied in lower concentrations which do not alter sensory perception of the product (Bastarrachea et al. 2015).

Active coatings can be an alternative to traditional cheese coatings in controlling the growth of microorganisms, loss of cheese weight, and drying of rind resulting in poor quality and economic losses (Youssef et al. 2017).

## 6.6 Intelligent Packaging for Dairy Products

Intelligent packaging (IP) solutions are developed to inform retailers and/or consumers of the quality and safety of food. They can monitor the conditions of the packed products and give information on temperature conditions in which the product was kept within the supply chain or the integrity of the packaging. Among IP, there are different indicators (time-temperature, gas/integrity, and freshness), data carriers (RFID, barcodes), and sensors (gas, temperature, biosensors, and nanosensors) (Davarcioglu 2017; Jafarizadeh-Malmiri et al. 2019; Tichoniuk et al. 2021). Intelligent packaging can be applied on the primary, secondary or even tertiary (transport) packaging in a form of labels or tags depending on the level at which the product and the conditions of its storage/transport in the supply chain are to be monitored.

Use of intelligent packaging can also help in reducing food waste as now much of the food wasted is due to overdue "use by" date printed on each food packaging. Retailers and consumers depend on that printed date and dispose of food when the date has expired. In some cases, such food can be still safe and of good quality, but in other it can be unconsumable even before the "use by" date, especially if there was abuse of storing or transport conditions within the supply chain, the integrity of a packaging was breached (Maskey et al. 2020).

### 6.6.1 Time-Temperature Indicators

Perishable food like dairy requires temperature control to minimize microbial growth and prolong shelf life. In retail, dairy products are usually stored at 0–4 °C and 4–10 °C at home refrigerators. These conditions, however, do not inhibit the growth of most fungi as they are psychrotrophic in general (Eshaghi et al. 2020). Temperature abuse can occur at any point in the dairy supply chain. That is why temperature monitoring as a quality and safety factors is crucial.

Time-temperature indicators (TTIs) are easy to read devices that are able to show full or partial temperature history of the packed product or it is thawing in case of frozen dairy products like ice-cream. In dairy products, TTIs can also be used in process control of pasteurization or high-temperature short-time pasteurization, sterilization or UHT treatments of milk to ensure that the desired temperature of the process was achieved. These indicators are based on several mechanisms such as mechanical, chemical, electrochemical, enzymatic or microbial irreversible reactions. Their response is visible usually in a color change or a color dye movement indicating temperature abuse above a set threshold (Tichoniuk et al. 2021).

TTIs are commercially available and well known in the US, Asia, and Australia. Examples with thermochromic ink are OnVu® or CoolVu® that can be used in time-temperature monitoring of dairy products (Drago et al. 2020). Their principle is based on an ink that is photosensitive. After activation with UV light, it is in a colored state that gradually returns to its initial colorless state at a temperature-dependent rate. It can be set at different length and temperature sensitivity by controlling the type of photochromic compound and UV light activation time (Taoukis 2010). Another example of an irreversible TTI label is Smart Dot® (Evigence, Israel) presented in Fig. 6.5.

These labels are activated upon placement at a packaging and their color changes gradually over time. This chemical reaction is irreversible and temperature-dependent, the higher the temperature the color changes faster (it does not change at all if kept frozen). Smart Dot® can be calibrated to match the shelf life of a given product in the range from hours to months.

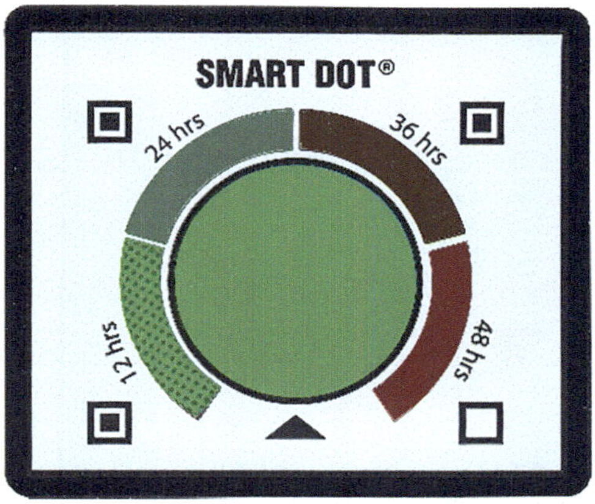


**Fig. 6.5** Evigence Smart Dot® label (Source: Image provided by Evigence (2021))

### 6.6.2 Freshness Indicators

Freshness indicators can give information on the product's quality as they are designed to react with microbial growth metabolites like organic acids, glucose, volatile nitrogen compounds, carbon dioxide, ethanol, biogenic amines, and sulfuric compounds. They detect changes in pH or gas composition (Mirza Alizadeh et al. 2020).

Eshaghi et al. (2020) used a smart label containing beetroot color and multilayers of polystyrene (PS) for milk freshness analysis. They determined the changing label color by calculating label color number. At the beginning of the experiment, the label was dark red and the visible color change was observed in the fifth day of the study. The color started to fade slowly until it turned light brown after 7 days. During the experiment, milk pH, acidity and total bacterial count (TC) were investigated. The label responded to bacterial activity, and changing acidity in pasteurized milk during refrigeration storage. They found a positive correlation between the label color changes and TC, but also pH (Roohi et al. 2018).

Mimica (London, UK) offers a temperature-sensitive indicator for food freshness (Fig. 6.6) which monitors the temperature conditions of a product. The label feels flat in touch when the product is stored at desired temperature, but when there is a temperature abuse it feels bumpy when swiping finger over it. This bumpy feel indicates that the product is no longer good for consumption.

### 6.6.3 Integrity Indicators

Integrity indicators are used to support aseptic or modified atmosphere packaging (MAP) to maintain their functionality. They can be in a form of labels, tablets or printing on the inner side of the packaging. They can either detect an undesired gaseous component in the packaging (headspace) present as a result of leakage or as a volatile compound of the product spoilage. Integrity indicators change color giving

**Fig. 6.6** Mimica freshness label (Source: Image provided by Mimica 2021)

a visible response easy to read in contact with the target gaseous compound. Unsealing of the packaging can lead to increase of oxygen concentration inside the packaging and as such can be detected with the use of an Ageless Eye oxygen indicator (Tichoniuk et al. 2021). Packed dairy products are susceptible to high oxygen levels which can contribute to product spoilage, undesired oxidation of lipids and micronutrients, growth of aerobic microbes, and also color, flavor, and odor changes. Maintaining low oxygen concentration in the packaging is crucial in MAP packaging. The above-mentioned Ageless Eye oxygen indicator is a colorimetric sensor in a form of a pink tablet that changes color to blue in the presence of oxygen ($\geq$ 0.5%) (Mirza Alizadeh et al. 2020).

### 6.6.4 Data Carriers

Data carriers are used in commerce and improve the purchasing process, inventory management, and traceability. According to Article 17 of the EU Regulation No 1935/2004, *the traceability of materials and articles shall be ensured at all stages in order to facilitate control, the recall of defective products, consumer information, and the attribution of responsibility*. Among traceability basic data are: producer's name and address, article number, and production date/identification of the product (Mania et al. 2018b).

#### 6.6.4.1 Graphic Codes

This type of intelligent packaging systems includes well-known GS1 identifiers barcodes such as EAN-8, EAN-13 or code-128 (Fig. 6.7). The use of such graphic codes creates a standardized global data format that is readable in a product's life cycle (Vigneshwaran et al. 2019). They can facilitate product tracking from farm to fork.

Quick response (QR) codes (as presented in Fig. 6.8) are in some ways better or more advanced than traditional barcodes, as they can store mode data and can be readable by means of smartphones or computers (Tan and Ngan 2020). Those two

**Fig. 6.7** An example of a Code-128 barcode (left) and EAN-13 code (right) (Source: generated at https://www.cognex.com/resources/interactive-tools/free-barcode-generator)

**Fig. 6.8** Examples of OR codes (Source: own work)

dimensional (2D) codes have the ability to store more data than barcodes of up to 1.1 kilobytes (Mirza Alizadeh et al. 2020).

QR codes can encode more data thanks to different modes such as numeric, alphanumeric, binary, and kanji (Drago et al. 2020).

#### 6.6.4.2 Radio Frequency Identification

Radio frequency identification (RFID) systems are an invisible reading mode in which data are decoded without human interaction (Tan and Ngan 2020). They are active tags that transfer data from a tag attached to a packaging to automatically track and trace it as well as identify it (Kuswandi 2016). RFID technology within the supply chain can improve inventory control, allow shelf life monitoring, and provide information on the temperature history of the product. RFID tags can also change the checkout method at the retail stores; because they do not require direct scanning of a readable barcode by a scanner. They simply need to be in the vicinity of checkout equipped with a RFID transducer. In the dairy sector, they are usually utilized to provide complete traceability of products (Mirza Alizadeh et al. 2020). Commercial applications of RFID solutions are brought by CAEN RFID (Viareggio, Italy) like an Easy2Log (A927Z), Easy2Log (RT0005) or qLog Temperature (RT0012) tags

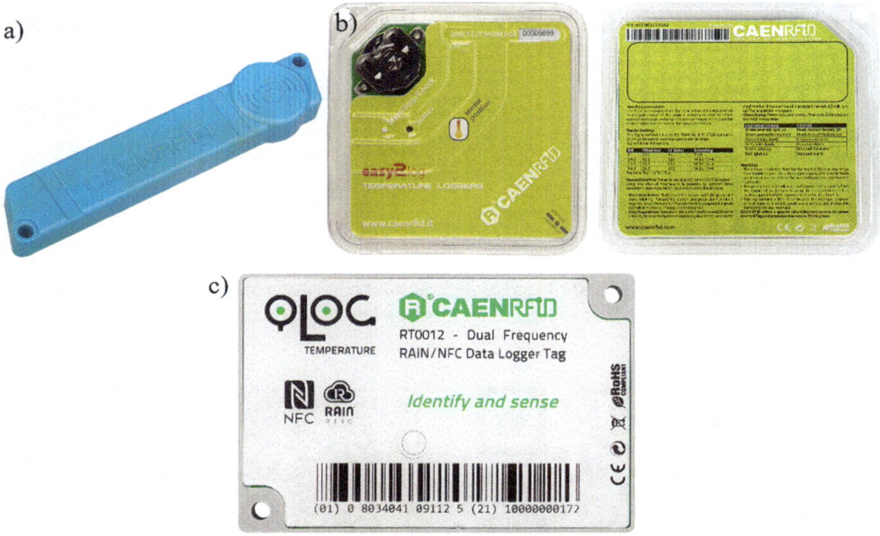


**Fig. 6.9** Examples of CAEN RFID tags: (**a**) Easy2Log - A927Z, (**b**) Easy2Log – RT0005, and (**c**) qLog Temperature (Source: Image provided by CAEN RFID (n.d.))

presented in Fig. 6.9. These tags can be applied in milk-based products. The A927Z is a low-cost, semipassive RAIN RFID logger tag with a rugged enclosure. It allows for temperature monitoring of perishable products. The tag can be used with standard RAIN RFID readers as it is compatible with EPCGlobal C1G2 and ISO 18000-63 standards. It can operate in the presence of strong vibrations and can store up to 8000 samples that can be preset in intervals from 8 s to 18 h. Alarms can be defined to control the temperature regime and when temperature is below or over a set threshold, user is informed.

The Easy2Log (RT0005) RAIN RFID tag is similar to the previously mentioned one. However, it can store up to 3958 samples and 16 temperature ranges with independent threshold alarms can be defined by user. It is also able to calculate the mean kinetic temperature thus enabling remaining shelf life configuration. The qLog Temperature tag is also applicable in milk-based products. It can store up to 4096 samples and their intervals can range from 5 s to 18 h. It has a rugged enclosure like the A927Z tag and can be used in air shipments. Moreover, it is equipped with an NFC interface which enables consumers to get information on the conditions of the product via smartphone (CAEN RFID n.d.).

### 6.6.4.3 Near-Field Communication

Near-field communication (NFC) is a technology that allows two NFC-enabled devices to communicate when held in close distance (ca. 4 cm). This technology evolved from radio frequency identification; here an NFC chip operates as one part of a wireless link. Once it is activated by another chip, small amounts of data between the two devices can be transferred. It does not require parring to connect the devices as it uses chips which are more power-efficient compared to other wireless types of communication. NFC smart packaging can be used to provide customers personalized experiences like with product information, promotional offers, how-to-guides, videos, and reordering reminders available with a simple tap of a smartphone. It can enhance brand loyalty as NFC tags are placed on each item, because they are small. This type of packaging has grown in popularity especially within the fast-moving consumer goods (FMCG). Yeo Valley Organic has introduced connected packaging across all products as part of the new "Put Nature First" brand platform. Consumers will be able to use their smartphones to scan the "Moo-R" QR codes which can be found on over 100 million products, linking to its "Yeokens" reward program, and bringing to life ongoing nature-led initiatives. Customers will have access to specific product information, brand content, and future promotions (AIPA 2020).

Maskey et al. (2020) designed a time-temperature history (TTH)-based wireless food label with an NFC antenna. The obtained label could be seamlessly integrated to the packaging. The label was designed by bridging a single Si-chip signal controller and transponder onto a roll-to-roll (R2R) gravure-printed antenna and thermistor. It was able to transfer logged temperature data to a smartphone using the ISO/IEC 15693 protocol. Such intelligent communication devices can inform retailers and/or consumers via smartphones of the product's quality in real-time.

#### 6.6.4.4 Blockchain Technology

Blockchain technology is a decentralized database that allows collecting and transferring information on the Internet network with peer-to-peer architecture. It is a distributed register of operations that are not handled by a single centralized server, but by computers connected together in a network. Information on these operations is stored here in batches (blocks) which are linked together by cryptography to form a blockchain. Blockchain is not governed by top-down authority and cannot be controlled by anyone. On the other hand, it is most often open source, which means that all its users have free access to it. More specifically, they can view the entire history of operations, but at the same time cannot edit any related data. What is more, every transaction that has been saved in the blockchain stays there forever. All these make blockchain an effective and secure way to store and transmit a variety of information.

This technology can be implemented to improve transparency and supply chain traceability "from farm to fork". In other words, it can enable access to information like origins of raw materials, finished products, the different processes it underwent thus ensuring food safety and quality. However, there are not many dedicated applications and to lower costs companies try to take advantage of existing integrated enterprise information systems like ERP. The most implemented traceability solutions so far are RFID and barcodes which are used for example by Walmart, TESCO, METRO Cach&Carry (Tan and Ngan 2020).

## 6.6.5 Sensors

There is a growing consumer demand for ensuring the safety of food products. They can only rely on their sensory evaluation and "use by" date to determine the quality of a purchased food item. Smart packaging is therefore a kind of safety indicator and determines/monitors the quality of the product. Sensors are more sophisticated devices than indicators as they are made of a receptor and a transducer. They should be reversible in their action to constantly (in real-time) monitor changes within the packaging headspace or the product itself. Sensors are able to detect, locate or quantify target substance and describe its physical or chemical characteristics (Park et al. 2015; Pereda et al. 2018; Tichoniuk et al. 2021). They can detect gases released during spoiling of food, monitor pH, temperature or color of the packed food (Kuswandi 2016; Basavegowda et al. 2020). Berna (2010) mentioned $SnO_2$ and other metal oxide semiconductor sensors to detect adulteration, contamination or off-flavor in milk and dairy products. Such sensors used in electronic noses that detect volatile compounds are also capable of determining aging/ripening or cheese type, but also origin of milk or caseinates. Moreover, sensors can be an alternative to traditional methods such as chromatographic techniques coupled with for example, mass spectrometry (expensive, time- consuming, and labor intensive) (Campanella and Tomassetti 2019).

### 6.6.5.1 Nanosensors

This type of sensors can be used to detect microbial contaminants, toxins, food-borne pathogenic bacteria, and facilitate food testing. Nanosensors have the advantage over traditional methods that they can rapidly detect analytes like microbes, contaminants within minutes up to days. They can inform consumers and retailers of the product's temperature, light or oxygen exposure history (Jafarizadeh-Malmiri et al. 2019; Basavegowda et al. 2020). In nanosensors, different particles in nanoscale can be used such as carbon tubes, nanoshells, dendrimers, quantum dots, nanorods, liposomes, and others. They are also able to detect allergens in dairy products (optical nanosensors). In addition, these sensors are capable of detecting gas content (oxygen, carbon dioxide) in the packaging headspace in a noninvasive and continuous manner, which is very important in MAP packaging systems to ensure packaging integrity (Kabariya and Ramani 2017).

### 6.6.5.2 Biosensors

First biosensors were designed in 1950s and since then they have been developed into more accurate and rapid solutions for monitoring food quality and safety. Their performance is based on their sensitivity, reproducibility, portability, high noise ratio, appropriate storage conditions, response time, and ease of use (Kabariya and Ramani 2017). Depending on the receptors used, there can be:

- Immunosensors
- DNA sensors

- Microbial sensors
- Enzymatic sensors (Chauhan et al. 2019)

They are comprised of protein, enzymes, oligonucleotide, cells or tissues that are used to generate a particular signal toward the objective analyte (Kabariya and Ramani 2017).

In dairy products like milk or cheese to detect pathogenic microorganism, a nanocomposite-associated smartphone-based immunosensor (pathogen—*S. enteritidis*) and UCNP-conjugated antibody biosensor (pathogens—*E. coli*, *S. aureus*) were studied. Moreover, to detect toxins in skimmed milk (botulinum neurotoxin A), a graphene oxide-based FRET biosensor and botulinum neurotoxin in milk and a graphene nanocomposite-based impedimetric immunosensor were investigated. There is also a commercially available sensor like Assurance® (Neogen Corp., USA) can detect enterohemorrhagic *E. coli* (EHEC) and *L. monocytogenes* in liquid milk (Chauhan et al. 2019).

Commercially available biosensor assay kit for determination of lactose levels in lactose-free or low-lactose dairy products is LactoSens®(DirectSens, Austria) shown in Fig. 6.10.

It consists of disposable test strips with an immobilized enzyme which oxidizes lactose in the sample and electrons are detected amperometrically by the detector as shown in Fig. 6.11 (Halbmayr-Jech et al. 2021).

Moreover, each strip has a QR label for sample tracking and lot-specific information. It is simple in its operation and requires only 1 mL of a liquid product sample or a homogenized solid sample and a portion (1 mL) of a LactoSens Buffer. Then the Reader is to be connected to a computer and a dedicated software changes the

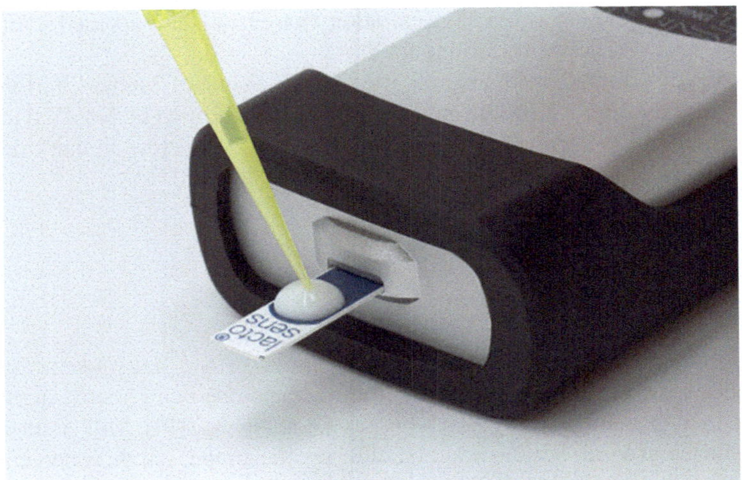


**Fig. 6.10** LactoSens® biosensor assay kit for lactose detection in dairy products (Source: Image provided by DirectSens (2021))

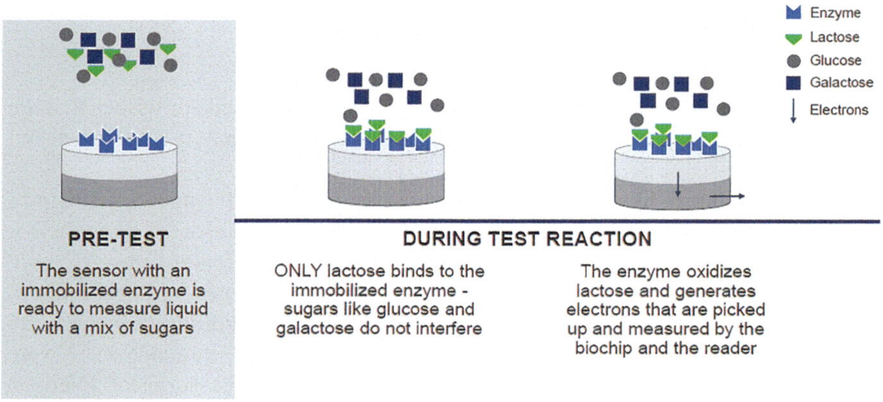


**Fig. 6.11** LactoSens® mode of action (Source: Image provided by DirectSens (2021))

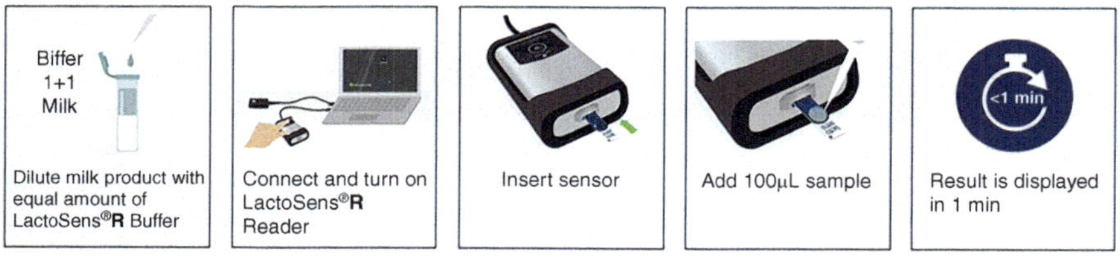


**Fig. 6.12** LactoSens® operating procedurę step-by-step (Source: Image provided by DirectSens (2021))

electrical signal into a lactose concentration (based on the specified calibration curve) (Fig. 6.12) (Halbmayr-Jech et al. 2021).

LactoSens® was validated according to AOAC Standard Method Performance Requirements (SMPR®) 2018.009 and the method was certified by NordVal in 2018. This is a rapid and selective method for determination of lactose in dairy products and products with dairy ingredients (Halbmayr-Jech et al. 2021).

## 6.7 Fraud in the Dairy Sector

Nowadays there can be observed two challenges in the dairy sector such as adulterations and counterfeit products and the second one being increasing the use of blockchain technology and traceability (Tan and Ngan 2020). Milk adulterations can take different forms such as addition of whey, melamine, starch, water, chlorine, formalin or mixing milk from different species. Moreover nondeclared substances can be added to milk and dairy products (e.g., urea, sodium bicarbonate, dicyandiamide) or mislabeling of geographical origin or as an organic food (Sharma et al. 2019; Hassoun et al. 2020).

Improved traceability and security (fraud control) of products, especially food products can be obtained through the use of global standard like GS1. The widely known and recognizable EAN-13 barcodes and the accompanying GTIN-13 numeric codes allow for better traceability of products in the supply chain and verification of their authenticity. The range of use covers 150 countries and about two million companies in the world. Each manufacturer has the rights to a specific GTIN-13 with which they can legally code their products (their brand). Unfortunately, with the growing popularity of e-commerce, the theft of these codes and impersonation of other companies under existing brands have increased. It is, however, possible to be picked up by GS1 and even taken to court. Therefore, from producers' perspective, better traceability allows for quicker response within the supply chain and products recall, but also fraud control enables maintaining the desired brand image, credibility in the eyes of consumers, and their safety. From consumers' point of view, those solutions/standards facilitate retail, especially in self-service stores and/or with self-service checkouts. These factors have become increasingly important in recent years (Mania et al. 2018b; Roohi et al. 2018). Traceability is crucial among other factors in dairy sector as it supports product quality and tacking in the entire life cycle of the product providing information even on rearing of milking cows. However, there was a necessity to develop a regional standard namely a Chinese Sensible Code that is adjusted to Chinese character environment (Li et al. 2015).

Also NFC tags provide product unique IDs that prevent counterfeiting and can support authentication of the product both by brand owner and consumer. Intelligent packaging is often utilized as tracking devices to ensure food security and avoid fake products (Davarcioglu 2017).

Development of tamper-evident packaging solutions is gaining speed. They are often additional packaging materials like special bottle liners or closures or changing color upon opening heat seals (Smirnova et al. 2020).

Among most counterfeited foods is milk, just after olive oil. This usually takes the form composition and quality falsification. Most common is diluting milk with water, milk whey, and addition of milk substitutes, preservatives, and N-containing substances. Milk fat is often replaced by cheaper plant analogs (Smirnova et al. 2020). In analysis of dairy products for adulterants like starch, urea, hydrogen peroxide, neutralizer, detergents, boric acid, melamine, mycotoxins or bacteria, different nanosensors can be used. Traditional methods are time-consuming, expensive, and require skilled personnel to conduct them. It is crucial to analyze the presence and amounts of those adulterants in dairy products and compare those concentrations with limits set by governmental food authorities around the world (Berna 2010; Kabariya and Ramani 2017). Examples of adulterants detected by means of nanosensors are presented in Table 6.3.

One of illegally added into dairy products substance is melamine which increases protein levels in the product. It is a small organic compound that is difficult to determine by a standard Kjeldahl method. In human bodies, melamine forms

**Table 6.3** Adulterants and nanosensors used to detect them in dairy products

| Dairy product adulterant | Nanosensor | Detection tool/technique |
|---|---|---|
| Melamine | Gold nanoparticle<br>Water-soluble CdTe quantum dots<br>Single-wall carbon nanotube | Colorimetric probe<br>Standard colorimetric card<br>Surface-enhanced Raman spectroscopy<br>Fluorescence probe<br>Electrochemical luminescence |
| Bacterial pathogens<br>*Salmonella*<br>*E. coli* O157:H7<br>*Staphylococcal enterotoxin B* | Carbon nanotubes<br>Magnetic nanoparticles and $TiO_2$ nanocrystals<br>Oligonucleotide-functionalized Au<br>Gold nanoparticles | Electrochemical sensor<br>Optical nanocrystal probes<br>Piezoelectric biosensor<br>Chemiluminescence |
| Toxins:<br>Brevotoxins<br>Aflatoxin<br>Aflatoxin B1<br>Mycotoxin<br>Ochratoxin A | Au NP-PAADs<br>Functionalized-gold nanoparticles<br>Silver core and gold shell<br>Nanostructured zinc oxide<br>Single-walled carbon nanotubes | Electrochemical immunosensor<br>immunoelectrode<br>immunodipstick assay<br>indium-tin-oxide glass plate<br>Fluorescent aptasensor |

Source: Based on (Kabariya and Ramani 2017)

insoluble melamine-cyanurate crystals in their kidneys which can lead to kidney damage or death. There was a melamine scandal in 2008 in China as it was found in milk and lead to several infant deaths and thousands suffered from kidney failures (Yang et al. 2018; Sharma et al. 2019; Shellaiah and Sun 2019; Montgomery et al. 2020). Melamine in raw milk can be detected by dopamine-stabilized silver nanoparticles as a colorimetric reader. It is a one-step assay that is simple and rapid, but also of high sensitivity (Poonia 2019). Other nanomaterial-based melamine detectors like quantum dots, nanocrystals, nanorods, and nanotubes were reviewed by Shellaiah and Sun (2019).

As an example of product authentication, it is worth mentioning cases of counterfeiting Mozarella di Bufala Campana Protected Designation of Origin. Novel analytical methods have been proposed to differentiate milk and cheese product from Protected Designation Area from those produced outside of it (Hassoun et al. 2020).

Despite development of novel adulterations detection methods, the battle against counterfeiting of milk and dairy products is still a challenge. One of the reasons is that there are cases where it is difficult to define the type of fraud or detect it. Aiding the monitoring and assessment of contamination issues are various country regulations, governing bodies, and companies that run their own tests. There are also databases that gather and provide information on food safety like in the EU Rapid Alert System for Food and Feed (RASFF) of Fera'sHorizonScan or the EU Food Frau Network (FFN) (Montgomery et al. 2020).

## 6.8 Conclusions

Food waste and losses can occur at any point in the supply chain especially within perishable goods like dairy products. Millions of tons of food waste can be avoided through the use of proper packaging. They no longer simply protect the product, but also can prolong its shelf life and ensure customers of the product's quality, safety or integrity. The use of active packaging systems offers a variety of solutions supporting shelf life and safety. On the other hand, intelligent packaging improves product's traceability and provides much crucial information on the product's conditions or possible adulterations. Novel edible coatings allow for improved protection of dairy products, can also act as active compounds or nutrients carriers with controlled release. Moreover, they can be an alternative to plastic packaging and help minimize the use of petroleum-based packaging materials. Those novel technologies enable producers to adapt to the constantly changing requirements on the dairy market. In addition, they can support inventory and supply chain management, enhance traceability, as well as be environmentally friendly and contribute to reducing food waste.

Innovations on the food packaging market and in the area of dairy products are dictated by both the changing and constantly growing requirements of consumers, market participants, as well as economic and environmental aspects. Growing consumer awareness of the quality, safety, and health benefits of dairy products drives innovation in this market segment. In turn, to meet these requirements and to extend the shelf life of products, ensure their authenticity, and improve logistics processes, appropriate packaging is necessary. Packaging innovation goes along with product development and is part of the fourth industrial revolution (Industry 4.0). Novel packaging materials and solutions add value in the food supply chain. In addition, it can help to reduce food waste and economic losses.

## References


Adeyeye OA, Sadiku ER, Babu Reddy A, et al (2019) The use of biopolymers in food packaging, pp 137–158. https://doi.org/10.1007/978-981-13-8063-1_6

Agrillo B, Balestrieri M, Gogliettino M et al (2019) Functionalized polymeric materials with bio-derived antimicrobial peptides for "active" packaging. Int J Mol Sci 20:601. https://doi.org/10.3390/ijms20030601

AIPA (2020) 100 million connected packs rolled out for UK dairy brand—News on smart, active & intelligent packaging. https://www.aipia.info/news-100-million-connected-packs-rolled-out-for-UK-dairy-Brand-1257.php. Accessed 21 Apr 2021

Alizadeh-Sani M, Ehsani A, Moghaddas Kia E, Khezerlou A (2019) Microbial gums: introducing a novel functional component of edible coatings and packaging. Appl Microbiol Biotechnol 103: 6853–6866. https://doi.org/10.1007/s00253-019-09966-x

Amjadi S, Emaminia S, Nazari M et al (2019) Application of reinforced ZnO nanoparticle-incorporated gelatin bionanocomposite film with chitosan nanofiber for packaging of chicken fillet and cheese as food models. Food Bioprocess Technol 12:1205–1219. https://doi.org/10.1007/s11947-019-02286-y

Anonymous (2019) Ampacet launches high-opacity masterbatch to protect dairy products from photodegradation. Addit Polym 2019:2. https://doi.org/10.1016/S0306-3747(19)30140-X

Ávila BP, da Rosa PP, Fernandes TA et al (2020) Analysis of the perception and behaviour of consumers regarding probiotic dairy products. Int Dairy J 106:104703. https://doi.org/10.1016/j.idairyj.2020.104703

Basavegowda N, Mandal TK, Baek KH (2020) Bimetallic and trimetallic nanoparticles for active food packaging applications: a review. Food Bioprocess Technol 13:30–44. https://doi.org/10.1007/s11947-019-02370-3

Bastarrachea LJ, Wong DE, Roman MJ et al (2015) Active packaging coatings. Coatings 5:771–791. https://doi.org/10.3390/coatings5040771

Berna A (2010) Metal oxide sensors for electronic noses and their application to food analysis. Sensors 10:3882–3910. https://doi.org/10.3390/s100403882

Bilska B, Kołozyn-Krajewska D (2019) Risk management of dairy product losses as a tool to improve the environment and food rescue. Foods 8:481. https://doi.org/10.3390/foods8100481

Braun S, Ilberg V, Blum U, Langowski HC (2020) Nanosilver in dairy applications—Antimicrobial effects on Streptococcus thermophilus and chemical interactions. Int J Dairy Technol 73:376–383. https://doi.org/10.1111/1471-0307.12686

Byberg L, Lemming EW (2020) Milk consumption for the prevention of fragility fractures. Nutrients 12:1–22. https://doi.org/10.3390/nu12092720

CAEN RFID (n.d.) Manuals & documents—RAIN RFID, UHF RFID readers—CAEN RFID. https://www.caenrfid.com/en/support/manuals-documents/. Accessed 22 Apr 2021

Campanella L, Tomassetti M (2019) The applications of sensors and biosensors in investigating drugs, foods, and nutraceuticals. Sensors (Switzerland) 19:3395. https://doi.org/10.3390/s19153395

Chadwicks (2021). https://chadwickslids.com/lidding-innovation/. Accessed 17 Apr 2021

Chauhan N, Jain U, Soni S (2019) Sensors for food quality monitoring. In: Nanoscience for sustainable agriculture. Springer International Publishing, pp 601–626. https://doi.org/10.1007/978-3-319-97852-9_23

Chávez-Martínez A, Reyes-Villagrana RA, Rentería-Monterrubio AL et al (2020) Low- and high-intensity ultrasound in dairy products: applications and effects on physicochemical and microbiological quality. Foods 9:1688. https://doi.org/10.3390/foods9111688

Clarke HJ, O'Sullivan MG, Kerry JP, Kilcawley KN (2020) Correlating volatile lipid oxidation compounds with consumer sensory data in dairy-based powders during storage. Antioxidants 9:338. https://doi.org/10.3390/antiox9040338

Crockett SL (2015) Allanblackia oil: phytochemistry and use as a functional food. Int J Mol Sci 16:22333–22349. https://doi.org/10.3390/ijms160922333

Davarcioglu B (2017) Nanotechnology applications in food packaging industry. In: Nanotechnology: food and environmental paradigm. Springer, Singapore, pp 87–113. https://doi.org/10.1007/978-981-10-4678-0_6

Dekker PJT, Koenders D, Bruins MJ (2019) Lactose-free dairy products: Market developments, production, nutrition and health benefits. Nutrients 11:551. https://doi.org/10.3390/nu11030551

Delorme MM, Guimarães JT, Coutinho NM et al (2020) Ultraviolet radiation: an interesting technology to preserve quality and safety of milk and dairy foods. Trends Food Sci Technol 102:146–154. https://doi.org/10.1016/j.tifs.2020.06.001

Dimitrellou D, Solomakou N, Kokkinomagoulos E, Kandylis P (2020) Yogurts supplemented with juices from grapes and berries. Foods 9:1158. https://doi.org/10.3390/foods9091158

Drago E, Campardelli R, Pettinato M, Perego P (2020) Innovations in smart packaging concepts for food: an extensive review. Foods 9:1628. https://doi.org/10.3390/foods9111628

Elopak (2020) #2 Elotalk news and views from Elopak's world of packaging. https://www.elopak.com/wp-content/uploads/2020/09/Elotalk-2-2020_WEB.pdf. Accessed 17 Apr 2021

Elsherif WM, el Hendy AHM, Elnisr NA, Zakaria IM (2020) Ameliorative effect of zeolite packaging on shelf life of milk. J Packag Technol Res 4:171–186. https://doi.org/10.1007/s41783-020-00093-x

Eshaghi R, Khalili Sadrabad E, Jebali A et al (2020) Application of pH indicator label based on beetroot color for determination of milk freshness. J Environ Health Sustain Dev 5:1053–1062. https://doi.org/10.18502/jehsd.v5i3.4277
Faccia M (2020a) The flavor of dairy products from grass-fed cows. Foods 9:1188. https://doi.org/10.3390/foods9091188
Faccia M (2020b) Chemical and technological characterization of dairy products. Foods 9:1475. https://doi.org/10.3390/foods9101475
Faccia M, D'Alessandro AG, Summer A, Hailu Y (2020) Milk products from minor dairy species: a review. Animals 10:1–25. https://doi.org/10.3390/ani10081260
Farag MA, Jomaa SA, El-wahed AA, El-seedi HR (2020) The many faces of kefir fermented dairy products: quality characteristics, flavour chemistry, nutritional value, health benefits, and safety. Nutrients 12:346. https://doi.org/10.3390/nu12020346
Garnier L, Valence F, Mounier J (2017) Diversity and control of spoilage fungi in dairy products: an update. Microorganisms 5:42. https://doi.org/10.3390/microorganisms5030042
Graça C, Raymundo A, Sousa I (2019) Wheat bread with dairy products-technology, nutritional, and sensory properties. Appl Sci 9:4101. https://doi.org/10.3390/app9194101
Guarnaschelli A, Salomone HE, Méndez CA (2020) A stochastic approach for integrated production and distribution planning in dairy supply chains. Comp Chem Eng 140:106966. https://doi.org/10.1016/j.compchemeng.2020.106966
Halbmayr-Jech E, Kittl R, Weinmann P et al (2021) Determination of lactose in lactose-free and low-lactose milk, milk products, and products containing dairy ingredients by the LactoSensVR R amperometry method: first action 2020.01. J AOAC Int 103:1534–1546. https://doi.org/10.1093/JAOACINT/QSAA080
Hassoun A, Måge I, Schmidt WF et al (2020) Fraud in animal origin food products: Advances in emerging spectroscopic detection methods over the past five years. Foods 9:1069. https://doi.org/10.3390/foods9081069
Hutchings C, Rajasekharan SK, Reifen R, Shemesh M (2020) Mitigating milk-associated bacteria through inducing zinc ions antibiofilm activity. Foods:9, 1094. https://doi.org/10.3390/foods9081094
International Dairy Federation (2018). https://www.fil-idf.org/wp-content/uploads/2018/02/Factsheet-001_Heat-treatment-1-1.pdf. Accessed 16 Apr 2021
Iriondo-Dehond M, Miguel E, del Castillo MD (2018) Food byproducts as sustainable ingredients for innovative and healthy dairy foods. Nutrients 10:1358. https://doi.org/10.3390/nu10101358
Irkin R, Esmer OK (2015) Novel food packaging systems with natural antimicrobial agents. J Food Sci Technol 52:6095–6111. https://doi.org/10.1007/s13197-015-1780-9
Jafarizadeh-Malmiri H, Sayyar Z, Anarjan N et al (2019) Nanobiotechnology in food packaging. In: Nanobiotechnology in food: concepts, applications and perspectives. Springer International Publishing, pp 69–79. https://doi.org/10.1007/978-3-030-05846-3_5
Jeske S, Zannini E, Arendt EK (2018) Past, present and future: the strength of plant-based dairy substitutes based on gluten-free raw materials. Food Res Int 110:42–51. https://doi.org/10.1016/j.foodres.2017.03.045
Jin TZ (2017) Current state of the art and recent innovations for antimicrobial food packaging. In: Microbial control and food preservation. Springer, New York, pp 349–372. https://doi.org/10.1007/978-1-4939-7556-3_16
Kabariya JH, Ramani VM (2017) Nanobiosensors, as a next-generation diagnostic device for quality & safety of food and dairy product. In: Nanotechnology: food and environmental paradigm. Springer, Singapore, pp 115–129. https://doi.org/10.1007/978-981-10-4678-0_7
Kuswandi B (2016) Nanotechnology in food packaging, pp 151–183. https://doi.org/10.1007/978-3-319-39303-2_6
Kuswandi B (2017) Environmental friendly food nano-packaging. Environ Chem Lett 15:205–221. https://doi.org/10.1007/s10311-017-0613-7
Kuswandi B, Moradi M (2019) Improvement of food packaging based on functional nanomaterial. In: Nanotechnology: applications in energy, drug and food. Springer International Publishing, pp 309–344. https://doi.org/10.1007/978-3-319-99602-8_16

Lai G, Pes M, Addis M, Pirisi A (2020) A cluster project approach to develop new functional dairy products from sheep and goat milk. Dairy 1:154–168. https://doi.org/10.3390/dairy1020010
Leialohilani A, de Boer A (2020) EU food legislation impacts innovation in the area of plant-based dairy alternatives. Trends Food Sci Technol 104:262–267. https://doi.org/10.1016/j.tifs.2020.07.021
Li T, Li L, Liu S (2015) Study of traceability system of dairy products based on Chinese-sensible code. In: LISS 2014. Springer Berlin Heidelberg, pp 771–776. https://doi.org/10.1007/978-3-662-43871-8_111
Ligaj M, Tichoniuk M, Cierpiszewski R, Foltynowicz Z (2020) Efficiency of novel antimicrobial coating based on iron nanoparticles for dairy products' packaging. Coatings 10:156. https://doi.org/10.3390/coatings10020156
Lipińska M, Tomaszewska M, Kołozyn-Krajewska D (2019) Identifying factors associated with food losses during transportation: potentials for social purposes. Sustainability 11:2046. https://doi.org/10.3390/su11072046
Mania I, Delgado AM, Barone C, et al (2018a) Food packaging materials in the cheesemaking field. In: Traceability in the dairy industry in Europe. Springer International Publishing, pp 141–145. https://doi.org/10.1007/978-3-030-00446-0_9
Mania I, Delgado AM, Barone C, et al (2018b) Food packaging and the mandatory traceability in Europe. In: Traceability in the dairy industry in Europe. Springer International Publishing, pp 129–139. https://doi.org/10.1007/978-3-030-00446-0_8
Maskey BB, Lee J, Majima Y et al (2020) A smart food label utilizing roll-to-roll gravure printed NFC antenna and thermistor to replace existing "use-by" date system. IEEE Sensors J 20:2106–2116. https://doi.org/10.1109/JSEN.2019.2948752
Mirabelli V, Majidi Salehi S, Angiolillo L et al (2018) Enzyme crystals and hydrogel composite membranes as new active food packaging material. Global Challenges 2:1700089. https://doi.org/10.1002/gch2.201700089
Mirza Alizadeh A, Masoomian M, Shakooie M, et al (2020) Trends and applications of intelligent packaging in dairy products: a review. Crit Rev Food Sci Nutr. 1–15. https://doi.org/10.1080/10408398.2020.1817847
Mohamed MSM, Elshaghabee FMF, Alharbi SA, El-Hussein A (2020) The prospective beneficial effects of red laser exposure on lactocaseibacilluscasei fermentation of skim milk. Biology 9:1–12. https://doi.org/10.3390/biology9090256
Montgomery H, Haughey SA, Elliott CT (2020) Recent food safety and fraud issues within the dairy supply chain (2015–2019). Global Food Security 26:100447. https://doi.org/10.1016/j.gfs.2020.100447
Munekata PES, Pateiro M, Rodríguez-Lázaro D et al (2020) The role of essential oils against pathogenic Escherichia coli in food products. Microorganisms 8:1–16. https://doi.org/10.3390/microorganisms8060924
Najib M, Hallab MW, Hallab K et al (2020) Thermal processing of milk as a main tool in the production of Qishta, Khoa and Kajmak. J Mater Environ Sci 2020:294–309
Newbold D, Koppel K (2018) Carbonated dairy beverages: challenges and opportunities. Beverages 4:66. https://doi.org/10.3390/beverages4030066
Odueke OB, Farag KW, Baines RN, Chadd SA (2016) Irradiation applications in dairy products: a review. Food Bioprocess Technol 9:751–767. https://doi.org/10.1007/s11947-016-1709-y
Omeni A, Daniel A (2020) Influence of packaging and brand name on the customer perception of dairy product quality in Lagos, Nigeria. 22:1–3. https://doi.org/10.9790/487X-2205030111
Owusu-Kwarteng J, Akabanda F, Agyei D, Jespersen L (2020) Microbial safety of milk production and fermented dairy products in Africa. Microorganisms 8:752. https://doi.org/10.3390/microorganisms8050752
Park YW, Kim SM, Lee JY, Jang W (2015) Application of biosensors in smart packaging. Mol Cell Toxicol 11:277–285. https://doi.org/10.1007/s13273-015-0027-1
Paul AA, Kumar S, Kumar V, Sharma R (2020) Milk analog: plant-based alternatives to conventional milk, production, potential and health concerns. Crit Rev Food Sci Nutr 60:3005–3023. https://doi.org/10.1080/10408398.2019.1674243

Pereda M, Marcovich NE, Ansorena MR (2018) Nanotechnology in food packaging applications: barrier materials, antimicrobial agents, sensors, and safety assessment. In: Handbook of ecomaterials. Springer International Publishing, pp 1–22. https://doi.org/10.1007/978-3-319-48281-1_30-1

Poonia A (2019) Recent trends in nanomaterials used in dairy industry. In: Nanoscience for sustainable agriculture. Springer International Publishing, pp 375–396. https://doi.org/10.1007/978-3-319-97852-9_17

Ramos M, Valdés A, Beltrán A, Garrigós M (2016) Gelatin-based films and coatings for food packaging applications. Coatings 6:41. https://doi.org/10.3390/coatings6040041

Ritota M, Manzi P (2020) Natural preservatives from plant in cheese making. Animals 10:749. https://doi.org/10.3390/ani10040749

Roohi SP, Bano K, Zaheer MR, Kuddus M (2018) Biodegradable smart biopolymers for food packaging: sustainable approach toward green environment. In: Bio-based materials for food packaging: green and sustainable advanced packaging materials. Springer, Singapore, pp 197–216. https://doi.org/10.1007/978-981-13-1909-9_9

Sajdakowska M, Gębski J, Guzek D et al (2020) Dairy products quality from a consumer point of view: study among Polish adults. Nutrients 12:1503. https://doi.org/10.3390/nu12051503

Sanyang ML, Sapuan SM (2015) Development of expert system for biobased polymer material selection: food packaging application. J Food Sci Technol 52:6445–6454. https://doi.org/10.1007/s13197-015-1759-6

Ščetar M, Barukčić I, Kurek M et al (2019) Packaging perspective of milk and dairy products. Mljekarstvo 69:3–20. https://doi.org/10.15567/mljekarstvo.2019.0101

Šerešová M, Kočí V (2020) Proposal of package-to-product indicator for carbon footprint assessment with focus on the Czech Republic. Sustainability 12:3034. https://doi.org/10.3390/su12073034

Sharma SD, Bhagat AR, Parisi S (2019) Food adulteration episodes. The impact of frauds in the american market of dairy raw materials, pp 43–54. https://doi.org/10.1007/978-3-030-14651-1_4

Shellaiah M, Sun KW (2019) Review on nanomaterial-based melamine detection. Chemosensors 7: 9. https://doi.org/10.3390/CHEMOSENSORS7010009

Sindi A, Badsha MB, Nielsen B, Ünlü G (2020) Antimicrobial activity of six international artisanal kefirs against bacillus cereus, listeria monocytogenes, salmonella enterica serovar enteritidis, and staphylococcus aureus. Microorganisms 8:849. https://doi.org/10.3390/microorganisms8060849

Singh A, Khamrai M, Samanta S et al (2018) Microbial, physicochemical, and sensory analyses-based shelf life appraisal of white fresh cheese packaged into pet waste-based active packaging film. J Packag Technol Res 2:125–147. https://doi.org/10.1007/s41783-018-0034-5

Singh R, Mann B, Sharma R, Singh S (2019) Application of nanotechnology in functional foods. In: Nanoscience for sustainable agriculture. Springer International Publishing, pp 547–579. https://doi.org/10.1007/978-3-319-97852-9_21

Smirnova A, Konoplev G, Mukhin N et al (2020) Milk as a complex multiphase polydisperse system: approaches for the quantitative and qualitative analysis. J Composites Sci 4:151. https://doi.org/10.3390/jcs4040151

Stora Enso (n.d.) The future of dairy packaging. https://www.storaenso.com/-/media/documents/download-center/documents/product-brochures/packaging-solutions/viewpoint-the-future-of-dairy-packaging-en.pdf. Accessed 22 Dec 2020

Tan A, Ngan PT (2020) A proposed framework model for dairy supply chain traceability. Sustainable Futures 2:100034. https://doi.org/10.1016/j.sftr.2020.100034

Taoukis PS (2010) Commercialization of time-temperature integrators for foods. In: Case studies in novel food processing technologies. Elsevier, pp 351–366. https://doi.org/10.1533/9780857090713.3.351

Tichoniuk M, Biegańska M, Cierpiszewski R (2021) Intelligent packaging. In: Sustainable food processing and engineering challenges. Elsevier, pp 279–313. https://doi.org/10.1016/B978-0-12-822714-5.00008-5

Tricarico JM, Kebreab E, Wattiaux MA (2020) MILK symposium review: sustainability of dairy production and consumption in low-income countries with emphasis on productivity and environmental impact. J Dairy Sci 103:9791–9802. https://doi.org/10.3168/jds.2020-18269

Velasco C, Spence C (2018) Multisensory product packaging: an introduction. In: Multisensory packaging: designing new product experiences. Springer International Publishing, pp 1–18. https://doi.org/10.1007/978-3-319-94977-2_1

Vigneshwaran N, Kadam DM, Patil S (2019) Nanomaterials for active and smart packaging of food. In: Nanoscience for sustainable agriculture. Springer International Publishing, pp 581–600. https://doi.org/10.1007/978-3-319-97852-9_22

Wedowati ER, Singgih ML, Gunarta IK (2020) Determination of modules in pleasurable design to fulfil customer requirements and provide a customized product in the food industry. Designs 4: 1–21. https://doi.org/10.3390/designs4010007

Wikström F, Verghese K, Auras R et al (2019) Packaging strategies that save food: a research agenda for 2030. J Indus Ecol 23:532–540. https://doi.org/10.1111/jiec.12769

Wipak (2021) Sustainable solutions. https://www.wipak.com/sustainability/sustainable-solutions. Accessed 12 Apr 2021

Yang H, Wang J, Wu Q et al (2018) Simple and label-free fluorescent detection of melamine based on melamine–thymine recognition. Sensors 18:2968. https://doi.org/10.3390/s18092968

Youssef AM, Assem FM, El-Sayed SM et al (2017) Utilization of edible films and coatings as packaging materials for preservation of cheeses. J Packag Technol Res 1:87–99. https://doi.org/10.1007/s41783-017-0012-3

# Functional Packaging Materials for Fishery Products Applications

7

C. O. Mohan, S. Remya, K. R. Sreelakshmi, and C. N. Ravishankar

**Abstract**

Demand for fishery products is on the rise as it is regarded as health food due to their easily digestible protein with all the essential amino acids, essential fatty acids, vitamins and minerals in good proportion. However, they are also highly perishable due to their intrinsic properties apart from other external factors which affect their quality. Suitable processing and storage will enhance the eating quality of fish for an extended period. Processing methods adopted to preserve the quality of fishery products have to be appropriately supplemented with the selection of better packaging materials. Conventional packaging materials used for food packaging offer limited advantages. Apart from this, the petroleum-based packaging materials pose environmental problems if not properly managed. As a result, the demand for biodegradable packaging materials is increasing worldwide. Biomaterials from the aquatic origin are gaining increased attention in developing biodegradable packaging materials, especially with certain functionality. This chapter gives a brief note on the functional packaging materials like oxygen scavenging film, fish freshness indicating film, thermochromatic indicating film, antimicrobial film, antioxidant film, pathogen indicating film, off-odour scalping, flavour emitting films, etc.

**Keywords**

Functional packaging film · $O_2$-scavenging film · Antimicrobial film · Freshness indicating film · Thermochromatic film

C. O. Mohan (✉) · S. Remya · K. R. Sreelakshmi · C. N. Ravishankar
Fish Processing Division, ICAR—Central Institute of Fisheries Technology, CIFT Junction, Willingdon Island, Kochi, Kerala, India



A. K. Shukla (ed.), *Food Packaging: The Smarter Way*,
https://doi.org/10.1007/978-981-16-7196-8_7

## 7.1 Introduction

The demand for very good quality fresh and minimally processed products is on the rise. Consumers do not want processing techniques adopted by the industry to affect both the nutritional and characteristic quality of the product. This has led to the development of minimal processing or novel food processing technologies which minimize the changes after processing and storage. Apart from processing, widespread awareness and advancements in hygienic handling, packaging and preservation have improved the quality of food products. The aim of consumer is to get best quality product for the price paid and at the same time food manufacturers strive to supply quality products with sustainable quality for an affordable price. Of the variety of options available for food, fish and fishery products have special importance. Historically, fish has long been an important part of human diets, and increasingly a major source of economic value. Globally, fish accounted 6.7% of all protein consumed by humans, as well as offering a rich source of omega-3 fatty acids, particularly EPA and DHA, vitamins, calcium, zinc and iron. There is a great demand for fish and shellfishes in international market due to its proven health benefits. Fish business not only provides foreign exchange it also helps in ensuring nutritious food, employment to millions of people, many of whom are below the poverty line. As per FAO (2018), the world fish production in 2016 has reached 171 million tonnes and 88% of this was used for direct human consumption. Of the total fish produced, aquaculture represented 47%. Global sale value of fisheries in 2016 was estimated at US$ 362 billion of which US$ 232 billion was the contribution from aquaculture production. The value of global fish exports in 2017 was US$ 152 billion, up from $8 billion in 1976 and 54% of this was originating from developing countries, indicating the contribution of seafood export to the building of nations. Nearly 57 million people are engaged in the primary fish production sector, a third of them in aquaculture. Of all the global merchandise trade, fishery products accounted for one per cent in terms of value, representing more than 9% of total agricultural exports. Export of fishery product is one of the major foreign exchange earners in developing countries which accounted to over US$ 80 billion in 2017, providing higher net trade revenues than meat, tobacco, rice and sugar combined. The per capita consumption during 2016 reached 20.3 kg and is expected to increase further. On an average, fish provides nearly 6.7% of all protein consumed by human beings. These indicate there is an ever-increasing demand for fish across the countries, which should be met by increasing the production.

Due to ever-increasing demand for fish, global requirement is increasing steadily. From the available data, it is observed that both fish production as well as fish consumption increased from 1975. Till 1990s, capture fisheries was the major contributor and later, the importance of aquaculture increased resulting in sizeable contribution to total production. As the years pass, the fish consumption level is also increasing. Per capita consumption was 9 kg in 1961 which increased to 17.2 kg in 2008 and further increased to 20.5 kg/year in 2018 (FAO 2020). Considering a stable consumption at 2008 level, the fish requirement estimated as nearly 140 and 152 million tonnes by 2025 and 2050, respectively. However, as the consumption is not stable and it is increasing steadily, the additional demand considering the rate of increase in consumption between 1975 and 2008 will be 164 and 232 million tonnes by 2025 and 2050, respectively. Meeting this huge demand is a herculin task, which creates huge pressure on both capture production as well as on aquaculture. As it is very important to increase the production of fish to meet the global demand, it is also equally important for its proper utilization without wasting by adopting responsible handling, advanced processing and packaging to reduce the post-harvest loss. It is estimated that, nearly 18–20% of global fish production is wasted as post-harvest loss resulting in huge loss of valuable nutrient-rich food commodity. This also affects the economy of the country negatively. For instance, in India alone, very huge economic loss is estimated annually, which is very huge loss to the economy as well as loss of nutritive food commodity. Proper handling of food including fish from its primary production centres and processing them into suitable forms, packaging and storage at desirable temperature will help in reducing the post-harvest losses. Proper handling for fishery products includes segregation of catch based on species and size. Evisceration of large size fishes and icing of fishes with appropriate fish to ice ratio is the first step immediately after catching fish for maintaining better quality. Avoiding more than 3 layers of stacking of fish and ice to avoid damage to fishes in the bottom layer is another good handling practice on board the fishing vessel. Upon arrival at landing centres, care should be taken to unload as early as possible. They should be placed on the raised platform/surface/floor which can be easily cleaned and chlorinated after the auction is over. Processing factories should receive only good quality fish which have maintained the minimum temperature of <4 °C. Once it is processed in suitable form viz., chilling, refrigerated, freezing, curing and drying, thermal processing, smoking, freeze-drying or developing value-added products, they have to be properly packed and stored at the proper temperature. Packaging material should be chosen wisely so that it will complement to the processed food product to extend the eating quality maintaining all the desirable quality.

Food packaging in simple words is wrapping or covering or placing in sealable materials to give protection from its external environment to preserve the quality for an extended duration or till it is utilized by consumers. Although all the activities including processing with advanced techniques are practised with utmost care and if the packaging techniques and packaging materials are not selected suitably it may not protect the food as it is expected. Hence, it assumes great importance to offer the required characteristics to the food till the end of shelf life. Proper packaging not only protects the food from chemical changes during storage, microbial contamination arising from its external environment but also prevents physical damage from the wear and tear during handling, transportation and storage. If the products are not packed properly, the packaging process can become a source of contamination. The main purpose of packaging is to provide protection from biological hazards (pathogens and spoilage organisms), pests, physical damage (shock or vibrations), should act as barrier to external gases, moisture and light, and should provide protection during storage, handing and distribution and transportation. Packaging also provides information about the product contained with ingredients, nutritional information, allergens if any, manufacturer details and also manufactured date. The concept of packaging is also fast changing as the modern packaging is also intended to provide security against tampering, counterfeit by providing authentic seals and helps in tracking and tracing the product.

Traditionally, food packaging is aimed for protection, communication, convenience and containment (Paine 1991; Robertson 2006). Packaging material intended to be used for food package should be able to withstand the internal pressure as in the case of vacuum packaging and modified atmosphere packaging, should be able to withstand the processing conditions like high temperature and pressure as in the case of thermal processing, should be able to withstand storage conditions like normal room temperature, chilled or refrigerated temperature, frozen temperature or elevated temperature as desired to maintain its quality and shelf life, should withstand process conditions and machinability. External factors which are considered for protecting food include but not limited to conditions like heat, light, presence or absence of moisture, pressure, gases, microbial contamination, etc. Convenience and time-saving attributes and different attractive shape and size of package to contain the products are other advantages offered by the packaging materials (Yam et al. 2005; Marsh and Bugusu 2007). The inertness of the packaging material was the key safety objective for traditional packaging materials.

## 7.2 Overview of Food Packaging Industry

Packaging industry is growing rapidly due to its ever-increasing importance. Globally packaging industry accounts to 700 billion US$ excluding turn-out for machineries. The United States of America is the major market for packaging materials accounting nearly 24% of the global market. European Union is the second largest market and Germany, France, Italy and UK dominate packaging industry in Europe. Among the different packaging materials, paper and board lead the sector with 36% of the world market followed by petroleum-based packaging materials. World food packaging market size is to the tune of over 300 billion US$ in 2019 and is expected to increase to more than 460 billion US$ by 2027 (https://www.fortunebusinessinsights.com/industry-reports/food-packaging-market-101941).
Glass, metal, paper & paperboard, wood and plastic packaging materials are widely used for a variety of food sectors like fruits and vegetables, meat, poultry and seafood, dairy products, bakery and confectionery, sauces, dressings, condiments, etc. Changing lifestyle, convenience, increasing purchase power, increased percentage of working women and crave for spending more valuable time with the family are the factors influencing the growth of packaged food and beverages apart from extended and stable shelf-life with improved safety and quality. Rising consumer awareness and demand for processed food is another factor influencing the growth. Multinational giants taking rapid strides in the food and beverages industry is another important factor influencing the growth of food packaging industry. As a result, the per capita consumption of packaging food is worth US$ 400, 260, 90, 46 and 9 for the USA, Europe, Latin America, Asia and India, respectively. Being one

of the fastest-growing industries, it is expected that the food packaging industry will grow to new heights.

## 7.3 Selection of Packaging Material

Appropriate packaging material should be chosen based on the composition of the food. The use of proper packaging will ensure the safety and durability of the product during its entire shelf life. Important aspects to be considered while selecting packaging materials are its moisture content, sensitiveness to oxygen, acidic or alkaline, alcohol content and fat content. A container that is appropriate for one type of food may not be suitable for another. It is very important to choose the proper packaging for the food intended to pack. The packaging materials/packages used in fish industry are both modern as well as traditional types, ranging from bamboo baskets, jute bags, leaf mats to corrugated fibre-board boxes, duplex board cartons, metal containers of aluminium and tinplate, plastic films and their laminates, thermoformed trays, polypropylene/high-density polythene crates, expanded polystyrene insulated boxes, glass bottles, etc. Each packaging material has its own merits and demerits. Paper & paperboards are very common packaging materials used for wide applications. These are made using pulp extracted from wood and other non-wood sources like straw, bamboos, etc. As wood pulp is very easily accessible, and commonly available its use is on the rise. Due to its natural and low price, it is one of the preferred packaging materials. They are suitable for small-scale, large-scale, retail and bulk packaging. All kinds of foods can be packed using paper and paperboard after slight modifications. They can be used as bags, pouches, wrapping, boxes, cartons, tetra packs, folding cartons, etc. The presence of cellulose fibres provides them good rigidity and high shock resistance. They are also light in weight, cheap and are widely available. However, they need to be laminated for packing foods with moisture content. They also need external protection during transportation to protect from damage. Apart from this, they have to be stored in proper facility to prevent damage by rodents and insects. Another important aspect to be considered is the selection of safe wood for the manufacturing of paper and paperboard which comes in contact with food directly. It is common practice to use chemical preservatives to protect wood. Care should be taken to select the preservative free wood. Due to its sustainability nature, the paperboard will be the most prevalent and attractive packaging material in the food industry in future due to increased investments leading to innovations in packaging design and digital printing.

Glass is commonly used as packaging material for a variety of foods due to its inertness. Glass containers are manufactured by heating the mixtures like silica (sand), calcium carbonate in the presence of sodium carbonate (melting agent) and alumina (hardner) at very high temperatures to melt the mixture, cast them into the mould to get suitable shape and size and finally cooling them (Keleş 1998; Marsh and Bugusu 2007). Surface coating and oiling is done during the production process to prevent the abrasion of the glass surface (Aday 2014). Normally soda-lime glass is used in the packaging industry for producing transparent, transluscent or coloured glasses. Also, to mask the colour emerging from minute quantity of impurities like iron and sulphur compounds especially during the manufacturing of colourless glass, nickel and cobalt are added (Keleş 1998). Glass is one of the highly preferred materials for food packaging due to its inertness, durable, transparent, touch as it withstands high temperature and pressure, impermeable to gas, odour, liquids and water vapour (MEB 2011). It has a very smooth and shiny surface, it will neither corrode nor degrade or erode over time. However, they are not devoid of drawbacks. Major drawback is its heaviness and very fragile leading to break. It requires extra care while transporting and distribution and hence are more costly than petroleum-based packaging materials. Being transparent, it allows light to pass through the glass resulting in light-induced oxidation and discolouration.

Metal is another important material for packaging applications. Unlike glass, metals have free and mobile electrons that absorb light energy. That is why they cannot be transparent and translucent as in the case of glass. There is a long history of packing food products in metal containers. Mainly tin plate or aluminium or stainless steel coated with the food-grade lacquer material is commonly used. Among these, tin-plate containers were very commonly used for many food applications. Tin plate was used to make containers for food over 120 years. Tin-plate containers are very attractive and have an appealing golden colour. Tin can is made up of 98% steel and 2% tin coated on both the sides of container differentially. The surface coming in contact with the food material is coated thicker layer of tin. Normally low carbon steel is used in the manufacturing of this type of can. They are light in weight, rigid, strong, corrosion-resistant and easy to handle. They are comparatively easy to fabricate and offer more shelf life to food products compared to other metal containers. The major drawback of this container is its limitation for packing acidic products. In the earlier days, 3-piece cans were used soldering with a lead:tin ratio of 98:2 percentage. Some lead from this solder may migrate to food depending on the type of product packed and on the amount of solder exposed. Apart from this, lead contamination may also originate as impurity from the tin coating. Due to its proven ill effects, regulatory agencies have stipulated a maximum level of 2 ppm for lead in canned food products. Apart from this, tin may also permeate into the food and higher levels of tin will implicate in the gastro-intestinal disturbance. A level of up to 250 ppm is generally permitted by regulatory authorities in canned foods. Invention of 2-piece, solder-free cans reduced the problem of lead in canned foods. Canned shrimp products from India were very well appreciated from many importing countries and this industry was flourishing until 1980s. Later on the canned seafood industry collapsed as it could not compete with the neighbouring countries in terms of its pricing. As India does not have tin resources, it has to import tin-plate for manufacturing cans and hence it became costlier. This has resulted in search for other alternative metal cans. Aluminium is one such metal used as an alternative to tin-plate cans. In India and also in many parts of the world, aluminium is very abundantly available and is light in weight. It is recognized as generally regarded as safe (GRAS) material by the US FDA. The use of 100% aluminium for manufacturing cans or containers results in weak material. To overcome this, small amount of magnesium, manganese, iron, zinc, copper and silicon is added while manufacturing. Aluminium-based containers can be used as cans for meat and fishes, beverages and as bottle tops or closures. Aluminium is easy to fabricate and one cannot make out the corrosion due to its colourless character. Metallic taste is not imparted when aluminium containers are used as compared to tin cans. Requirement of very high energy for the production, tendency to bleach with some pigmented foods and impossibility of soldering are the drawbacks of this container. Apart from this, aluminium is implicated in Alzheimer's disease up on chronic exposure from water and food. World Health Organization (WHO) has stipulated the limit of 1 mg $kg^{-1}$ body weight per day. Tin-free steel is another metal container used for a wide range of food products like fruits, vegetables, meat and seafood. TFS cans are made by electrolytic coating of a thin layer of chromium and chromium oxide on the steel base material. The presence of chromium helps in preventing rust formation and corrosion. It can withstand very high internal pressure and is suitable for attractive printing. The major problem with this container is it cannot be recycled or reused and is not suitable for soldering or welding.

Petroleum-based plastic packaging material is another widespread material used for many applications including food packing. Plastic is the material that is obtained when the bonds between carbon and hydrogen, oxygen, nitrogen and other organic or inorganic elements are broken from their simple structured molecule, known as 'monomer' and formed into the long and chained structure, known as 'polymers' (Durusoy and Karababa 2011; MEB 2011). With the help of polymer science and engineering, many variety of plastics with varying characteristics can be manufactured so that almost all the food products with varying processing, storage and transportation conditions can be packed with plastic materials. Their light-weight, transparent/translucent with low or high gas and moisture barrier properties, requiring less storage space and easy to use, seal, open and dispose makes one of the most preferred packaging materials for food products (Kızılırmak 1997). They can be made either as thin films or as rigid containers depending on the requirements. Different packaging materials with their code and applications are given in Table 7.1. Codes from 1 to 7 are given for easy identification during its recycling. Different types of packaging material either single-layered or multi-layered packaging materials have to be selected appropriately depending on the category of food products which is given in Table 7.2. Among the different packaging materials, not all the materials are suitable for direct food contact application. The use of plastics and plastic-based materials are increasing across the food industry including fisheries.

**Table 7.1** Different packaging materials and their applications

| Code | Name | Applications |
|---|---|---|
| 1 PET | Polyethylene terephthalate | Water and soft drink bottles |
| 2 HDPE | High-density polyethylene | Bottles for milk, juice, water, yoghurt, cosmetics, shampoo, dish and laundry detergent, bleach bottles, ice boxes |
| 3 V | Polyvinyl chloride (PVC) | Bottles packaging sheet, pipes and fittings from hard PVC, wire and cable insulation, film and sheet, floor coverings synthetic leather products, coatings, blood bags, medical tubing etc. from soft PVC |
| 4 LDPE | Low-density polyethylene | Frozen food packaging and for squeezable bottles |
| 5 PP | Polypropylene | Retortable pouches |
| 6 PS | Polystyrene | Lids, cups, bottles and trays, food service applications, meat trays, egg cartons |
| 7 OTHER | Other | Multi-layer combination, plastic baby bottles, clear plastic "sippy" cups, sports water bottles, metal food can liners, some juice and ketchup containers, compact discs, cell phones, computers |

During the manufacturing of plastic packaging materials, along with basic polymer materials, other non-polymeric compounds are added either inherent or deliberately to achieve desired properties. These are categorized into three different classes viz., polymerization residues (residual monomers, catalyst remnants, polymerization

**Table 7.2** Selection of packaging materials depending on the processing method for different food category

| Food categories | Packaging material |
|---|---|
| Fresh chilled | HDPE, PP, HM-HDPE |
| Fresh frozen | Laminates or co-extruded pouches, LDPE, BOPP |
| Dried | HDPE woven gusseted sack, laminates |
| MAP | Nylon/suryln laminates, PVC moulded trays laminated with polyethylene, polyester/LDPE film, EVOH |
| Thermal processed | Metal cans, retort pouches, HIPP trays |
| Surimi | LDPE/LLDPE/HMHDPE with waxed duplex carton & 5/7 ply CFB |
| Sausage | PVDC or natural casings |
| Breaded & battered products | Thermoformed trays of PVC, HIPP & HDPE |
| Pickles | Glass bottles or PEST/LDPE-HDPE co-extruded film |

**Table 7.3** Limits of monomer and heavy metals in plastics

| Country | Monomer | Heavy metals |
|---|---|---|
| BIS-India | VCM in PVC-1 ppm; in food migration-10 ppb, styrene in polystyrene-2000 ppm | Lead 1 ppm and others 0.01 ppm in PVC |
| EEC-Europe | VCM in PVC-1 ppm | Nil |
| EPF-UK | VCM in PVC-1 ppm, styrene in PS-5000 ppm | Nil |
| Japan | VCM in PVC-1 ppm, volatile component in polystyrene-5000 ppm<br>Vinylidine chloride in PVDC-6 ppm, Caprolactum in Nylon-15 ppm | 1. Lead, Cadmium & Barium 100 ppm each in PVDC<br>2. 0.05 ppm antimony & 0.1 ppm germanium in PET |
| FDA-USA | VCM not specified styrene in PS-10000 ppm, acrylonitrile in ABS plastics-11 ppm | Nil |

*VCM* Vinylchloride Monomer, *PVC* Polyvinylchloride, *PVDC* Polyvinylidenechloride

solvent, etc.); processing aids (plasticizers, stabilizers, antioxidants, slip agents, lubricants, antistatic agents, etc.) and end-use additives (antioxidants, brighteners, blowing agents, mould release agents, colourants, UV stabilizers, etc.). Among these, the polymerization residues compounds are unavoidable whereas processing aids and end-use additives are deliberately added to the polymer either during manufacture or subsequently to achieve the desired end properties of the finished plastic material. If we are not adding these additives, desirable properties of the plastic packaging materials will not be noticed in the end product. Polymers by their nature like very high molecular weight and low solubility in both aqueous and fatty systems may make them inert. The other additives, which are non-polymeric in nature, are lower molecular weight which may leach out from these plastic packaging materials and dissolve in the food system. These may pose threat to human health due to their risk factor and the awareness in this matter has led the national and international regulatory authorities in the formation of guidelines for the proper use of plastics for food packaging application. Such guidelines are necessary to restrict

**Table 7.4** Limits of heavy metals in colours used in plastic manufacture

| Heavy metals | Limit (ppm) |
|---|---|
| Lead | 0.01 |
| Arsenic | 0.005 |
| Mercury | 0.005 |
| Cadmium | 0.20 |
| Selenium | 0.20 |
| Barium | 0.01 |

the indiscriminate use or abuse of plastics in food packaging. The residual monomer content and heavy metal content in different plastics specified by different countries and limits of heavy metals in colours used are presented in Tables 7.3 and 7.4 respectively. The other regulations on food packaging materials comprise of regulations for adjuvants (antioxidants, colourants, plasticizers, etc.) used in food packaging materials. Only permitted materials within their allowable limits can be used in the manufacturing of plastic packaging material.

Packaging protects, but it is also a potential source of risk. Plastics, paper, cardboard, and other packaging materials that come into contact with the food can react with it and affect its safety. The plastic packaging material may contain impurities such as leftover monomer residues, additives, stabilizers, odorous adjuvants, colorants and antioxidants. At different stages of converting, preparing and storing, chemical substances may migrate to the food. If the packaging is made from wrong materials which include printing substrates like, inks, varnishes and all the auxiliaries used in the printing process, it may pose a risk to the packed product, thus to human health. In paper, the group of potential contaminants include 1,2-benzisothiazoline-3-one (BIT); 2-(thiocyanomethylthio) benzothiazole (TCMTBT); 2,4,5,6-tetrachloro-isophthalonitrile (TPN); 2,4,6-trichlorophenol (TCP); pentachlorophenol (PCP); 4,4′-bis (dimethylamino)-benzophenone; 4,4′-bis (diethylamino)benzophenone; 4-(dimethylamino) benzophenone (DMAB) and bisphenol A (BPA), which in chlorinated form (BPAs) may be found in effluent from wastepaper processing plants. Sometimes even slimicides and fungicides used in both virgin and recycled paper may pose risk to consumers. Pentachlorophenol is used for wood conservation, therefore if the conserved wood or slimicides containing this substance are used in paper production there is a risk of contamination of packed food. Leaching of all these chemicals from packaging material into food affects the quality of food and increases the risk to consumer. Hence the materials coming in contact with food product need to comply with strict regulations in many countries. It is the manufacturer's sole responsibility to deliver safe packaging materials which comply with all legal requirements. Due to this as well as due to the impact of these packaging materials on the environment and biota after disposal, there is an increased priority worldwide for the development of biodegradable packaging material which will not impact any health risk to consumers.

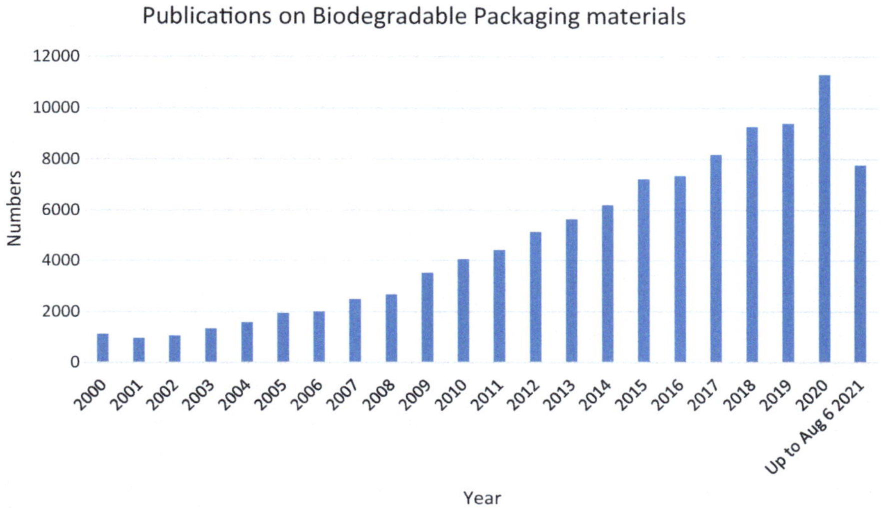


**Fig. 7.1** Number of publications on Biodegradable packaging materials over the period 2000 to Aug 2021

## 7.4 Biodegradable Packaging Materials

Biodegradable packaging materials are the materials derived from natural biological materials. This can be either plant- or animal-based packaging materials. The research interest on the development of biodegradable packaging material is on the rise constantly from 2000 onwards as indicated from the publications (Fig. 7.1). In 2000, the publications were only 1120 which have crossed 11,000 in 2020. The growing interest is basically due to its eco-friendly, sustainable nature and will not create any health or environmental related issues. These biodegradable packaging materials can be either protein-based or carbohydrate-based materials in addition to a plasticizer. They can be in the form of gels, film, bag and box. As the biodegradable packaging materials are made using eco-friendly materials, their recycle will be easier. They require less energy to produce and are non-toxic in nature. Their carbon emission will be very less. Long-term and increased dependence on the biodegradable packaging material may result in requirement of more plant- or animal-based matter for their production. If any additives or petroleum-based polymers are mixed in compounded packaging materials, they require special facility for composting. The physical barrier properties are very poor compared to petroleum-based materials and not all the biodegradable packaging materials are sealable. Bioactive ingredients derived from aquatic sources which can be used for biodegradable packaging material production are fish and shellfish proteins, protein from Surimi waste, blanched and cooked water protein extract from heat processing, protein extract from clam shuck water, chitosan extracted from shrimp, crab, lobster and squid pen, collagen and gelatin from fish scale and skin, squid and cephalopod skin and a variety of seaweeds. Although biodegradable packaging materials have various advantages, functionalizing these packaging materials will increase their applications and enhance the quality of food products.

## 7.5 Smart Functional Packaging Materials

Globalization and dynamism in the exchange of food products, along with reduced time for selection/cooking with fresh ingredients, and the growing interest in health safety and environment are the main challenges which trigger the development of new improved packaging concepts. Traditionally, the functions of the packages include protection, containment, communication with the user, ergonomics and marketing. However, in recent years additional functions are incorporated into the packaging materials to meet the growing global demand on the safety of food products. One of the main objectives of food law is the safety of the food products. Controlling the quality and assuring the safety of the product at all the stages of food supply is very much essential to enhance the market reach for a product. This can be achieved by using active and intelligent packaging technology, which is also known as smart packaging. Active packaging involves altering the surrounding environment of food suitably with favourable gases, antimicrobial and antioxidant agents, flavour bearing compounds, etc. whereas intelligent packaging monitors interaction between the food, the packaging, and the environment and provides information like the conditions of the package, freshness, leakage and presence of pathogen and informs to the manufacturer and consumer. Thus, the smart packaging provides a specific functionality beyond function physical barrier between the food product and the surrounding environment. Knowing information about the product quality, the packaging or the environment establishes a bond of responsibility throughout the food supply chain (storage, transport, distribution and sale). The global market for active and intelligent packaging will double between 2011 and 2021, growing at an annual rate of 8% until 2016, reaching US$ 17,230 million, and later at an annual rate of 7, 7%, reaching US$ 24,650 million in 2021. Among the smart packaging technologies, the development of intelligent packaging material which can detect the presence of pathogen assumes great importance as it can assure the safe and wholesome food to consumers without any microbial pathogen contamination. The technology of developing smart pathogen indicating film can be used in all the food products to monitor the quality and safety of food from the producers to the consumers. The technology provides an on-line quality control and safety for the consumers beyond the existing conventional technologies which are helpful for the authorities and food producers as well.

## 7.6 Active Packaging

Active packaging is an advanced concept that can be defined as '*a type of packaging that changes the condition of the packaging and maintains these conditions throughout the storage period to extend shelf-life or to improve safety or sensory properties while maintaining the quality of packaged food*' (Vermeiren et al. 1999). Unlike other conventional packaging methods wherein packaging material or packaging system will not alter and remain passive in nature apart from their inertness. Any gases formed or introduced in the packaging atmosphere may escape or may enter from external atmosphere depending on the permeability of the packaging material resulting in alteration of the package atmosphere which affects the quality and shelf life of the product. In the case of active packaging, the packaging provides extra protection by interacting with the package atmosphere and modify the conditions of the package suitably to help the retention of the quality. This can be achieved either by removing undesirable constituents like oxygen, $CO_2$, off-odour, ethylene or by releasing suitable constituents like nitrogen, carbon dioxide, antimicrobials, antioxidants, ethanol, pesticide, etc., depending on the food products. Major active packaging techniques are concerned with substances that absorb oxygen, ethylene, moisture, carbon dioxide, flavours/odours and those which release carbon dioxide, antimicrobial agents, antioxidants and flavours. The most important active packaging concepts for fishery products include $O_2$ scavenging, $CO_2$ emitters, moisture regulators, antimicrobial packaging concepts, antioxidant release and release or absorption of flavours and odours.

### 7.6.1 $O_2$- Scavenger

Oxygen is a lifeline for all the animals when they are in living condition but becomes detrimental as soon as life is lost leading to spoilage and decay. Many food products including fish are highly sensitive to the presence of oxygen as it leads to the growth of aerobic microorganisms and oxidation which causes undesirable colour changes (e.g. discolouration of pigments such as myoglobin, carotenoids), off-odours and flavours (e.g. rancidity as a result of lipid oxidation) and leads to loss of nutrients (e.g. oxidation of vitamin E, β-carotene, ascorbic acid) which adversely affects the quality. Hence control of oxygen assumes its importance to control the spoilage and enhance the shelf life. This can be achieved by packing in high barrier films like EVOH, polyester-laminated with aluminium foil or laminated packaging materials. Further advanced packaging technologies like vacuum packaging and modified atmosphere packaging may be adopted. However, these techniques may not remove oxygen completely and there is no control on the permeation of oxygen through packaging materials. In such cases, the use of oxygen scavenger is highly suitable and very effective in reducing the oxygen level to very low level like less than 0.05% within short time. $O_2$ scavengers were first commercialized in the late 1970s by Japan's Mitsubishi Gas Chemical Company (Ageless®). $O_2$ scavenging concepts are mainly based on iron powder oxidation, ascorbic acid oxidation, photosensitive dye oxidation, enzymatic oxidation (e.g. glucose oxidase and alcohol oxidase), unsaturated fatty acids (e.g. oleic or linolenic acid), rice extract or immobilized yeast on a solid substrate. These oxygen scavengers are useful in preventing discolouration of fresh and cured fish, rancidity problems, mould spoilage of intermediate and high moisture products or oxidative flavour changes. Apart from fish, they find its applications in most of the dried food items like grains, snacks, bakery items, fish and meat, dry fruits, ready meal, etc.

$O_2$-scavenging film is developed either incorporating any oxygen scavenging systems mentioned above or by using the principle of photosensitive dye oxidation. This involves sealing of a small coil of an ethyl cellulose film containing a dissolved photosensitive dye and a singlet $O_2$-acceptor in the headspace of a transparent package. Due to illumination of the film with the light of the appropriate wavelength, excited dye molecules sensitize $O_2^-$ molecules, which have diffused into the polymer, to the singlet state. These singlet $O_2^-$ molecules react with acceptor molecules and are consumed (Rooney 1985):

$$\text{Photon} + \text{dye} \rightarrow \text{Dye}*$$

$$\text{Dye} * + O_2 \rightarrow \text{Dye} + O_2*$$

$$O_2 * + \text{Acceptor} \rightarrow \text{Acceptor oxide}$$

$$O_2* \rightarrow O_2$$

$O_2$ acceptors can be tetraphenyl porphine (TPP), dimethyl anthracine (DMA), dioctyl thallate (DOT), etc. This can be used for both wet and dry fish products as it does not require any moisture for its activation. Most $O_2$ scavengers in commercial use today are iron-based systems and only limited reports are available on film based $O_2$ scavenger.

### 7.6.2 Antimicrobial Packaging

Antimicrobial packaging is the second commonly used active packaging system after oxygen scavenger. The concept of antimicrobial packaging was developed due to increased awareness on the antimicrobial agents used directly by the food manufacturer to retain the quality of food. As microbial spoilage is the major cause for food deterioration, various chemicals were used for treating food to reduce both food spoilage and pathogenic microorganisms. The classes of these antimicrobial agents include but not limited to alcohol (ethanol), ammonium compound (quaternary ammonium salts), antibiotics (natamycin), antimicrobial peptides (leucocin, sakacin, enterocin), antioxidant phenolic compounds (butylatedhydroxyanisole (BHA), butylatedhydroxytoluene (BHT), tertiary butylhydroquinone (TBHQ), grape seed extract, pomegranate peel and seed extract), bacteriocin (bavaricin, lacticin, nisin, pediocin), chelators (citric acid, EDTA, lactoferrin, polyphosphate), enzymes (chitinase, ethanol oxidase, glucose oxidase, glucosidase, lysozyme, lactoperoxidase, hydrolase), fatty acids (lauric acid, palmitoleic acid), fungicide (benomyl, imazalil, sulphur dioxide), metals (copper, silver), natural phenols (catechin, hydroquinones), organic acid (acetic acid, benzoic acid, citric acid, lactic acid, propionic acid, sorbic acid, tartaric acid), organic acid salt (potassium sorbate, sodium benzoate, acetic acid, propionic acid, benzoic acid, sorbic acid, calcium sorbate, benzoic anhydride, propionic acid, propyl paraben), paraben (ethyl, methyl and propyl paraben), plant based volatile compounds (allylisothiocyanate, cinnamaldehyde, eugenol, terpineol, thymol), polysaccharide (chitosan, carrageenan), etc. Recent advancements in the nanotechnology research have expanded the list of antimicrobial compounds with silver nanoparticle as one of the highly potential antimicrobial agents. The principle action of antimicrobial films is based on the release of antimicrobial entities into the food which extends the lag phase and reduces the growth phase of microorganisms in order to prolong shelf life and to maintain product quality and safety. To confer antimicrobial activity, antimicrobial agents may be coated, incorporated, immobilized or surface modified onto package materials. However, there is a growing concern both by consumers and by the regulatory agencies on the very high levels of these antimicrobial compounds in the food products if they are used directly to treat with these antimicrobial agents. The residual levels will be very high leading to increased concerns which can be overcome with the adoption of active antimicrobial packaging techniques. The antimicrobial compound embedded into the polymer acts by two different kinds of mechanisms. In the first method, the preservative is covalently immobilized into the polymer matrix and acts directly from the film when the food is brought in contact with the active material. Regarding the latter, the preservative is embedded into the matrix in the dry state. When the active material is brought in contact with a moist food or a liquid-like food, the preservative is released from the material and acts directly. In both cases the aim of the system is to extend the shelf life of the packaged foodstuff, inhibiting the microbial growth and preserving its properties. The classes of antimicrobials listed range from acid anhydride, alcohol, bacteriocins, chelators, enzymes, organic acids and polysaccharides. Several compounds tested for antimicrobial activity in food packaging including organic acids such as potassium sorbate, sorbic acid, propionate and benzoate or their respective acid anhydrides, bacteriocins, e.g. nisin and pediocin, enzymes such as lysozyme, metals, fungicides such as benomyl and imazalil, Ag- zeolite. Antimicrobial agents are selected based on their ability to withstand the processing or extrusion condition of the packaging material and its compatibility. Ethanol is another commonly used antimicrobial agent for surface immobilization with the polymer films. At a concentration of 60–70%, v/v, it exerts its effectiveness. Even at relatively low concentrations (4–12%), ethanol has proved effective in controlling growth of several moulds and bacteria (De Kruijf et al. 2002). Spraying ethanol onto fresh or dry fish prior to packaging can be adopted, but another option is to use sachets generating ethanol vapour. This contains food-grade ethanol embedded or absorbed or encapsulated in a carrier material. This is meant for either slow or rapid release depending on the requirement and is adjusted by varying permeability of the sachet material to water vapour. Its application mainly depends on the food products intended and the sensitivity of the food products to ethanol. Apart from these, various plant derivatives can be incorporated into the packaging system as antimicrobials.

### 7.6.3 Antioxidant Release

Antioxidants are another class of compounds used extensively in most of the oxygen-sensitive foods to improve the oxidation stability. Fish being highly rich in polyunsaturated fatty acid, are prone to oxidation. Common antioxidants used for direct food applications were BHA, BHT, etc. whose use is now not permitted in most of the food products due to its ill effects by accumulating in adipose tissue. Incorporation of natural plant extracts are also practised however it requires to use more quantity leading to influence the sensory properties. Direct incorporation of antioxidants into the packaging films and use as antioxidant is gaining more attention worldwide. However, the antioxidant compound should be able to withstand the extrusion conditions for its applications. Vitamins E and C are the common natural antioxidants, and their incorporation in polymer films to exert antioxidative effects is still at the experimental stage. Vitamin E is stable under processing conditions and has an excellent solubility in polyolefins. However, it is confirmed that, vitamin E is a less mobile antioxidant in low-density polyethylene (LDPE) than BHT, as vitamin E is a larger molecule (Wessling et al. 1998). Apart from these, natural antioxidants extracted from plant and animal substances and their use as antioxidant packaging are under experimental stages.

### 7.6.4 Release or Absorption of Flavours and Odours

Fish and shellfish have a typical fresh flavour, which is distinct from any food systems. Food packaging materials, particularly some plastics, may interact with these flavours, resulting in loss of flavour known as '*flavour scalping*' which affects the quality of the product. Furthermore, flavours are usually lost or degraded during various stages of processing at different temperatures or after packaging. Therefore, there is a need to replace the lost flavour constituents when scalping or degradation occurs. Flavour incorporation in packaging material might be used to minimize flavour scalping. Consumers always like to smell good flavours when they first open a food package. The applications of suitable flavour enriched packaging materials have the potential to improve the organoleptic quality of the product by emitting desirable flavours into the food and to encapsulate pleasant aromas that are released upon opening. Flavour release may also provide a means to mask off odours coming from the fish or the packaging. It is of utmost importance that the aforementioned technology should not be misused to mask the development of microbial off-odours thereby concealing the marketing of products that are below standard or even dangerous for the consumer (Nielsen 1997).

Flavour scalping, i.e. sorption of food flavours by polymeric packaging materials, may result in loss of flavour. Generally, flavour scalping is detrimental to food quality, but it could be used in a positive way to selectively absorb unwanted odours or flavours. In contrast to flavour releasing systems, flavour absorbers scavenge undesirable flavours, aromas and odour present in the package headspace. The formation of off-flavour and off-odour in fish products are mainly from the oxidation of fat and oils, leading to the formation of peroxides and their decomposition products like aldehydes, breakdown of fish proteins into amines forming alkaline compounds. Flavour absorbing systems employ cellulose triacetate, acetylated paper, citric acid, ferrous salt/ascorbate and activated carbon/clays/zeolites to absorb off-odour and off-flavour. Removal of aldehydes from package headspaces can be achieved by means of the layer Bynel IXP101 which is a HDPE master batch. Amines can be removed by reacting with acidic compounds, e.g. citric acid incorporated in polymers.

### 7.6.5 Time-Temperature Indicators

Food and pharmaceutical products are highly sensitive to the temperature of storage. They need to be maintained at specified temperature throughout their shelf-life to ensure its quality and safety. Frozen food products have relatively very long shelf life ranging from 6 months to 2 years. They should be stored at $-18 \pm 2$ °C from the point of production till they are consumed. However, cold chain is broken due to many reasons like improper cold chain facilities at all the stages of food distribution, frequent power failure and regular opening and closure of freezer door at retail outlets. Failure to maintain specified temperature will indirectly affect the quality and safety of the product. It is difficult to find out whether the frozen food product has thawed at some point and then refrozen. As there is no device/method to monitor such temperature abuses, consumers will end up buying inferior quality products. At the same the producers will get bad reputation due to breakdown of cold chain which is not in their hand to maintain. The principle behind this is that as the temperature increases, biochemical and microbial reactions take place at rapid rate leading to deterioration of food. TTIs are based on either chemical, electrochemical, enzymatic, microbial or mechanical reaction leading to visible response. TTIs find its applications not only in food products, but also in pharmaceutical products to ensure better quality product to consumers. Recent research on TTI indicates that the active compound, sensitive to temperature fluctuation can be incorporated into the polymer/biodegradable film that can change the colour as temperature is abused. Indicator dyes, plant and fruit extracts containing specific pigments responsible for giving colour change with the fluctuation in the temperature are used in this.

### 7.6.6 Freshness Indicators

Fish production as well as fish consumption is on the rise both in the domestic and international market. Understanding the ever-increasing demand for fish, retail marketing and online marketing is flourishing across India and the price of fish is also increasing. Due to recently reported adulteration menace, consumers are always at doubt while purchasing fish. Fish being highly perishable, undergoes spoilage leading to formation of various chemicals (oxidation products and amines) which may affect the health of the consumers. The fish quality is either ensured by sensory attributes or by analytical methods. However, the analytical methods are time consuming, costly and are not real-time in nature. This has resulted in relying on sensory quality assessment to judge the quality of fish being marketed. However, sensory quality analysis is qualitative and it can be biased and hence, quality control requires rapid methods for measuring fish freshness in real-time. An intelligent packaging technology will be beneficial for this purpose. This can be achieved by using freshness indicator or spoilage indicators. Commonly developed freshness indicators are based on the reaction between the volatile compounds or other constituents generated during the process of spoilage with the indicating material. Freshness indicators provide direct product quality information resulting from microbial growth or chemical changes within a food product. Microbiological quality may be determined through reactions between indicators included within the package and microbial growth metabolites (Smolander, 2003). Normally, the freshness indicators are incorporated into the packaging film, which reacts with volatile amines and other indicating agents produced during the storage of fish and other seafoods, and the freshness is indicated by a colour change.

## 7.7 Studies on Functional Packaging Materials

Smart packaging includes both active and intelligent packaging systems. Active packaging refers to alternation of package atmosphere/incorporation of suitable compounds to enhance the quality and shelf life, whereas intelligent system monitors the condition of packaged food to give information regarding the quality of food. Research on development of smart packaging devices for perishables including fish is very limited in India. Many researchers are carrying out work on various aspects of smart functional packaging materials. Chemical combinations for $O_2$ scavenger, $CO_2$ emitter is optimized using GRAS chemicals which can reduce the $O_2$ level to 0.01% and increase the $CO_2$ level to >80% within 24 h, respectively (Mohan 2008). Dual-action sachet, which combines the scavenging of $O_2$ and emits $CO_2$, is also developed and found to extend the shelf life of fatty fish up to 25 days (Mohan 2008). These developed active packaging systems follows first-order reaction and extends the shelf life of fishery products significantly (Mohan 2008). Studies indicated that $O_2$ scavenger was very efficient in reducing oxygen concentration by 99.58% within 24 h inside the packages and found to extend the catfish steaks shelf-life up to 20 days, compared to 10 days in control air packs (Mohan et al. 2008). Studies on Seer fish indicated a shelf life extension of 20 days under $O_2$ scavenger compared to only 12 days for air packs and inhibited the formation of biogenic amines, especially histamine by inhibiting bacterial enzyme activity (Mohan et al. 2009a). The use of $O_2$ scavenger positively extended the shelf life by inhibited the formation of volatile bases, inhibiting the nucleotide degradation resulting in delayed formation of hypoxanthine, which is associated with the spoilage of fish (Mohan et al. 2009b). The use of $O_2$ scavenger improved the shelf life of barracuda steaks by 20 days (Remya et al. 2018) and Indian oil sardine by 15 days (Mohan et al. 2019a, b) under chilled storage. A delay in the growth of microorganisms including specific spoilage flora like *Pseudomonas* spp. and $H_2S$ forming bacteria was observed in fishes packed with $O_2$ absorber by extending the lag phase which is mainly due to the effect of altered atmosphere (Mohan et al. 2010a). A shelf life of 9–10 days was observed for long tail tuna (*Thunnus tonggol*) packed under $O_2$ scavenger under chilled stored (Mohan et al. 2014).

Active antimicrobial packaging films prepared using chitosan incorporating ginger (*Zingiber officinale*) essential oil (GEO) was effective against foodborne pathogens (Remya et al. 2016). Keeping quality of steaks of barracuda (*Sphyraena jello*) fish improved significantly in the chitosan films with GEO (Remya et al. 2016). Antimicrobial packaging film incorporating silver nanoparticles synthesized using low and high molecular weight and other chemicals can be used effectively to control the growth of foodborne pathogens (Pankaj et al. 2017). Combination of $O_2$ scavenger and antimicrobial film incorporating essential oil resulted in enhanced quality retention and reduced oxidation and extended the shelf life up to 30 days in chilled storage condition (Remya et al. 2017). Combination of curry leaf essential oil and $O_2$ scavenger resulted in increased lag phase and reduced oxidation in *Rachycentroncanadum* and extended shelf life up to 30 days (Remya et al. 2014). Antimicrobial coating with chitosan resulted in reduced microbial growth, volatile formation, oxidation, drip loss and improved water holding capacity and improved the texture of Indian oil sardine (Mohan et al. 2012; Renuka et al. 2016). The formation of total volatile base nitrogen and trimethylamine nitrogen was less by 14.9–32.7 and 26.1–49% for different concentrations of chitosan-treated samples (Mohan et al. 2012). Biodegradable antioxidant packaging film developed using rosemary essential oil resulted in improved DPPH activity and total phenolic content (Mohan et al. 2018). Moisture absorbing system developed using aquatic weed, water hyacinth to absorb the drip formed during the storage of fish.

Similar to active packaging systems, ICAR-CIFT has also developed various intelligent packaging systems. A simple, easy to use and cheap (Only Rs 0.4 per pack equivalent to US$ 0.0055) freshness indicator is developed for indicating quality of fish and shellfishes. The effectiveness of freshness indicator is evaluated with fishes of freshwater, marine and brackish water origin and found effective, except for freshwater fishes. Studies on nanoparticle-based intelligent packaging system to develop temperature history sensor were developed in collaboration with University of Wisconsin-Madison, USA (Wang et al. 2017). Nanocomposite of chitosan and gold nanoparticles (AuNPs) was used to develop sensors that can indicate the frozen state and thermal history of foods and other temperature-sensitive products based on the visual colour change (Wang et al. 2017). A greener method used for the synthesis of Gold nanoparticle using chitosan to develop temperature history indicator to ensure the quality and safety of frozen stored perishable food and pharma products during shipment and transportation (Mohan et al. 2019a, b). Developed and characterized gold, silver and copper nanoparticles using different chemicals and biological sources of marine origin which finds application as biogenic amine and heavy metal detecting sensor. Developed paper-based colorimetric nano-biosensor strip for detection of foodborne pathogens including *E. coli* 0157:H7 and *E. coli* which reduced the detection time (Nadella et al. 2019). A detailed review on active and intelligent packaging systems is documented (Mohan et al. 2009a, b, 2010b, 2018; Biji et al. 2016).

Apart from this, the institute is also steer heading the research on developing biodegradable smart packaging materials with improved properties. Chitosan, collagen, chitosan-collagen-based films were developed for its application as wrap. PLA-based biodegradable packaging material with improved mechanical properties and heat sealability is developed. Seaweed based functional and edible films developed exhibited good sealing and antioxidant properties and can be used as novel packaging material in food industry as a sachet/pouch/bag for seasoning powder for instant noodles, instant coffee/tea, etc. Continued research and development is needed for enhancing the efficiency of smart packaging systems developed. Collaboration with other research institutes is the need of the hour to fine tune and upscale the developed smart packaging technologies, validate properties of smart packaging devices developed in field condition and commercialization.

## 7.8 Conclusion

Apart from proper handling, there is a need to adopt advanced packaging technologies, particularly cost-effective smart packaging to overcome this problem. Although advanced packaging technologies like vacuum and modified atmosphere packaging technologies are available, their adoption in middle- and low-income countries are very insignificant due to its high cost and its maintenance. Apart from reducing post-harvest losses, providing quality and safe fish products without the use of any chemical preservative is a challenge world is facing today. Advanced, low-cost packaging options to enhance the quality, shelf-life and safety are the need of the hour. Advancements in the biodegradable packaging material with functional attributes will continue to progress which brings new concepts and opportunities to enhance the quality and safety of perishable commodities. Adoption of these functional packaging materials will bring new avenues to entrepreneurs and industry persons to enhance their marketing capabilities as the consumers will get the trust on such products which use functional packaging materials to show that these products are better compared to conventional ones.

## References


Aday MS (2014) Ambalaj Materyalleri. Turktarım, Gıda Tarım ve Hayvancılık Bakanlığı Dergisi (220):16–21

Biji KB, Ravishankar CN, Venkateswarlu R, Mohan CO, SrinivasaGopal TK (2016) Biogenic amines in seafood: a review. J Food Sci Technol 53(5):2210–2218

De Kruijf N, Van Beest M, Rijk R, Sipilainen-Malm T, Losada PP, De Meulenaer B (2002) Active and intelligent packaging: application and regulatory aspects. Food Addit Contam 19:144–162

Durusoy, Karababa AO (2011) Plastik Gıda Ambalajları ve Sağlık. TAF Prev Med Bull 10(1):87–96

FAO (2018) The State of World Fisheries and Aquaculture 2018 – Meeting the sustainable development goals. Rome. Licence: CC BY-NC-SA 3.0 IGO

FAO (2020) The State of World Fisheries and Aquaculture 2020. Sustainability in action, Rome. https://doi.org/10.4060/ca9229en

Keleş (1998) Gıda Ambalajlama İlkeleri. Ataturk Universitesi Ziraat Fakultesi Yayınları No: 189.2. Baskı, Ataturk Universitesi Ziraat Fakultesi Ofset Tesisi, Erzurum

Kishore P, Mohan CO, Sreelakshmi KR, Panda SK (2017) Antimicrobial activity of silver nanoparticles (AgNPs) against human significant pathogens. Fish Tech Rep 3(2):33–34

Kızılırmak O (1997) Gıda Sanayiinde Kullanılan Plastik Ambalajlar ve Migrasyon. TMMOB Gıda Muhendisliği Dergisi 4:19–21

https://www.fortunebusinessinsights.com/industry-reports/food-packaging-market-101941. Accessed 26 July 2021

Marsh K, Bugusu B (2007) Food packaging: roles, materials, and environmental issues. J Food Sci 72:39–55

MEB, Ministry of Eduction (2011) Cevre Sağlığı - Gıda Maddelerinde Kullanılan Ambalajlar. T.C. Milli Eğitim Bakanlığı, Ankara

Mohan CO (2008) Shelf-life extension of seer fish (*Scomberomorus commerson*) steaks using $O_2$-scavenger and $CO_2$ emitters in chilled condition. Mumbai, ICAR-CIFT, p 234

Mohan CO, Ravi Shankar CN, SrinivasaGopal TK (2008) Effect of $O_2$ scavenger on the shelf life of catfish (*Pangasius sutchi*) steaks during chilled storage. J Sci Food Agric 88(3):442–448

Mohan CO, Ravishankar CN, SrinivasaGopal TK, Ashok Kumar K, Lalitha KV (2009a) Biogenic amines formation in Seer fish (*Scomberomorus commerson*) steaks packed with $O_2$-scavenger during chilled storage. Food Res Int 42:411–416

Mohan CO, Ravishankar CN, SrinivasaGopal TK, Ashok Kumar K (2009b) Nucleotide breakdown products of seer fish (*Scomberomorus commerson*) steaks stored in $O_2$ scavenger packs during chilled storage. Innov Food Sci Emerg Technol 10:272–278

Mohan CO, Ravishankar CN, SrinivasaGopal TK, Lalitha KV (2010a) Effect of reduced oxygen atmosphere and sodium acetate treatment on the microbial quality changes of Seer fish (*Scomberomorus commerson*) steaks stored in ice. Food Microbiol 27:526–534

Mohan CO, Ravishankar CN, SrinivasaGopal TK (2010b) Active packaging of fishery products- a review. Fish Technol 47(1):1–18

Mohan CO, Ravishankar CN, SrinivasaGopal TK, Lalitha KV (2012) Effect of chitosan edible coating on the quality of double filleted Indian oil sardine (*Sardinella longiceps*) during chilled storage. Food Hydrocoll 26(1):167–174

Mohan CO, Ravishankar CN, SrinivasaGopal TK (2014) Quality of longtail tuna (*Thunnus tonggol*) chunks packed under reduced oxygen atmosphere in chilled storage condition. In: International symposium "MECOS-2 marine ecosystems—challenges and opportunities", The Marine Biological Association of India at Cochin during 2–5 Dec 2014

Mohan CO, Ashitha VA, Kishore P, Panda SK, Ravishankar CN (2018) Influence of mono- and multilayered packaging material on the quality of Seer fish (*Scomberomorus commerson*) during chilled storage. J Packag Technol Res 2(1):67–76

Mohan CO, Abin J, Kishore P, Panda SK, Ravishankar CN (2019a) Effect of vacuum and active packaging on the biochemical and microbial quality of Indian oil sardine (Sardinella longiceps) during iced storage. J Packag Technol Res:1–13

Mohan CO, Gunasekaran S, Ravishankar CN (2019b) Chitosan-capped gold nanoparticles for indicating temperature abuse in frozen stored products. npj Sci Food 3(2):1–6

Nadella RK, Mohan CO, Kishore P, Panda SK, Nilavan E, Prasad MM, Ravishankar CN (2019) Development of colorimetric nano-biosensor strip for detection of E. coli O157:H7 and E. coli. In: National conference Nano India 2019 organized by Mahatma Gandhi University, Kottayam, Kerala on April 26 & 27th, 2019

Nielsen T (1997) Active packaging-the literature review. SIK-rapport no 631, p 20

Paine FA (1991) The packaging users handbook. AVI, Van Nostrand Reinhold, New York

Remya S, Mohan CO, Bindu J, Sivaraman GK, Ravishankar CN (2014) A new generation active packaging system with antimicrobial film and O2 scavenger for fish preservation. In: The 10th Indian fisheries & aquaculture forum (10ifaf), NBFGR, NBFGB, Lucknow, 12–15 Nov 2014 organized by Asian Fisheries Society, Indian Branch (AFSIB), in collaboration with NBFGR, Lucknow

Remya S, Mohan CO, Bindu J, Sivaraman GK, Venkateshwarlu G, Ravishankar CN (2016) Effect of chitosan based active packaging film on the keeping quality of chilled stored barracuda fish. J Food Sci Technol 53(1):685–693

Remya S, Mohan CO, Venkateshwarlu G, Sivaraman GK, Ravishankar CN (2017) Combined effect of $O_2$ scavenger and antimicrobial film on shelf life of fresh cobia (*Rachycentron canadum*) fish steaks stored at 2°C. Food Control 71:71–78

Remya S, Mohan CO, Ravishankar CN, Sivaraman GK, Jha AK, Venkateshwarlu G (2018) Effect of active packaging atmosphere on the shelf life of chilled stored steaks of barracuda *Sphyraena jello*. Indian J Fish 65(4):109–115

Renuka V, Mohan CO, Kirplani Y, Sivaraman GK, Ravishankar CN (2016) Effect of chitosan edible coating on the microbial quality of Ribbonfish, *Lepturacanthus savala* (Cuvier, 1929) steaks. Fish Technol 53(2):146–150

Robertson G (2006) Food packaging principles and practices. Taylor & Francis, Boca Raton, FL

Rooney ML (1985) Oxygen scavenging from air in package headspaces by singlet oxygen reactions in polymer media. J Food Sci 47(291–294):298

Vermeiren L, Devlieghere F, Van Beest M, de Kruijf N, Debevere J (1999) Developments in the active packaging of foods. Trends Food Sci Technol 10:77–86

Wang Y-C, Mohan CO, Jiehao G, Ravishankar CN, Gunasekaran S (2017) Chitosan and gold nanoparticles-based thermal history indicators and frozen indicators for perishable and temperature-sensitive products. Food Control 85:186–193

Wessling C, Nielsen T, Leufven A, Jagerstad M (1998) Mobility of tocopherol and BHT in LDPE in contact with fatty food simulants. Food Addit Contam 15:709–715

Yam KL, Takhistov PT, Miltz J (2005) Intelligent packaging: concepts and applications. J Food Sci 70:1–10

# Edible Films and Coatings: Major Challenges and Potential Applications in Food Packaging. A Review

8

Aitboulahsen Mohamed, Ouiam El Galiou, Said Zantar, Abdelhay Arakrak, Amin Laglaoui, and Mounir Hassani Zerrouk

**Abstract**

Petroleum-based plastics, largely used in food packaging, have raised a great number of health and environmental issues mainly due to their non-renewable source. Biopolymers-based films and coatings designed to preserve the food quality have attracted the attention of many researchers since they are made from renewable resources, and could in addition carry bioactive compounds like antimicrobials and antioxidants that lead to prolong the shelf life of food. However, to be adequately used in food packaging, they must meet a certain number of functional properties like mechanical resistance and gas permeability. This work systematically discusses the most significant functional characteristics of edible films and coatings like mechanical, gas barrier and water vapor permeability, solubility, and organoleptic properties as well as their bioactive characteristics. It also highlights their potential applications and limits in fruits and vegetables, meat products, cheese, and sea foods. The challenges faced so far, and the prospects for their progress and development are also discussed.

A. Mohamed · M. H. Zerrouk (✉)
Department of Biology, Polydisciplinary, Faculty of Larache, Abdelmalek Essaâdi University, Larache, Morocco

O. El Galiou · A. Arakrak · A. Laglaoui
Department of Biology, Faculty of Science and Technology, Abdelmalek Essaâdi University, Tangier, Morocco

S. Zantar
Research Unit of Nuclear Technology, Environment and Quality, National Institute of Agronomic Research, Tangier, Morocco


A. K. Shukla (ed.), *Food Packaging: The Smarter Way*,
https://doi.org/10.1007/978-981-16-7196-8_8

## 8.1 Introduction

Food preservation aims to increase the storage life of food and preserve their quality. Several methods have therefore been developed. Classical empirical techniques such as cooking, drying, or salting, and others such as freezing, canning, pasteurization, and packaging have been commonly used (Augusto et al. 2018). The role of food packaging is to contain food and protect it from external contaminations. It constitutes a physical barrier against mechanical shocks, transfers of matter and energy, and against microorganisms (Berk 2018).

In the early twentieth century, with polyethylene's invention, food transportation and storage was revolutionized. Plastic packaging has since occupied an important place in the food industry because of its practicability, low cost, and functionality. Therefore, global production of plastic packaging exceeded 150 million in 2015 (Groh et al. 2019). However, a number of environmental concerns have been raised regarding the continued use of plastic materials in food packaging, considering the fact that they are non-biodegradable, almost non-recyclable, and have a very limited shelf life, resulting in large amounts of waste that have huge environmental impacts (Hahladakis et al. 2018). On the other side, toxic and undesirable substances such as monomers, synthetic antioxidants, chemicals, etc., could be transmitted once in contact with food (Bolívar-Monsalve et al. 2019). For these reasons, there is a major interest in the development of new packaging materials, biodegradable, non-contaminating, made from renewable and recyclable resources.

In this context, films and coatings made from biopolymers have attracted the attention of many researchers as an alternative to petroleum-based plastics (Menezes n.d.; Umaraw and Verma 2017; Rojas-Graü et al. 2009; Han 2005). Biodegradable films and coatings are described as thin films or coating solutions (Fig. 8.1) made from biopolymers (proteins, polysaccharides, etc.) and are applied as primary food packaging providing a barrier to factors such as moisture, gases, and vapors, and improving therefore their quality (Huber and Embuscado 2009).

In addition, these biopolymers-based films and coatings (BFC) could also carry bioactive compounds, such as antimicrobial agents and antioxidants, contributing in this case to improve the food shelf life (Quezada-Gallo 2009; Salgado et al. 2015; Raybaudi-Massilia et al. 2016; Atarés and Chiralt 2016; Ganiari et al. 2017).

However, to be used effectively in food packaging, BFC must fulfill several functional properties such as mechanical, water, and gas permeability, as well as organoleptic and antimicrobial properties (Fig. 8.2).

This work is an overview aiming to present and analyze the main findings of recent studies with regard to the functional properties of BFC and their potential food applications.

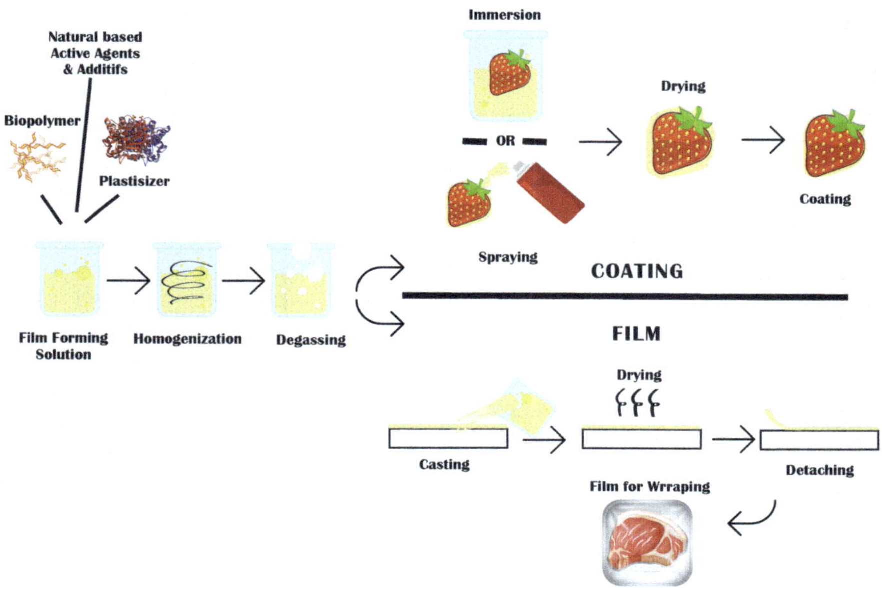


**Fig. 8.1** Development of biopolymers based films and coatings

## 8.2 Functional Properties of Biopolymers-Based Films and Coatings

### 8.2.1 Mechanical Properties

To be efficient, BFC used in food packaging must be mechanically strong enough to protect the product from any abrasion and maintain its integrity during manipulation, storage, and transportation (Wihodo and Moraru 2013). The mechanical resistance of films is determined mainly by their tensile strength (TS) and elongation at break (EB) values.

The improvement of films' mechanical properties depends on the biopolymer nature itself, the additional use of plasticizers, and the control of film manufacturing parameters such as the type of process used, environmental humidity, speed, and temperature of drying or cooling (Debeaufort et al. 1998).

According to Jiménez et al. (2012), the solvent evaporation step for films manufactured by wet process is as important as the cooling step in the dry process. Cho and Rhee (2002) investigated the impact of the environmental humidity on the films' mechanical properties and showed that hydrophilic films tend to absorb moisture more easily, thereby increasing the plasticizing effect of water leading to reduced TS.

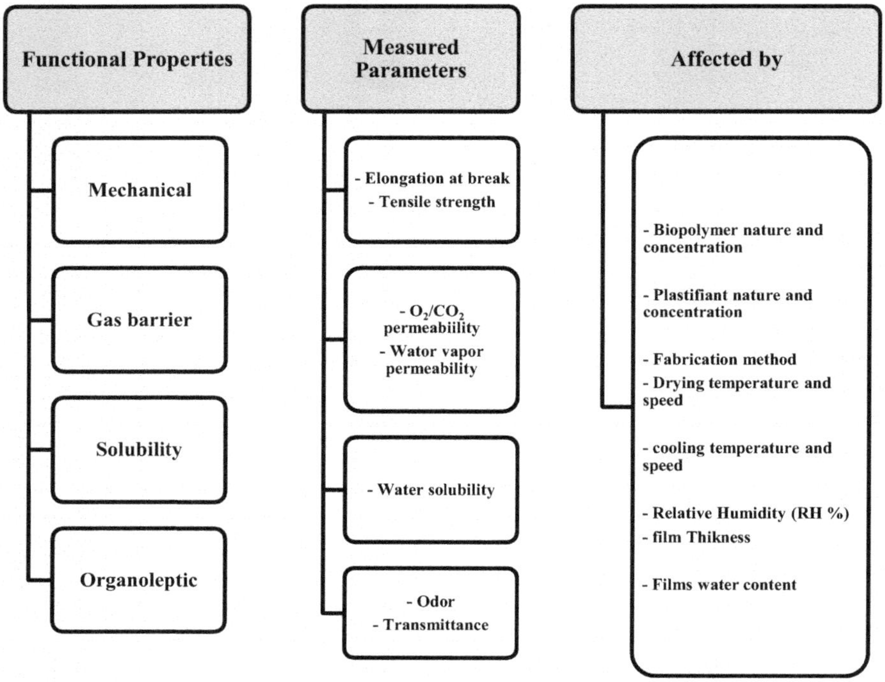


**Fig. 8.2** Functional properties of biopolymers-based films and coatings

The use of plasticizers has a considerable influence on the mechanical properties of films. The presence of a plasticizer on the film matrix increases the mobility of polymer chains by increasing the intermolecular space, leading to higher EB and lower TS levels (Vieira et al. 2011).

It was demonstrated that a double increase of glycerol results in a large increase of elongation at break (76% instead of 17%) and a threefold decrease of TS values (Butler et al. 1996). It should be noted that mechanical properties are also affected by the type of plasticizer and molar mass modifications of biopolymers (Cao et al. 2018). Sothornvit and Krochta (2005) revealed that a decrease in molar mass creates more terminal groups and free volumes in polymers, giving films with higher elongation at break levels. In this sense, Aitboulahsen et al. (2020a) showed that films made from sorbitol as a plasticizer displayed higher TS than films made from glycerol for three different biopolymer-based films (gelatin, pectin, and starch).

Table 8.1 compares TS and EB values of films made with bio- and synthetic polymers. Synthetic polymers have comparable TS levels to that of films made with polysaccharides but higher values of elongation at break (up to 1000 %). However, since many factors are involved, results cannot be compared easily.

**Table 8.1** Mechanical properties of edible and/or biodegradable films (~25°C, 50% HR)

| Polymer | Tensile strength (MPa) | Elongation at break (%) | Reference |
|---|---|---|---|
| Cellulose | 44–65 | 10–50 | Saremnezhad et al. (2011) |
| Chitosan | 10–100 | 20–80 | Saremnezhad et al. (2011) |
| Type A gelatin | 25–85 | 7–22 | Mark (1999) |
| Type B gelatin | 28–140 | 7–12 | Mark (1999) |
| Starch | 35–46 | 1.7–3.4 | Mark (1999); Saremnezhad et al. (2011) |
| Soy protein | 3.7–4.5 | 152–160 | Saremnezhad et al. (2011) |
| Whey protein | 2.5–3 | 15–18 | Saremnezhad et al. (2011) |
| Corn zein | 3–4 | 50–120 | Padua and Wang (2002) |
| Low-density Polyethylene (LDPE) | 9–20 | 100–1200 | Mark (1999) |
| High-density polyethylene (HDPE) | 10–60 | 400–1800 | Mark (1999) |

### 8.2.2 Gas Barrier and Water Vapor Properties

Oxygen fixation in food products is irreversible and may lead to oxidative rancidity; its absence, however, could result in anaerobiosis (Yano 1990). Gas barrier properties of BFC are therefore of great importance. They should act by protecting the quality of food susceptible to oxidation (rancidity, loss of vitamins) and controlling respiratory exchanges of fruits and vegetables (Zahedi 2019).

Permeability is defined as permeants transmission through materials (Han 2005). The fundamental principle of permeation is based on the adsorption/dissolution-diffusion/desorption mechanism.

In absence of pores, defects, and membrane perforations, the permeability P is equal to the product of diffusion coefficient D (representing permeants mobility in the polymer) and of solubility coefficient (representing the concentration of permeants in the film at equilibrium with the external pressure).

$P = D*S$ (Colak 2014)

In this process, gases such as $O_2$ and $CO_2$ cross the biopolymer matrix and are therefore transferred to the external environment or the reverse and lead to changes in food safety and organoleptic characteristics of food (Han 2005).

According to Debeaufort et al. (1998), edible films made with proteins or polysaccharides have generally good oxygen barrier properties particularly under conditions of low humidity; nevertheless, they are often lower to those of conventional synthetic films such as polyethylene.

Due to their linear structure and high percentage of amino acid groups, films made with corn zein, wheat gluten, soy, or wheat protein are characterized with higher permeability to oxygen than the ones made with polysaccharides (Wu et al. 2002). The same authors reported that gas permeability properties of starch or lipid films affected the shelf life of the packed food.

The plasticizer nature and concentration also affected the oxygen permeability of BFC.

Water vapor permeability (WVP) can be defined as the rate of water vapor transmission by unit area of a giving material with known thickness and by difference in vapor pressure between two surfaces, under known temperature and humidity conditions. It is directly related to environmental conditions and to the amount of –OH groups in the molecule (Henrique et al. 2007).

Water vapor permeability (WVP) of BFC was investigated by several authors (Gennadios et al. 1994; Debeaufort and Voilley 1994; McHugh et al. 1994; Bertuzzi et al. 2007; Al-Hassan and Norziah 2012; Aitboulahsen et al. 2020a). It depends on various factors such as molecules solubility on the film matrix, hydrophobic/hydrophilic nature of biopolymers, their structure, polymeric chain mobility, interactions between polymers functional groups, hydrophobic/hydrophilic and crystalline/amorphous ratios, type and concentration of plasticizer, film thickness and ambient conditions (temperature, relative humidity. . .) (Debeaufort and Voilley 1994; McHugh et al. 1994; Pérez-Gago and Krochta 1999).

The addition of lipidic, hydrophobic compounds can be an effective approach to improve WVP of BFC. The low polarity of lipids and their ability to form dense and well-structured molecular matrices reduces water mobility (Debeaufort and Voilley 2009). Regarding the plasticizer effect, the use of polyols, especially glycerol and sorbitol, improves films flexibility to the detriment of water vapor barrier (Haq et al. 2014). The same authors observed that WVP increases with glycerol concentration increase for gum-based films. In contrast, sorbitol was proved better than glycerol for obtaining less permeable films to water vapor (Razavi et al. 2015).

On the other hand, incorporation of hydrophobic substances in BFC matrix such as essential oils (EOs) would improve water vapor barrier properties. Basil seed gum incorporated with oregano EO allowed the decrease of WVP values due to the interference of oil globules on the transmission of water molecules in the coating matrix (Gahruie et al. 2017).

Table 8.2 shows gas and water vapor property values of edible and synthetic films and coatings. Starch and protein films are generally highly permeable to water vapor but less permeable to gases. It is therefore necessary to optimize the barrier properties according to food product requirements (Wihodo and Moraru 2013).

### 8.2.3 Thickness

Mechanical and barrier properties of BFC are directly influenced by their thickness (Zahedi 2019). Biopolymers-based films and coatings thickness depends on many factors such as the production process and the concentration and type of components, particularly, plasticizers (Zahedi 2019). Gheribi et al. (2018) showed that glycerol addition as a plasticizer can absorb a great deal of water resulting in inflated films with high thickness.

Razavi et al. (2015) observed that sage seed gum films supplemented with sorbitol were thicker than those containing added glycerol. Density difference

**Table 8.2** Water vapor and gas permeability values of some bio- and synthetic polymers (Krochta and de Mulder-Johnson 1997; Embuscado and Huber 2009)

| Materials | WVP ($10^{-11}$ g/m.s. Pa) ~38 °C, 90/0% HR | $PO_2$ ($10^{-17}$ g/m.s. Pa) ~25 °C, 0–50% HR |
|---|---|---|
| Methylcellulose | 8.7 | 15–150 |
| Starch | 17.5–25.7 | 15–150 |
| Gelatin | 16 | 1.5–15 |
| Chitosan (2%) | 3.6–4.9 | 15–150 |
| Zein | 53-89 | 15–150 |
| Gluten | 0.1–10 | 1.5–15 |
| Whey protein | 71 | 1.5–15 |
| High-density polyethylene HDPE | 0.02 | 4 |
| Low-density polyethylene LDPE | 0.09 | 14.3 |
| Polyethylene terephthalate PET | 0.28 | 0.08 |

between the different phases of the films and emulsified components interactions may modify thickness (Razavi et al. 2015).

According to Gahruie et al. (2017), thyme EO incorporation in basil seed gum films increases thickness. This may be explained by the presence of lipid molecules between polysaccharide chains which inhibits the development of ordered and compact structure, dilating film matrix (Haq et al. 2016).

### 8.2.4 Solubility

The majority of BFC are soluble in water (Menezes n.d.). This property is of great interest through its applicability on foodstuffs with different compositions and for other technological purposes. Films solubility in water should be controlled depending on food product and according to the intended technological purpose. Indeed, ready-to-eat semi-finished food products require films with high solubility. In contrast, for packaging of high water activity products, films should be less soluble. Inversely, packaging that should be dissolved before product consumption (cooking, infusion) should be highly soluble. To optimize packages' water solubility, the use of insoluble lipids and proteins and some other hydrophobic components was investigated in order to preserve films from inflation or disaggregation (Debeaufort and Voilley 2007). Composite films allow better resistance to water and polymers crosslinking. Vachon et al. (2000) have successfully reduced films solubility to more than 75% favoring reticulation by irradiation treatment.

Rompothi et al. (2017) observed that concentration increase of plasticizer increases solubility, whereas concentration increase of the biopolymer (starch) decreases solubility. They also noted that glycerol gives less soluble films than sorbitol. According to the same authors, this is due to plasticizer migration which was difficult when film matrix was more compact with high proportion of cohesive forces between starch chains.

### 8.2.5 Organoleptic Properties

Organoleptic neutrality is also important for BFC. Transparency/opacity, smell, taste, brightness, etc. are important organoleptic properties that should be controlled. Furthermore, compatibility of these properties with food matrix is a parameter that must be taken into account (Fernández-Pan et al. 2011).

Generally, protein or polysaccharide films are more neutral than lipid ones, which are contrarily more opaque and sliding with light wax taste (Krochta 2010). Additives addition, especially EOs which plays a major role as antimicrobial agent may influence significantly the smell and taste of films and consequently those of the food (Atarés and Chiralt 2016). Khwaldia et al. (2004) showed that edible films made with milk proteins result in a bad taste during storage due to lipid oxidation products, Maillard reaction, and vitamins degradation.

Films' optical properties are also of paramount importance as they have direct impact on product appearance and acceptability by the consumer (Ozdemir and Floros 2008).

Kaya et al. (2018) have shown that incorporation of seed oil and dried fruits extract in film matrix made with chitosan decreases film transparency.

Nisar et al. (2018) also observed a decrease in films' transparency with clove EO addition in pectin films and reported that this may be due to phenolic components which can absorb light at different wavelengths with a natural yellowish color attributed to lipids or EOs.

### 8.2.6 Antimicrobial Properties

Antimicrobial substances can be added to BFC to confer them antimicrobial properties inhibiting microbes' growth on food surfaces. These incorporated antimicrobial agents are released in a progressive and controlled manner maintaining low microbial counts throughout the storage period (Bolívar-Monsalve et al. 2019; Sánchez-González et al. 2011; Janes and Dai 2012; Espitia et al. 2016; Avila-Sosa et al. 2016; López et al. 2007).

Table 8.3 shows the inhibitory effect of some antimicrobial films on pathogens. Lysozyme incorporation in chitosan-based films was effective to reduce *Escherichia coli* and *Streptococcus faecalis* growth levels (Park et al. 2004). Chitosan-based coatings supplemented with potassium sorbate and nisin showed inhibitory action against *Escherichia coli*, *Staphylococcus aureus*, *Salmonella typhimurium*, *Listeria monocytogenes,* and *Bacillus cereus* (Pranoto et al. 2005).

Hettiarachchy and Satchithanandam (n.d.) incorporated organic acids in soy proteins films and demonstrated their antimicrobial effect against *Listeria monocytogens*, *Salmonella gaminara*, and *Escherichia coli O157:H7*.

The antimicrobial effect of BFC containing EOs was widely investigated (Avila-Sosa et al. 2016). Rojas-Graü et al. (2007a, b) examined the antimicrobial properties of films incorporated with various EOs. The most effective was carvacrol followed by oregano, lemon, lemongrass, and cinnamon. Ravishankar et al. (2009) have

**Table 8.3** Antimicrobial activity of biopolymers-based films and coatings containing antimicrobial agents

| Composition of the film | Antimicrobial agent | Targeted microorganisms | References |
|---|---|---|---|
| Whey proteins isolates | Essential oils: Oregano/rosemary/ garlic | *Escherichia coli* O157 : H7<br>*Staphylococcus Aureus*<br>*Salmonella* enteritidis<br>*Listeria monocytogenes* | Seydim and Sarikus (2006) |
| Chitosan | Nisin<br>Potassium sorbate | *Escherichia coli*<br>*Salmonella typhimurium*<br>*Staphylococcus aureus*<br>*Bacillus cereus*<br>*Listeria monocytogenes* | Pranoto et al. (2005) |
| Pectin—apple puree | Essential oils: Oregano/ lemongrass/ cinnamon | *Escherichia coli* O157 : H7 | Rojas-Graü et al. (2007a, b) |
| Whey proteins isolates | Lactoperoxidase | *Salmonella enterica and Escherichia coli O157:H7* | Min et al. (2006) |
| Sodium alginate | Castor oil | *Staphylococcus aureus Bacillus Subtilis, Escherichia coli, and Salmonella typhimurium* | Abdel Aziz Mohamed et al. (2018) |
| Starch | Methanolic extracts of *Hibiscus sabdariffa* | *Listeria monocytogenes* | Cruz-Gálvez et al. (2018) |

shown that chicken breast fillet and ham wrapped with pectin-apple puree films supplemented with different concentrations of cinnamaldehyde or carvacrol have significantly reduced *Salmonella enteritidis*, *Escherichia coli O157: H7,* and *Listeria monocytogenes* growth levels.

Basil seed gum films containing oregano EO produced by Gahruie et al. (2017) showed significant antimicrobial activity against *Escherichia coli, Salmonella typhimurium, Pseudomonas aeruginosa, Staphylococcus aureus,* and *Bacillus cereus*.

Lin and Zhao (2007) and Ozdemir and Floros (2008) showed that EOs diffusion on films matrix is influenced by many factors such as biopolymer nature, additives, manufacturing process, pH and water activity of the food and storage time and temperature.

Due to its film-forming and antimicrobial properties, the chitosan effect on food shelf life was widely investigated. Coma et al. (2002) and Ponce et al. (2008) reported its inhibitory effect against *Listeria monocytogenes* and Maqbool et al. (2011) proved the efficacy of chitosan-based coatings in slowing the growth of *Fusarium* spp., *Colletotrichum musae,* and *Lasiodiplodia theobromae* when preserving bananas. It has been proven that the use of these films and coatings improved fruits and vegetables shelf life by inhibiting microorganisms' growth, reducing ethylene production, increasing carbon dioxide concentration, and by reducing oxygen levels (Geraldine et al. 2008; Lazaridou and Biliaderis 2002).

## 8.3 Films and Coatings Applications on Food Products

During storage and distribution, food products are subject to dynamic changes, mainly due to interactions between them and their surrounding environment that adversely affect their quality and lead to deterioration. Several formulations of BFC made from different biopolymers, additives, and their mixtures have been tested and applied to food products such as fruits and vegetables, cereals and dried fruits, meats, poultry or seafood, cheese, fresh food, salted, frozen and processed, etc...

In the following section, an overview on the various applications of BFC on food products will be addressed.

### 8.3.1 Applications in Fruits and Vegetables

Fruits and vegetables are usually altered after harvest, during ripening and storage because of gas exchange, respiration, transpiration, and microorganisms' attacks. Edible films and coatings applied to fruit and vegetables aim to reduce the respiratory rate, water loss, microbial degradation, and thus help to preserve their quality (Yousuf and Qadri 2020). Besides, the incorporation of antimicrobial agents such as plant extracts and EOs prevents harmful microbial growth, particularly molds, and helps improve the shelf life of fruits and vegetables (Saxena et al. 2020). In this context, some proteins and polysaccharides-based coatings have been developed to control the ripening of fruits and vegetables. Their selective permeability creates a changed internal atmosphere that helps to sustain a reduced respiratory rate and thus slows down the metabolism of fruit and vegetables after harvesting, thus enhancing their shelf life (Yousuf and Qadri 2020). In this regard, a recent study has shown that the application of alginate-based coating with rhubarb extract has contributed to a substantial reduction in the respiration rate of peach fruit (Li et al. 2019). Similarly, a polysaccharide-based coating applied to potatoes was an efficient oxygen barrier and decreased respiration during storage (Wu 2019).

Applying films and coatings to fruit and vegetables should retain firmness after harvesting and minimize water and weight loss. Accordingly, Correa-Pacheco et al. (2021) were able to preserve the firmness of green pepper fruits, during 12 days of storage at $10 \pm 1$ °C, using chitosan nanoparticles-based coating.

Aitboulahsen et al. (2018) and Pinzon et al. (2020) also demonstrated the positive effect of the coating on the texture of strawberries during storage, thus addressing the problem of cell wall degradation of strawberries during storage. Kozlu and Elmacı (2020) obtained similar results for mandarin fruits coated with a quince seed mucilage-based coating.

The incorporation of active substances, such as antioxidants, helps to delay postharvest browning of fruits and vegetables, by chemical reduction of quinones to colorless orthodiphenol reducing agents, acting in the enzymatic browning process, which consists of the oxidation of phenolic substrates to colored orthoquinones (Saxena et al. 2020).

Browning can also be inhibited in an atmosphere containing less than 1 kPa of oxygen (Gorny et al. 2002), for this, packing fresh fruits and vegetables with BFC, whether or not containing antioxidants can be a better way to prevent oxygen diffusion due to mechanical damage during handling, transport, or storage. These films and coatings improve the oxygen and light barrier, prevent polyphenol oxidase (PPO) contact with oxygen, and maintain a stable amount of ascorbic acid (Lin and Zhao 2007).

El-Mogy et al. (2020) concluded that artichoke bottoms coated with a Cordiamyxa gum coating incorporated with calcium dichloride ($CaCl_2$ and ascorbic acid prolong the shelf life of fresh artichoke bottoms by delaying their browning and improving their texture compared to control fruits. Wang and Gao (2013) obtained the same results on strawberries coated with chitosan. The authors reported a decrease in the weight loss and the activity of PPO, as well as a stable content of anthocyanins, flavonoids, and total phenolics. A recent study carried out on freshly cut lettuce claimed that coating with chitosan is an effective method to prevent enzymatic browning and therefore extend the shelf life of fruits and vegetables (Li et al. 2021).

On the other hand, the damage of fruit, vegetable, or plant tissue establishes ideal conditions for microbial growth (Chen et al. 2021). The applications of BFC containing antimicrobial agents have been largely used to overcome this problem (Azarakhsh et al. 2014; Shahbazi 2018; Hu et al. 2020; Meindrawan et al. 2020; Arabpoor et al. 2021).

Maqbool et al. (2011) demonstrated that the incorporation of lemongrass and cinnamon EO in Arabic gum-based coatings extended the banana's shelf life up to 33 days, and allowed to control anthracnose caused by *Colletotrichum spp.* Figure 8.3 shows the evolution of the hygienic quality of strawberries coated with gelatin incorporated with pennyroyal EO. In this study, Aitboulahsen et al. (2018) showed that the bioactive coating used decreased fungal contamination of strawberries during storage and thus increased their shelf life. Table 8.4 summarizes the most relevant applications of BFCon fruits and vegetables.

### 8.3.2 Applications in Cheese

Cheese is an ancient food product that makes part of humans' regular diet, due to its composition characterized by high amount of protein, calcium, minerals, and vitamins. The intensive growth of yeasts, molds, and undesirable bacteria on cheese surfaces can considerably reduce its quality. Biopolymers-based films and coatings have been largely used for preserving the quality of cheeses (Duan et al. 2007; Kuorwel et al. 2013; Di Pierro et al. 2011; Cerqueira et al. 2010; Oliveira et al. 2017).

**Fig. 8.3** Appearance change of strawberries coated with a gelatin coating incorporated by pennyroyal essential oil (MEO) or not during 13 days of storage at 4 °C. (T1: Control; T2: Gelatin coating + 0% MEO; T3: Gelatin Coating + 0.5% MEO; T4: Gelatin Coating + 1% MEO) (Aitboulahsen et al. 2018)

Low moisture Mozzarella slices coated with chitosan incorporated with lysozyme allowed a delay of microbial growth of *Escherichia coli, Pseudomonas fluorescens, and Listeria monocytogenes* after 14 days of storage at 10 °C (Duan et al. 2007). For ricotta without the rind, the coating based on whey and chitosan extended cheese shelf life up to 30 days (Di Pierro et al. 2011). The same type of cheese was coated with a galactomannan-based coating containing nisin and showed a significant growth reduction of *Listeria monocytogenes* inoculated intentionally on the cheese surface, compared to uncoated control (Martins et al. 2010). According to Kuorwel et al. (2013), the starch-based coating incorporated with linalool, carvacrol, and thymol was effective against the development of *Aspergillus niger* on cheddar's surface during storage. However, the use of EOs can produce strong odors and aromas influencing the sensorial quality of the cheese (Yangilar 2016). More recent studies summarized in Table 8.5 also showed the effects of bioactive films and coatings on sensory characteristics of cheeses such as color, odor, flavor, and taste, as well as on the development of harmful microorganisms during the ripening process.

### 8.3.3 Applications in Meat

Several studies have demonstrated the effectiveness of BFC in preserving the quality of meat and poultry products (Guo et al. 2014; Shin et al. 2017; Vital et al. 2018; Saricaoglu et al. 2018; Souza et al. 2019; Bolívar-Monsalve et al. 2019). Gallego

**Table 8.4** Recent applications of biopolymers-based films and coatings on fruits and vegetables

| Fruit/ vegetable | Biopolymers | Additives/active agents | Application method | Storage conditions | Major findings | References |
|---|---|---|---|---|---|---|
| Peach fruit | Chitosan | Chlorogenic acid | Immersion for 2 min | Packed in unsealed polyethylene bags and stored for 8days at 20 °C with 80–90% RH | Antioxidant activity<br>Preservation of physicochemical characteristics | Jiao et al. (2019) |
| Papaya fruit | Chitosan | *Rutagraveolens L* essential oil | Immersion for 2 min, and air-drying for 1 h at room temperature | Stored at 20 °C ± 1 °C and 88% RH for 12 days | Antifungal activity (inhibition of *Colletotrichum gloeosporioides* growth) | Peralta-Ruiz et al. (2020) |
| Guava fruit | Chayotextle starch and resistant starch | Ascorbic acid | Dipping for 15 s, draining and drying at room temperature | Packed in polypropylene plastic trays with a perforated cover at 4 °C and 65% RH | Delay ofmaturation changes<br>Preservation of organoleptic characteristics (firmness and color) | Martínez-Ortiz et al. (2019) |
| Orange fruit | Aloe vera gel | Salicylic acid | Immersion for 5 min at 22 °C, drying at room temperature for 2 h | 4 ± 1 °C and 80 ± 5% RH and transfer of 4 samples to 25 °C for 1 day | Antimicrobial activity<br>Amelioration of chilling injury | Rasouli et al. (2019) |
| Strawberry | Pullulan | Cinnamon essential oil | Immersion for 60 s, drying at room temperature for 4 h | 20 °C ± 2 °C and 70–75% RH for 6 day | Antimicrobial activity<br>Prolongation of the shelf life | Chu et al. (2020) |
| Cherry fruit | Nano-chitosan | *Eryngium campestre L.* essential oil | Immersion for 3 min and drying at room T°C | Packed in disposable polypropylene containers and stored in dark at a 4 ± 1 °C/70% RH | Antimicrobial and antioxidant activity<br>Improvement of physicochemical characteristics | Arabpoor et al. (2021) |
| Mandarin fruit | Carboxymethyl cellulose | | | 5 ± 1 °C for 30 days | Preservation of membrane integrity | Ali et al. (2021) |

(continued)

**Table 8.4** (continued)

| Fruit/ vegetable | Biopolymers | Additives/active agents | Application method | Storage conditions | Major findings | References |
|---|---|---|---|---|---|---|
| | | | Dipping for 3 min, drying at room temperature | | Improvement of sensory characteristics | |
| Potato | Cactus *Opuntia dillenii* | | Dipping for 3 min and draining the excess | Packed in racks at 5 °C for 5 days | Extension of shelf life (suppression of browning, weight loss, and microbial growth)<br>Preservation of total sugars content | Wu (2019) |
| Artichoke | *Cordiamyxa*gum | Calcium dichloride ($CaCl_2$) and ascorbic acid | Dipping for 5 min, drying under sterile laminar flow hood at ambient conditions | Packed in polypropylene trays, thermally sealed by the stretch film and stored at 2 °C and 95% RH for 9 days | Antimicrobial activity and antibrowning activity | El-Mogy et al. (2020) |
| Cherry tomato | Protein hydrolysate | | Dipping for 3 min and air-drying at room temperature | Packed in plastic trays at 21 °C and chilled stored 4 °C for 45 days | Inhibition of the proliferation of molds and yeasts | de da Costa et al. (2020) |
| Persimmon fruit | Gumarabic | | Dipping for 3 min, drying for 2 h at 25 °C | Packed in clamshell clear PET boxes, and kept at 20 ± 1 °C with 80 ± 2% RH for 20 days | Reduction of weight loss, membrane leakage, and fruit softening<br>Preservation of physicochemical characteristics | Saleem et al. (2020) |
| Tomato | Rice bran wax | | Dipping 2–3 min, drying under the fan for about 3–5 h | 32.7–34.4 °C and RH of 57.5–88.3% | Controlling spoilage of tomatoes until 27 days | Abhirami et al. (2020) |
| Pomegranate fruit | Chitosan | Malic acid (MA) and oxalic acid (OA) | Immersion for 2 min and drying at 20 ± 1 °C | Cold storage at 2 ± 1 °C with 80–90% RH for 120 days | Antioxidant and antibrowning activity | Ehteshami et al. (2020) |

| | | | | | | |
|---|---|---|---|---|---|---|
| Apple | Fenugreek and flax polysaccharides | | Dipping for 1 min, twisting and air-drying for 20 min | 20 °C ± 5 with relative humidity 80–85% | Delaying the respiration rate and inhibiting the ethylene production<br>Preservation of quality attributes | Rashid et al. (2020) |
| Carrots | Carboxymethyl cellulose/ candelilla wax | Ascorbic acid | Dipping for 1 min, and air-drying for 1 h | Placed in sterile plastic vessels overwrapped with a perforated PE foil, and stored at 5 ± 1 °C and ~70% RH for 21 days | Preservation of fresh characteristics up to 21 days of cold storage | Kowalczyk et al. (2020) |
| Mango | Chitosan | Lactoperoxidase | Dipping | 18 °C and 60% RH | Antifungal activity | Mohamed et al. (2020a, b) |
| Banana | Cassava starch/ carboxy methyl cellulose | Lactic acid bacteria | Wrapping with the developed film | 30 °C and 40% RH for 7 days | Antibrowning activity | Li et al. (2020) |
| Lettuce | Chitosan | | Sprayed by both sides air-drying for 20 min at 17 °C | Packing into low-density polyethylene bags and stored at 4 °C for 15 days | Improving the hygienic quality preservation of chlorophyll levels, increasing the total phenolics, flavonoids content, and antioxidant activity | Fasciglione et al. (2020) |
| Cucumber | Chitosan, Carnauba Wax | Oregano essential oil | Manual application with brushes and drying with forced air at 25 °C for 10 min | Stored on polypropylene trays at 10 °C and 90% RH for 15 days | Antimicrobial activity<br>Reduction of weight loss | Gutiérrez-Pacheco et al. (2020) |

(continued)

**Table 8.4** (continued)

| Fruit/ vegetable | Biopolymers | Additives/active agents | Application method | Storage conditions | Major findings | References |
|---|---|---|---|---|---|---|
| Eggplant | Chitosan | | Immersion for 5 min and air drying | Packed in perforated plastic baskets and covered with low-density polyethylene film, and stored at 10 ± 2 °C | Preservation of overall quality attributes and sensory acceptance | Sharma et al. (2021) |
| Avocado | Sodium alginate | *Meyerozyma caribbica* yeast | Dipping for 1 min and drying for 1 h (25 °C, 45% RH) | 15 days storage | Prevention of anthracnose reduction of weight loss | Iñiguez-Moreno et al. (2020) |

**Table 8.5** Recent applications of biopolymers-based films and coatings on meat products

| Meat product | Biopolymers | Additives/active agents | Application method | Storage conditions | Major findings | References |
|---|---|---|---|---|---|---|
| Chevon sausages | Calcium alginate/ Maltodextrin | *Terminalia arjuna* | Individually enclosing | 6 units packed in low-density polyethylene pouches and stored at (4 ± 1 °C) | Antimicrobial activity control of lipid oxidation and storage changes | Kalem et al. (2018) |
| Beefloins | Chitosan | Cumin essential oil | Wrapping | Packed in polyethylene pouches, stored at 3 ± 0.2 °C for 21 days | Shelf life extension control of foodborne pathogens | Dini et al. (2020) |
| Raw beef meat | Pectin/fish gelatin | Olive antioxidants hydroxytyrosol and 3,4-dihydroxy phenyl glycol | Wrapping | Stored at 4 °C for 6–7 days | Delaying the lipid oxidation | Bermúdez-Oria et al. (2019) |
| Chicken breast | Semi-refined κ-carrageenan | Water extract of germinated fenugreek seeds | Wrapping and heat sealing | Stored at 5 °C for 7 days | Shelf life extension 6 days light-barrier films | Farhan and Hani (2020) |
| Fresh pork | Chitosan/gelatin | Nisin and grape seed extract | Immersion in coating solutions for 30 s, and air-drying | Packed in plastic trays at 4 °C, with light exposure | Shelf life extension Prevention of lipid and protein oxidation inhibition of microbial growth | Xiong et al. (2020) |
| Chilled black pork tender loin | Curdlan/polyvinyl alcohol/ | Thyme essential oil | Wrapping | Stored at 4 °C for 20 days | Shelf life extension for up to 10 days | Zhang et al. (2020) |
| Strips of beef loins | Chitosan | | Immersion for 5 s | Stored at 4 ± 1 °C | Inhibition of the growth of *Staphylococcus aureus* | Duran and Kahve (2020) |

(continued)

**Table 8.5** (continued)

| Meat product | Biopolymers | Additives/active agents | Application method | Storage conditions | Major findings | References |
|---|---|---|---|---|---|---|
| Chicken nuggets | *Manihotesculenta* starch/ carrageenan | Anise, caraway, and nutmeg essential oils | Overwrapping | Stored at 4 ± 1 °C | Antioxidant and antimicrobial activity with well sensorial acceptance | Bharti et al. (2020) |
| Pork patties | Chitosan | Clove essential oil and nisin | Wrapping | Placed in plastic boxes and storedat 4 ± 2 °C for up to 15 days | Shelf life extension for 12 days instead of 6 days antioxidant and antimicrobial activity | Venkatachalam and Lekjing (2020) |
| Turkey meat | Gelatin/chitosan | *Ferulago angulate* essential oil | Wrapping | Stored at 4 °C for 15 days | Inhibition of total viable count, *Staphylococcus aureus* and coliform count | Naseri et al. (2020) |
| Lamb meat | Alginate | Thyme and garlic essential oils | Coating | Packed in an individual polystyrene tray over-wrapped with permeable oxygen film and stored between 2 °C and 4 °C under light exposure | Reducing lipid oxidation color maintenance | Guerrero et al. (2020) |
| Ground beef | Starch | Red cabbage extract and sweet whey | Wrapping | Stored at 4 °C ± 1 °C for 4 days | Preservation of the overall quality | Sanches et al. (2021) |
| Harbin red sausage | Chitosan | | Immersion for 30 s and drying for ½ h | 12 h at room temperature | Improving storage stability slowing down the growth of aerobic bacteria and lactic acid bacteria | Dong et al. (2020) |
| Fresh ostrich meat | QodumeShirazi seed mucilage- | Lavender essential oil | Immersion for 1 min | Refrigerated storage | Shelf life extension preservation of sensorial characteristics | Heydari et al. (2020) |

| | | | | | | |
|---|---|---|---|---|---|---|
| Turkey filets | Chitosan | *Zataria Multiflora Boiss* and *Bunium persicum Boiss* essential oils | Immersion | Kept in sterile bags during the storage at 4 ± 1 °C for 18 days | Prolongation of shelf life to 9 days<br>reduction of about 3 log CFU/g in *S. enteritidis* population and about 2 log CFU/g in *L. monocytogenes* | Keykhosravy et al. (2020) |

et al. (2020) showed that gelatin-based coating enriched with antioxidants can reduce water loss and improve the texture of pork meat. Antimicrobial BFC has been widely studied to extend the shelf life of meat and poultry products. They are more efficient than EOs used alone (Cutter 2006). Potato starch-based films incorporated with thyme EO as an antimicrobial agent applied to chilled pork meat reduced *Escherichia coli* and *Staphylococcus aureus* growth and thus extended the shelf life by eight days longer than unpackaged meat (Yuan et al. 2021). Naseri et al. (2020) observed inhibition of the growth of total flora, coliforms, and *Staphylococcus aureus* on turkey meat wrapped with composite films based on gelatin and chitosan, incorporated with *Ferulago angulate* EO. Coating based on sodium alginate, rosemary EO and nisin was effective in reducing the growth of *Listeria monocytogenes* in poultry products during cold storage (Raeisi et al. 2016). The incorporation of antioxidant agents such as garlic or ascorbic acid in carrageenan-based coatings has been reported to improve the shelf life of poultry products (Ustunol 2009).

A comparative study on the antioxidant efficacy of different antioxidant agents (oregano and chili pepper EOs and their combination) incorporated into milk protein-based edible films for the protection of beef slices was conducted by Oussalah et al. (2004). Results showed that films incorporated with chili pepper EO were more effective than oregano in inhibiting lipid oxidation. Several other studies (Table 8.6) from different meat origins have reported the effectiveness of films and coatings in preserving the sensory and hygienic quality, as well as improving the shelf life of meat products.

### 8.3.4 Applications in Seafood

Due to their rich and complex composition, fish and seafood, in general, are highly perishable (Tahergorabi et al. 2015). The addition of preservatives such as phosphates is often used as a technique to increase water retention capacity, reduce oxidation and extend the shelf life of seafood products (Gram and Huss 1996). Also, freezing and deep-freezing remain the most widely used techniques to control or reduce biochemical changes in seafood products (Ghaly 2010). Biopolymers-based coatings and films have been largely used during storage to enhance the quality of seafood (Table 8.7). Several authors developed chitosan-based films and evaluated their effects on lipid oxidation during the storage of various seafood products (Jeon et al. 2002; Abdollahi et al. 2014; Gómez-Estaca et al. 2010; Günlü and Koyun 2013). The results obtained by Sathivel (2005) using chitosan and soy protein concentrate-based coating showed significant efficacy in delaying lipid oxidation for frozen salmon fillets.

Kim et al. (2012) developed an antioxidant edible film based on defatted mustard flour that was able to reduce lipid oxidation for smoked salmon without affecting the sensorial quality. A coating based on gelatin and cinnamon EO was applied to fresh rainbow trout fillets, reducing bacterial growth during 15 days of cold storage (Andevari and Rezaei 2011). *Listeria monocytogenes*, considered to be among the

**Table 8.6** Recent applications of biopolymers based films and coatings on cheese

| Cheese product | Biopolymers | Additives/Active agents | Application method | Storage conditions | Major findings | References |
|---|---|---|---|---|---|---|
| Kashar cheese | Natamycin | Oregano and rosemary essential oils | Dipping twice for 60 s and drying for 2 h | Stored at 4 °C and 70–80% RH | Inhibition of mold growth during ripening without any adverse effects on sensorial quality | Yangilar (2017) |
| Commercial Tybo cheese and handcraft goat cheese | Agar | Enterocins synthesized by *Enterococcus avium* | Immersion for 10 s and drying for 2 h at 30 °C and 24 h at 25 °C | Stored at 4 °C to 8 °C | Anti-*L. monocytogenes* activity, with a decrease of at least 1 log cycle in the viability of the pathogen | Guitián et al. (2019) |
| White brined cheese | Chitosan | Zinc oxide nanoparticle | Dipping for 1 min and drying in a laminar flow hood for 2 h at room temperature | Vacuum-packaged in plastic bags and stored at 4 or 10 °C for 28 days | Antimicrobial activity<br>Reducing growth of f E. coli O157:H7<br>Preservation of color attributes | Al-Nabulsi et al. (2020) |
| Gouda cheese | κ-Carrageenan | Black carob extracts | Coating on both sides | Placed in the dark at 4 °C for 20 days | Antioxidant activity (reduced peroxide value) | Pérez et al. (2020) |
| Sliced Prato cheese | Gelatin/chitosan | Boldo extract | Covering both surfaces | Placed in Styrofoam trays, wrapped with a plastic film and stored at 4 °C for 10 days | Antioxidant and antimicrobial activity (reducing the growth of psychrotrophic and coliform counts, and reducing peroxide value) | Bonilla and Sobral (2019) |

(continued)

**Table 8.6** (continued)

| Cheese product | Biopolymers | Additives/Active agents | Application method | Storage conditions | Major findings | References |
|---|---|---|---|---|---|---|
| Queso Blanco Cheese | Whey protein isolate | Oregano essential oil | Individually and completely wrapping | Stored at 8 °C with exposure to fluorescent light | Enhancing barrier properties against pro-oxidant agents<br>Antifungal properties | Gurdian et al. (2017) |
| Cheddar cheese | Guargum/ Tragacanthgum | | Three-laired coating by Gentle brushing approach, drying of 1 h between brushing layers and finally drying for 4 h at 24 °C | Kept at 8 °C for 90 days | Antimicrobial activity<br>Enhancing the rate of lipolysis | Pourmolaie et al. (2018) |
| Goat'smilkcheese | Chitosan | Oregano and rosemary essential oils | Immersion for 30 s and drying, 2 to 3 times coating | Stored at $10 \pm 5$ °C, $8 \pm 5\%$ RH for 15 days | Preventing weight loss during ripening<br>Improving microbial safety<br>Reducing lipolytic and proteolytic activity | Embuena et al. (2017) |
| "Coalho" cheese | Galactomannan of *Caesalpinia pulcherrima* | *Cymbopogoncitratus* essential oil | Aseptically coating | Placed in polystyrene trays, covered with polyvinyl chloride and stored at $4 \pm 1$ °C | Reduction in microbial growth, especially of total coliforms<br>Preservation of sensorial characteristics | Lima et al. (2021) |

| | | | | | | |
|---|---|---|---|---|---|---|
| Home-made cheese | Sodium alginate | *Menthapulegium L*, *Artemisiaherba alba Asso*, *Ocimumbasilicum L*, *Rosmarinus officinalis L* essential oils | Immersion for 1 min, and drying at 30 °C and 60% RH for 60 min | Stored at 5–7 °C for 10 days | Prevention against weight loss, hardness, discoloration, loss off-flavor, texture, and microbial spoilage | Mahcene et al. (2020) |
| Roquefort cheese | Sodium alginate/ methylcellulose/ hydroxypropy methylcellulose/ sodium caseinate/ gellan gum/κ-carrageenan/ whey protein isolate / xanthan gum | Oleic acid, beeswax , potassium sorbate, Gallic Tannin and natamycin | Immersion for 1 min | Stored at 4 °C and 60% RH | Effective mass loss control during the ripening period<br>Antifungal activity | Ordoñez et al. (2021) |
| Ras cheese | Chitosan/poly vinyl alcohol | Titanium dioxide nanoparticles | Brushing on the different faces of the cheese piece and drying for 2 h at room temperature | Stored at 12 ± 2 °C and about 80% RH for 3 months | Prevention of mold growth on the cheese surface during ripening | Youssef et al. (2019) |

**Table 8.7** Recent applications of edible films and coatings on seafood products

| Seafood product | Biopolymers | Additives/Active agents | Application method | Storage conditions | Major findings | References |
|---|---|---|---|---|---|---|
| Salmon sushi | Furcellaran/ gelatin | Green and pu-erh tea extracts | Spraying technique (5 cm spraying distance) | Packed in sterile PET trays with a cover and stored at 4 °C for 8 days | Shelf life extension | Kulawik et al. (2019) |
| Fish burger rainbow trout | Chitosan | Seaweeds extracts | Aseptically coating | Were placed in trays cover with aluminum foil and stored at 4 °C | Reducing oxidation improvement in microbial spoilage control | Albertos et al. (2019) |
| Shrimp | Microalgalexo polysaccharides | Red seaweed extract | Dipping and draining for 5 min | Placed into open sterile Petri dishes and stored at 4 °C for 8 days | Inhibition of lipid peroxidation, protein decomposition, and bacterial proliferation<br>Preservation of all sensory attributes | Balti et al. (2020) |
| Grass carp fillets | κ-Carrageenan/ gelatin/zein | Nisin/curcumin | Soaking for 30 s and airdrying at room temperature | Stored at 4 °C | Detecting the change of freshness<br>Shelf life extension | He et al. (2020) |
| Rainbow trout fillets | Quinoa | Sage and lemon essential oils | Wrapping on the whole surface | Stored in refrigerator conditions for 15 days | Preventing lipid oxidation<br>Antimicrobial activity | Alak et al. (2019) |
| Cold-smoked sunshine bass | Corn zein | Lemongrass essential oil/nisin | Spraying and drying for 30 min, for 2 times | Placed on sanitized plastic trays and 4 ± 1 °C | Reducing some spoilage bacteria individually (*Listeria spp.* and other Gram-positive bacteria) | Hager et al. (2019) |

| | | | | | | |
|---|---|---|---|---|---|---|
| Bonito fish fillets | Chitosan | Peppermint essential oil | Wrapping | Packed with flexible and low-density polyethylene films and coated using pyrolytic coating devices, and stored at +4 °C for 12 days | Antimicrobial activity<br>Shelf life extension of 6 days | Kalkan et al. (2019) |
| Trout fillet | Alginate coarse | *ZatariamultifloraBoiss* essential oil | Sunking for 1 min, and draining for 2 min | Stored at 4 °C for 16 days | Preservation against the growth of spoilage microflora<br>Shelf life extension of 4 days | Khanzadi et al. (2020) |
| Tilapia fillets | Fish gelatin/ pectin | *Menthapulegium* essential oil | Wrapping | Stored at 4 °C for 15 days | Antimicrobial activity<br>Preservation of physicochemical quality<br>Shelf life extension | Aitboulahsen et al. (2020b) |
| Golden Pomfret | Fish gelatin | Tea polyphenol | Immersing for 30 min and drying for 60 min | Packed in high-quality polyethylene ziplock bags and stored at 4 ± 1 °C | Reducing weight loss<br>Reducing microbial growth slowing down the myofibril degradation | Feng et al. (2017) |
| Boiled-dried anchovy | Fish gelatin | | Coating and drying with cold-air blast for 13–15 h | Packed in a zipper film bag and stored at 20 °C | Suppression lipid oxidation (lowering oxidant factors) | Kim et al. (2017) |
| Sierra fish fillets | Chitosan | Tomato plant extract | Immersion for 2 min | Placed into plastic bags prior to storage-on-ice for 15 days | Shelf life extension of 5 days | Ramírez-Guerra et al. (2018) |
| Spanish Mackerel fish | Whey protein | Transglutaminase enzyme | Dipping for 3 min | Stored at 7 °C for 15 days | Preservation of physicochemical quality | Karina and Setiadi (2020) |
| Rainbow trout fish patties | Chitosan/ carboxymethyl cellulose | Glutaraldehyde and cinnamon essential oil | Separation by paper sheets | Kept in stainless steel trays at 4 ± 1 °C for 16 days | Shelf life extension antibacterial and antioxidant activity | Valizadeh et al. (2020) |

bacteria of primary importance to be monitored by the seafood industry, has been retained at low levels in cold-smoked salmon coated with chitosan film containing sodium lactate and potassium sorbate (Jiang et al. 2011). Heydari et al. (2015) found a 1.5 $\log_{10}$ CFU/g reduction in the microbial population in carp fillets packed with sodium alginate films with mint EO. This film resulted in a considerable reduction in total volatile basic nitrogen (TVBN) levels during the storage of the wrapped fillets compared to the control fillets. In another study, Gómez-Estaca et al. (2010) evaluated the antimicrobial effect of different composite films based on gelatin and chitosan incorporated with eight EOs (clove, fennel, cypress, lavender, thyme, verbena, pine, and rosemary) against six types of bacteria responsible for fish spoilage. The results showed that the films incorporated with the clove EO had the best antimicrobial activity and therefore was applied on fish fillets and monitored during storage. The authors found that composite films containing this EO allow a controlled release of the active substances of EO and maintain a sufficient concentration over a longer period, and therefore a considerable reduction in bacterial growth during the storage of fish fillets.

## 8.4 Limits of Application

Edible films and coatings are among the most promising preservation methods for foodstuffs. Their main drawback, however, is their specificity. For instance, for fresh or freshly cut fruits and vegetables, coatings should be optimized for each species and each cultivar (Porta 2013). Therefore, to obtain the desired results, the biopolymer type, concentration, and/or the coating formulation must be well controlled. Besides, storage conditions could influence the behavior of packaged or coated foods. Thick films or coatings could cause anaerobic conditions that could lead to undesirable flavors. Films and coatings with high WVP result in weight loss and therefore, loss of appearance and texture of the product. Conversely, films and coatings with low gas permeability will prevent the transfer of the amount of oxygen required for the fruit's ripening phase. Also, the risks of microbial proliferation often increase, depending on the relative humidity of the environment, as well as the solubility of BFC, which can dissolve during storage and promote microbial proliferation (Menezes n.d.). On the other hand, food application methods for BFC must be enhanced and adapted to industrial-scale applications and ensure homogeneous and uniform application on the product surface with effective drying (Lin and Zhao 2007).

Several authors have focused on the constraints and limits encountered when applying an edible and bioactive film or coating as a food packaging. Hager et al. (2019) reported that nisin is not commonly used as an active agent in the formulation of bioactive films, despite its proven antimicrobial effects, due to its high cost. Also, the production and marketing of such bioactive film-coated products worldwide are limited, due to the variety of regulations and standards that differ between countries. Sensorial quality could also limit their application in foods in particular, for films and coatings incorporated with EOs. In this sense, Cano Embuena et al. (2017) applied one, two, and three layers of coating incorporated with EOs for goat cheese preservation and found that the cheese coated with a triple layer coating received the lowest sensory scores, because the smell, flavor, and taste of cheese were affected by the EOs.

## 8.5 Conclusions

The research and studies conducted on BFC have been highlighted in this chapter. Edible films and coatings could be an important alternative method toward replacing the unnecessary use of plastic-based, hazardous, and non-biodegradable synthetic packaging. The various functions of BFC based on their potential applications for food preservation were also explained. Composite films combining the use of several biopolymers in the matrix of BFC offer numerous advantages, such as the possibility of taking advantage of the different properties of each compound to obtain a film or coating with the desired qualities. Moreover, the incorporation of antimicrobial agents like EOs is an efficient way of enhancing the hygienic quality of food products. Efforts should be made to resolve, on the one hand, the challenges of industrial development and marketing and, on the other, the regulatory constraints of implementation, which differ between countries.

**Acknowledgments** The authors thank CNRST Morocco (PPR/2015/10) for financial support.

## References

Abdel Aziz Mohamed S, Salama HE, Sabaa MW (2018) Biobased alginate/castor oil edible films for active food packaging. LWT—Food Sci Technol 96(October):455–460. https://doi.org/10.1016/j.lwt.2018.05.049

Abdollahi M, Rezaei M, Farzi G (2014) Influence of chitosan/clay functional bionanocomposite activated with rosemary essential oil on the shelf life of fresh silver carp. Int J Food Sci Technol 49(3):811–818. https://doi.org/10.1111/ijfs.12369

Abhirami P, Modupalli N, Natarajan V (2020) Novel postharvest intervention using rice bran wax edible coating for shelflife enhancement of *Solanum lycopersicum* fruit. J Food Process Preserv 44(12):e14989. https://doi.org/10.1111/jfpp.14989

Aitboulahsen M, Zantar S, Laglaoui A, Chairi H, Arakrak A, Bakkali M, Hassani Zerrouk M (2018) Gelatin-based edible coating combined with *Mentha pulegium* essential oil as bioactive packaging for strawberries. J Food Qual 2018(July):1–7. https://doi.org/10.1155/2018/8408915

Aitboulahsen M, El Galiou O, Laglaoui A, Bakkali M, Hassani Zerrouk M (2020a) Effect of plasticizer type and essential oils on mechanical, physicochemical, and antimicrobial characteristics of gelatin, starch, and pectin-based films. J Food Process Preserv 44:e14480. https://doi.org/10.1111/jfpp.14480

Aitboulahsen M, Chairi H, Laglaoui A, Zantar S, Abdelhay A, Bakkali M, Zerrouk MH (2020b) Gelatin/pectin-based film incorporated with essential oils: functional characteristics and shelf life extension of tilapia fillets under refrigeration. J Food Safety 40(3):e12774. https://doi.org/10.1111/jfs.12774

Alak G, Guler K, Ucar A, Parlak V, Kocaman EM, Yanık T, Atamanalp M (2019) Quinoa as polymer in edible films with essential oil: effects on rainbow trout fillets shelf life. J Food Process Preserv 43(12):e14268. https://doi.org/10.1111/jfpp.14268

Albertos I, Martin-Diana AB, Burón M, Rico D (2019) Development of functional bio-based seaweed (*Himanthalia Elongata* and *Palmaria Palmata*) edible films for extending the shelf life of fresh fish burgers. Food Packag Shelf Life 22(December):100382. https://doi.org/10.1016/j.fpsl.2019.100382

Al-Hassan AA, Norziah MH (2012) Starch–gelatin edible films: water vapor permeability and mechanical properties as affected by plasticizers. Food Hydrocoll 26(1):108–117. https://doi.org/10.1016/j.foodhyd.2011.04.015

Ali S, Anjum MA, Ejaz S, Hussain S, Ercisli S, Saleem MS, Sardar H (2021) Carboxymethyl cellulose coating delays chilling injury development and maintains eating quality of "Kinnow" mandarin fruits during low temperature storage. Int J Biol Macromol 168(January):77–85. https://doi.org/10.1016/j.ijbiomac.2020.12.028

Al-Nabulsi A, Osaili T, Sawalha A, Olaimat AN, Albiss BA, Mehyar G, Ayyash M, Holley R (2020) Antimicrobial activity of chitosan coating containing ZnO nanoparticles against *E. coli* O157:H7 on the surface of white brined cheese. Int J Food Microbiol 334(December):108838. https://doi.org/10.1016/j.ijfoodmicro.2020.108838

Andevari GT, Rezaei M (2011) Effect of gelatin coating incorporated with cinnamon oil on the quality of fresh rainbow trout in cold storage: preservation of trout with gelatin coating. Int J Food Sci Technol 46(11):2305–2311. https://doi.org/10.1111/j.1365-2621.2011.02750.x

Arabpoor B, Yousefi S, Weisany W, Ghasemlou M (2021) Multifunctional coating composed of Eryngium Campestre L. essential oil encapsulated in nano-chitosan to prolong the shelflife of fresh cherry fruits. Food Hydrocoll 111(February):106394. https://doi.org/10.1016/j.foodhyd.2020.106394

Atarés L, Chiralt A (2016) Essential oils as additives in biodegradable films and coatings for active food packaging. Trends Food Sci Technol 48(February):51–62. https://doi.org/10.1016/j.tifs.2015.12.001

Augusto PED, Soares BMC, Castanha N (2018) Conventional technologies of food preservation. In: Innovative technologies for food preservation. Elsevier, pp 3–23. https://doi.org/10.1016/B978-0-12-811031-7.00001-7

Avila-Sosa R, Palou E, López-Malo A (2016) Essential oils added to edible films. In: Essential oils in food preservation, flavor and safety. Elsevier, pp 149–54. https://doi.org/10.1016/B978-0-12-416641-7.00015-8

Azarakhsh N, Osman A, Ghazali HM, Tan CP, Adzahan NM (2014) Lemongrass essential oil incorporated into alginate-based edible coating for shelflife extension and quality retention of fresh-cut pineapple. Postharvest Biol Technol 88(February):1–7. https://doi.org/10.1016/j.postharvbio.2013.09.004

Balti R, Mansour MB, Zayoud N, Le Balc'h R, Brodu N, Arhaliass A, Massé A (2020) Active exopolysaccharides based edible coatings enriched with red seaweed (*Gracilaria Gracilis*) extract to improve shrimp preservation during refrigerated storage. Food Biosci 34 (April):100522. https://doi.org/10.1016/j.fbio.2019.100522

Berk Z (2018) Food packaging. In: Food process engineering and technology. Elsevier, pp 625–41. https://doi.org/10.1016/B978-0-12-812018-7.00027-0

Bermúdez-Oria A, Rodríguez-Gutiérrez G, Rubio-Senent F, Fernández-Prior Á, Fernández-Bolaños J (2019) Effect of edible pectin-fish gelatin films containing the olive antioxidants hydroxytyrosol and 3,4-dihydroxyphenylglycol on beef meat during refrigerated storage. Meat Sci 148(February):213–218. https://doi.org/10.1016/j.meatsci.2018.07.003

Bertuzzi MA, Castro Vidaurre EF, Armada M, Gottifredi JC (2007) Water vapor permeability of edible starch based films. J Food Eng 80(3):972–978. https://doi.org/10.1016/j.jfoodeng.2006.07.016

Bharti SK, Pathak V, Alam T, Arya A, Singh VK, Verma AK, Rajkumar V (2020) Materialization of novel composite bio-based active edible film functionalized with essential oils on

antimicrobial and antioxidative aspect of chicken nuggets during extended storage. J Food Sci 85(9):2857–2865. https://doi.org/10.1111/1750-3841.15365
Bolívar-Monsalve J, Ramírez-Toro C, Bolívar G, Ceballos-González C (2019) Mechanisms of action of novel ingredients used in edible films to preserve microbial quality and oxidative stability in sausages—a review. Trends Food Sci Technol 89(July):100–109. https://doi.org/10.1016/j.tifs.2019.05.011
Bonilla J, Sobral PJA (2019) Gelatin-chitosan edible film activated with Boldo extract for improving microbiological and antioxidant stability of sliced prato cheese. Int J Food Sci Technol 54(5):1617–1624. https://doi.org/10.1111/ijfs.14032
Butler BL, Vergano PJ, Testin RF, Bunn JM, Wiles JL (1996) Mechanical and barrier properties of edible chitosan films as affected by composition and storage. J Food Sci 61(5):953–956. https://doi.org/10.1111/j.1365-2621.1996.tb10909.x
Cao L, Liu W, Wang L (2018) Developing a green and edible film from cassia gum: the effects of glycerol and sorbitol. J Clean Prod 175(February):276–282. https://doi.org/10.1016/j.jclepro.2017.12.064
Cerqueira MA, Sousa-Gallagher MJ, Macedo I, Rodriguez-Aguilera R, Souza BWS, Teixeira JA, Vicente AA (2010) Use of galactomannan edible coating application and storage temperature for prolonging shelflife of "Regional" cheese. J Food Eng 97(1):87–94. https://doi.org/10.1016/j.jfoodeng.2009.09.019
Chen T, Ji D, Zhang Z, Li B, Qin G, Tian S (2021) Advances and strategies for controlling the quality and safety of postharvest fruit. Engineering 7(8):1177–1184. https://doi.org/10.1016/j.eng.2020.07.029
Cho SY, Rhee C (2002) Sorption characteristics of soy protein films and their relation to mechanical properties. LWT—Food Sci Technol 35(2):151–157. https://doi.org/10.1006/fstl.2001.0829
Chu Y, Gao CC, Liu X, Zhang N, Xu T, Feng X, Yang Y, Shen X, Tang X (2020) Improvement of storage quality of strawberries by pullulan coatings incorporated with cinnamon essential oil nanoemulsion. LWT 122(March):109054. https://doi.org/10.1016/j.lwt.2020.109054
Colak BY (2014) Utilisation du caséinate de sodium pour la fabrication de films actifs pour l'emballage alimentaire: étude des propriétés barrières aux gaz, de l'activité antimicrobienne et de la biodégradabilité. Polymères. Université Jean Monnet - Saint-Etienne. https://tel.archives-ouvertes.fr/tel-01311112
Coma V, Martial-Gros A, Garreau S, Copinet A, Salin F, Deschamps A (2002) Edible antimicrobial films based on chitosan matrix. J Food Sci 67(3):1162–1169. https://doi.org/10.1111/j.1365-2621.2002.tb09470.x
Correa-Pacheco ZN, Corona-Rangel ML, Bautista-Baños S, Ventura-Aguilar RI (2021) Application of natural-based nanocoatings for extending the shelf life of green bell pepper fruit. J Food Sci 86:95–102. https://doi.org/10.1111/1750-3841.15542
Cruz-Gálvez AM, Castro-Rosas J, Rodríguez-Marín ML, Cadena-Ramírez A, Tellez-Jurado A, Tovar-Jiménez X, Chavez-Urbiola EA, Abreu-Corona A, Gómez-Aldapa CA (2018) Antimicrobial activity and physicochemical characterization of a potato starch-based film containing acetonic and methanolic extracts of Hibiscus Sabdariffa for use in sausage. LWT 93 (July):300–305. https://doi.org/10.1016/j.lwt.2018.02.064
Cutter CN (2006) opportunities for bio-based packaging technologies to improve the quality and safety of fresh and further processed muscle foods. Meat Sci 74(1):131–142. https://doi.org/10.1016/j.meatsci.2006.04.023
da Costa de Quadros C, Lima KO, Bueno CHL, dos Santos Fogaça FH, Rocha M, Prentice C (2020) Effect of the edible coating with protein hydrolysate on cherry tomatoes shelf life. J Food Process Preserv 44(10):e14760. https://doi.org/10.1111/jfpp.14760
Debeaufort F, Voilley A (1994) Aroma compound and water vapor permeability of edible films and polymeric packagings. J Agric Food Chem 42(12):2871–2875. https://doi.org/10.1021/jf00048a041

Debeaufort F, Voilley A (2007) Effect of surfactants and drying rate on barrier properties of emulsified edible films. Int J Food Sci Technol 30(2):183–190. https://doi.org/10.1111/j.1365-2621.1995.tb01370.x

Debeaufort F, Voilley A (2009) Lipid-based edible films and coatings. In: Huber KC, Embuscado ME (eds) Edible films and coatings for food applications. Springer New York, New York, pp 135–168. https://doi.org/10.1007/978-0-387-92824-1_5

Debeaufort F, Quezada-Gallo J-A, Voilley A (1998) Edible films and coatings: tomorrow's packagings: a review. Crit Rev Food Sci Nutr 38(4):299–313. https://doi.org/10.1080/10408699891274219

Dini H, Fallah AA, Bonyadian M, Abbasvali M, Soleimani M (2020) Effect of edible composite film based on chitosan and cumin essential oil-loaded nanoemulsion combined with low-dose gamma irradiation on microbiological safety and quality of beef loins during refrigerated storage. Int J Biol Macromol 164(December):1501–1509. https://doi.org/10.1016/j.ijbiomac.2020.07.215

Di Pierro P, Sorrentino A, Mariniello L, Giosafatto CVL, Porta R (2011) Chitosan/whey protein film as active coating to extend Ricotta cheese shelf-life. LWT Food Sci Technol 44(10):2324–2327. https://doi.org/10.1016/j.lwt.2010.11.031

Dong LM, Quyen NTT, Thuy DTK (2020) Effect of edible coating and antifungal emulsion system on *Colletotrichum Acutatum* and shelf life of strawberries: effect of edible coating and antifungal emulsion system on *Colletotrichum Acutatum* and shelf life of strawberries. Vietnam J Chem 58(2):237–244. https://doi.org/10.1002/vjch.201900169

Duan J, Park S-I, Daeschel MA, Zhao Y (2007) Antimicrobial chitosan-lysozyme (CL) films and coatings for enhancing microbial safety of Mozzarella cheese. J Food Sci 72(9):M355–M362. https://doi.org/10.1111/j.1750-3841.2007.00556.x

Duran A, Kahve HI (2020) The effect of chitosan coating and vacuum packaging on the microbiological and chemical properties of beef. Meat Sci 162(April):107961. https://doi.org/10.1016/j.meatsci.2019.107961

Ehteshami S, Abdollahi F, Ramezanian A, Dastjerdi AM (2020) Maintenance of quality and bioactive compounds in pomegranate fruit (*Punica Granatum* L.) by combined application of organic acids and chitosan edible coating. J Food Biochem 44(9):e13393. https://doi.org/10.1111/jfbc.13393

Embuscado EM, Huber KC (2009) Edible films and coatings for food applications. Springer, Dordrecht

El-Mogy MM, Parmar A, Ali MR, Abdel-Aziz ME, Abdeldaym EA (2020) Improving postharvest storage of fresh artichoke bottoms by an edible coating of cordia myxa gum. Postharvest Biol Technol 163(May):111143. https://doi.org/10.1016/j.postharvbio.2020.111143

Embuena C, Isabel A, Nácher MC, Boix AC, Pilar Molina Pons M, Llopis MB, Martínez MCB, Martínez CG (2017) Quality of goat's milk cheese as affected by coating with edible chitosan-essential oil films. Int J Dairy Technol 70(1):68–76. https://doi.org/10.1111/1471-0307.12306

Espitia PJP, Batista RA, Azeredo HMC, Otoni CG (2016) Probiotics and their potential applications in active edible films and coatings. Food Res Int 90(December):42–52. https://doi.org/10.1016/j.foodres.2016.10.026

Farhan A, Hani NM (2020) Active edible films based on semi-refined κ-carrageenan: antioxidant and color properties and application in chicken breast packaging. Food Packag Shelf Life 24 (June):100476. https://doi.org/10.1016/j.fpsl.2020.100476

Fasciglione G, Goñi MG, Yommi AK, Perez-Bravo JJ, Ortueta R, Scampini A, Buffa L, Andreu AB, Creus CM (2020) Revaluation of waste from fishing industry through generation of chitosan coatings to improve quality and extend shelflife of minimally processed lettuce. Postharvest Biol Technol 170(December):111310. https://doi.org/10.1016/j.postharvbio.2020.111310

Feng X, Ng VK, Mikš-Krajnik M, Yang H (2017) Effects of fish gelatin and tea polyphenol coating on the spoilage and degradation of myofibril in fish fillet during cold storage. Food Bioprocess Technol 10(1):89–102. https://doi.org/10.1007/s11947-016-1798-7

Fernández-Pan I, Ignacio J, Caballero M (2011) Biopolymers for edible films and coatings in food applications. In: Plackett D (ed) Biopolymers—new materials for sustainable films and coatings. Wiley, Chichester, pp 233–254. https://doi.org/10.1002/9781119994312.ch11

Gahruie H, Hadi EZ, Eskandari MH, Hosseini SMH (2017) Characterization of basil seed gum-based edible films incorporated with Zataria multiflora essential oil nanoemulsion. Carbohydr Polym 166(June):93–103. https://doi.org/10.1016/j.carbpol.2017.02.103

Gallego M, Arnal M, Talens P, Toldrá F, Mora L (2020) Effect of gelatin coating enriched with antioxidant tomato by-products on the quality of pork meat. Polymers 12(5):1032. https://doi.org/10.3390/polym12051032

Ganiari S, Choulitoudi E, Oreopoulou V (2017) Edible and active films and coatings as carriers of natural antioxidants for lipid food. Trends Food Sci Technol 68(October):70–82. https://doi.org/10.1016/j.tifs.2017.08.009

Gennadios A, Weller CL, Gooding CH (1994) Measurement errors in water vapor permeability of highly permeable, hydrophilic edible films. J Food Eng 21(4):395–409. https://doi.org/10.1016/0260-8774(94)90062-0

Geraldine RM, de Fátima Ferreira Soares N, Botrel DA, de Almeida Gonçalves L (2008) Characterization and effect of edible coatings on minimally processed garlic quality. Carbohydr Polym 72(3):403–409. https://doi.org/10.1016/j.carbpol.2007.09.012

Ghaly (2010) Fish spoilage mechanisms and preservation techniques: review. Am J Appl Sci 7(7):859–877. https://doi.org/10.3844/ajassp.2010.859.877

Gheribi R, Puchot L, Verge P, Jaoued-Grayaa N, Mezni M, Habibi Y, Khwaldia K (2018) Development of plasticized edible films from opuntia ficus-indica mucilage: a comparative study of various polyol plasticizers. Carbohydr Polym 190(June):204–211. https://doi.org/10.1016/j.carbpol.2018.02.085

Gómez-Estaca J, López de Lacey A, López-Caballero ME, Gómez-Guillén MC, Montero P (2010) Biodegradable gelatin–chitosan films incorporated with essential oils as antimicrobial agents for fish preservation. Food Microbiol 27(7):889–896. https://doi.org/10.1016/j.fm.2010.05.012

Gorny JR, Hess-Pierce B, Cifuentes RA, Kader AA (2002) Quality changes in fresh-cut pear slices as affected by controlled atmospheres and chemical preservatives. Postharvest Biol Technol 24(3):271–278. https://doi.org/10.1016/S0925-5214(01)00139-9

Gram L, Huss HH (1996) Microbiological spoilage of fish and fish products. Int J Food Microbiol 33(1):121–137. https://doi.org/10.1016/0168-1605(96)01134-8

Groh KJ, Backhaus T, Carney-Almroth B, Geueke B, Inostroza PA, Lennquist A, Heather A. Leslie, et al. (2019) Overview of known plastic packaging-associated chemicals and their hazards. Sci Total Environ 651(February):3253–3268. https://doi.org/10.1016/j.scitotenv.2018.10.015

Guerrero A, Ferrero S, Barahona M, Boito B, Lisbinski E, Maggi F, Sañudo C (2020) Effects of active edible coating based on thyme and garlic essential oils on lamb meat shelf life after long-term frozen storage. J Sci Food Agric 100(2):656–664. https://doi.org/10.1002/jsfa.10061

Guitián MV, Carolina I, Cecilia Soria M, Hovanyecz P, Banchio C, Carina Audisio M (2019) Anti-listeria monocytogenes effect of bacteriocin-incorporated agar edible coatings applied on cheese. Int Dairy J 97(October):92–98. https://doi.org/10.1016/j.idairyj.2019.05.016

Günlü A, Koyun E (2013) Effects of vacuum packaging and wrapping with chitosan-based edible film on the extension of the shelf life of sea bass (Dicentrarchus Labrax) fillets in cold storage (4 °C). Food Bioprocess Technol 6(7):1713–1719. https://doi.org/10.1007/s11947-012-0833-6

Guo M, Jin TZ, Wang L, Joseph Scullen O, Sommers CH (2014) Antimicrobial films and coatings for inactivation of Listeria innocua on ready-to-eat deli turkey meat. Food Control 40 (June):64–70. https://doi.org/10.1016/j.foodcont.2013.11.018

Gurdian C, Chouljenko A, Solval KM, Boeneke C, King JM, Sathivel S (2017) Application of edible films containing Oregano (*Origanum Vulgare* ) essential oil on Queso Blanco cheese prepared with flaxseed (*Linum Usitatissimum* ) oil: oregano edible films on cheese with oil.... J Food Sci 82(6):1395–1401. https://doi.org/10.1111/1750-3841.13733

Gutiérrez-Pacheco MM, Ortega-Ramírez LA, Silva-Espinoza BA, Cruz-Valenzuela MR, González-Aguilar GA, Lizardi-Mendoza J, Miranda R, Ayala-Zavala JF (2020) Individual and combined coatings of chitosan and carnauba wax with oregano essential oil to avoid water loss and microbial decay of fresh cucumber. Coatings 10(7):614. https://doi.org/10.3390/coatings10070614

Hager JV, Rawles SD, Xiong YL, Newman MC, Thompson KR, Webster CD (2019) *Listeria monocytogenes* is inhibited on fillets of cold-smoked sunshine bass, *Morone Chrysops* × *Morone Saxatilis* , with an edible corn zein-based coating incorporated with lemongrass essential oil or nisin. J World Aquac Soc 50(3):575–92. https://doi.org/10.1111/jwas.12573

Hahladakis JN, Velis CA, Weber R, Iacovidou E, Purnell P (2018) An overview of chemical additives present in plastics: migration, release, fate and environmental impact during their use, disposal and recycling. J Hazard Mater 344(February):179–199. https://doi.org/10.1016/j.jhazmat.2017.10.014

Han JH (Ed) (2005) Innovations in food packaging. Food science and technology international series. Elsevier, Amsterdam

Haq MA, Hasnain A, Azam M (2014) Characterization of edible gum cordia film: effects of plasticizers. LWT—Food Sci Technol 55(1):163–169. https://doi.org/10.1016/j.lwt.2013.09.027

Haq MA, Hasnain A, Jafri FA, Akbar MF, Khan A (2016) Characterization of edible gum cordia film: effects of beeswax. LWT—Food Sci Technol 68(May):674–680. https://doi.org/10.1016/j.lwt.2016.01.011

He F, Kong Q, Jin Z, Mou H (2020) Developing a unidirectionally permeable edible film based on κ-carrageenan and gelatin for visually detecting the freshness of grass carp fillets. Carbohydr Polym 241(August):116336. https://doi.org/10.1016/j.carbpol.2020.116336

Henrique CM, Teófilo RF, Sabino L, Ferreira MMC, Cereda MP (2007) Classification of Cassava starch films by physicochemical properties and water vapor permeability quantification by FTIR and PLS. J Food Sci 72(4):E184–E189. https://doi.org/10.1111/j.1750-3841.2007.00342.x

Hettiarachchy NS, Satchithanandam E (n.d.) Organic acids incorporated edible antimicrobial films, p 22

Heydari R, Bavandi S, Javadian SR (2015) Effect of sodium alginate coating enriched with horsemint (*Mentha longifolia*) essential oil on the quality of bighead carp fillets during storage at 4°C. Food Sci Nutr 3(3):188–194. https://doi.org/10.1002/fsn3.202

Heydari S, Jooyandeh H, Behbahani BA, Noshad M (2020) The impact of Qodume shirazi seed mucilage-based edible coating containing lavender essential oil on the quality enhancement and shelf life improvement of fresh ostrich meat: an experimental and modeling study. Food Sci Nutr 8(12):6497–6512. https://doi.org/10.1002/fsn3.1940

Hu X, Saravanakumar K, Sathiyaseelan A, Wang M-H (2020) Chitosan nanoparticles as edible surface coating agent to preserve the fresh-cut bell pepper (Capsicum Annuum L. Var. Grossum (L.) Sendt). Int J Biol Macromol 165(December):948–957. https://doi.org/10.1016/j.ijbiomac.2020.09.176

Huber KC, Embuscado ME (eds) (2009) Edible films and coatings for food applications. Springer New York, New York. https://doi.org/10.1007/978-0-387-92824-1

Iñiguez-Moreno M, Ragazzo-Sánchez JA, Barros-Castillo JC, Sandoval-Contreras T, Calderón-Santoyo M (2020) Sodium alginate coatings added with Meyerozyma Caribbica: postharvest biocontrol of *Colletotrichum Gloeosporioides* in Avocado (Persea Americana Mill. Cv. Hass). Postharvest Biol Technol 163(May):111123. https://doi.org/10.1016/j.postharvbio.2020.111123

Janes ME, Dai Y (2012) Edible films for meat, poultry and seafood. In: Advances in meat, poultry and seafood packaging. Elsevier, pp 504–21. https://doi.org/10.1533/9780857095718.4.504

Jeon Y-J, Kamil JYVA, Shahidi F (2002) Chitosan as an edible invisible film for quality preservation of herring and Atlantic cod. J Agric Food Chem 50(18):5167–5178. https://doi.org/10.1021/jf0116931

Jiang Z, Neetoo H, Chen H (2011) Control of *Listeria monocytogenes* on cold-smoked salmon using chitosan-based antimicrobial coatings and films. J Food Sci 76(1):M22–M26. https://doi.org/10.1111/j.1750-3841.2010.01925.x

Jiao W, Shu C, Li X, Cao J, Fan X, Jiang W (2019) Preparation of a chitosan-chlorogenic acid conjugate and its application as edible coating in postharvest preservation of peach fruit. Postharvest Biol Technol 154(August):129–136. https://doi.org/10.1016/j.postharvbio.2019.05.003

Jiménez A, Fabra MJ, Talens P, Chiralt A (2012) Edible and biodegradable starch films: a review. Food Bioprocess Technol 5(6):2058–2076. https://doi.org/10.1007/s11947-012-0835-4

Kalem IK, Bhat ZF, Kumar S, Noor S, Desai A (2018) The effects of bioactive edible film containing Terminalia Arjuna on the stability of some quality attributes of chevon sausages. Meat Sci 140(June):38–43. https://doi.org/10.1016/j.meatsci.2018.02.011

Kalkan S, Akben SB, Çanga D (2019) Modeling the microbiological effect of essential peppermint oil-supplemented chitosan film on Bonito fish (*Sarda Sarda*) fillets and analysis via image processing methods. J Food Process Eng 42(6). https://doi.org/10.1111/jfpe.13232

Karina S, Setiadi (2020) Influence of transglutaminase enzyme incorporated into protein based edible coating for preservation of spanish mackerel fish (Scomberomorus Commersoni). IOP Conf Series Mater Sci Eng 722(January):012081. https://doi.org/10.1088/1757-899X/722/1/012081

Kaya M, Ravikumar P, Ilk S, Mujtaba M, Akyuz L, Labidi J, Salaberria AM, Cakmak YS, Erkul SK (2018) Production and characterization of chitosan based edible films from berberis crataegina's fruit extract and seed oil. Innov Food Sci Emerg Technol 45(February):287–297. https://doi.org/10.1016/j.ifset.2017.11.013

Keykhosravy K, Khanzadi S, Hashemi M, Azizzadeh M (2020) Chitosan-loaded nanoemulsion containing Zataria multiflora Boiss and Bunium persicum Boiss essential oils as edible coatings: its impact on microbial quality of Turkey meat and fate of inoculated pathogens. Int J Biol Macromol 150(May):904–913. https://doi.org/10.1016/j.ijbiomac.2020.02.092

Khanzadi S, Keykhosravy K, Hashemi M, Azizzadeh M (2020) Alginate coarse/nanoemulsions containing *Zataria Multiflora* Boiss essential oil as edible coatings and the impact on microbial quality of trout fillet. Aquacult Res 51(3):873–881. https://doi.org/10.1111/are.14418

Khwaldia K, Perez C, Banon S, Desobry S, Hardy J (2004) Milk proteins for edible films and coatings. Crit Rev Food Sci Nutr 44(4):239–251. https://doi.org/10.1080/10408690490464906

Kim I-H, Yang H-J, Noh B-S, Chung S-J, Min SC (2012) Development of a defatted mustard meal-based composite film and its application to smoked salmon to retard lipid oxidation. Food Chem 133(4):1501–1509. https://doi.org/10.1016/j.foodchem.2012.02.040

Kim HJ, Son KT, Lee SG, Park SY, Heu MS, Kim J-S (2017) Suppression of lipid deterioration in boiled-dried anchovy by coating with fish gelatin hydrolysates. J Food Biochem 41(2):e12331. https://doi.org/10.1111/jfbc.12331

Kowalczyk D, Skrzypek T, Łupina K (2020) Effect of carboxymethyl cellulose/Candelilla wax edible coating incorporated with ascorbic acid on the physicochemical and sensory qualities of prepackaged minimally processed carrots (*Daucus Carota* L.) during cold storage. J Food Process Preserv 44(9):e14713. https://doi.org/10.1111/jfpp.14713

Kozlu A, Elmacı Y (2020) Quince seed mucilage as edible coating for mandarin fruit; determination of the quality characteristics during storage. J Food Process Preserv 44(11):e14854. https://doi.org/10.1111/jfpp.14854

Krochta JM (2010) Films, edible. In: Kirk-Othmer encyclopedia of chemical technology. American Cancer Society, pp 1–12. https://doi.org/10.1002/0471238961.filmkroc.a01

Krochta JM, de Mulder-Johnson C (1997) Edible and biodegradable polymer films: challenges and opportunities. Food Technol 51:59–65

Kulawik P, Jamróz E, Zając M, Guzik P, Tkaczewska J (2019) The Effect of furcellaran-gelatin edible coatings with green and Pu-Erh tea extracts on the microbiological, physicochemical and sensory changes of salmon sushi stored at 4 °C. Food Control 100(June):83–91. https://doi.org/10.1016/j.foodcont.2019.01.004

Kuorwel KK, Cran MJ, Sonneveld K, Miltz J, Bigger SW (2013) Migration of antimicrobial agents from starch-based films into a food simulant. LWT—Food Sci Technol 50(2):432–438. https://doi.org/10.1016/j.lwt.2012.08.023

Lazaridou A, Biliaderis CG (2002) Thermophysical properties of chitosan, chitosan ± starch and chitosan ± pullulan ®lms near the glass transition. Carbohydr Polym 48:179–190

Li X-Y, Xiao-Long D, Liu Y, Tong L-J, Wang Q, Li J-L (2019) Rhubarb extract incorporated into an alginate-based edible coating for peach preservation. Sci Horticult 257(November):108685. https://doi.org/10.1016/j.scienta.2019.108685

Li S, Ma Y, Ji T, Sameen DE, Ahmed S, Qin W, Dai J, Li S, Liu Y (2020) Cassava starch/carboxymethylcellulose edible films embedded with lactic acid bacteria to extend the shelf life of banana. Carbohydr Polym 248(November):116805. https://doi.org/10.1016/j.carbpol.2020.116805

Li L, Yi P, Li C, Xin M, Sun J, He X, Sheng J et al (2021) Influence of polysaccharide-based edible coatings on enzymatic browning and oxidative senescence of fresh-cut lettuce. Food Sci Nutr 9(2):888–899. https://doi.org/10.1002/fsn3.2052

Lima AEF, Andrade PL, Lemos TLG, de Andrade Uchoa DE, Siqueira MCA, Egito AS, Braga RC, Costa JN, Sá DMAT (2021) Development and application of galactomannan and essential oil-based edible coatings applied to " *Coalho* " cheese. J Food Process Preserv 45:e15091. https://doi.org/10.1111/jfpp.15091

Lin D, Zhao Y (2007) Innovations in the development and application of edible coatings for fresh and minimally processed fruits and vegetables. Comp Rev Food Sci Food Safety 6(3):60–75. https://doi.org/10.1111/j.1541-4337.2007.00018.x

López P, Sánchez C, Batlle R, Nerín C (2007) Development of flexible antimicrobial films using essential oils as active agents. J Agric Food Chem 55(21):8814–8824. https://doi.org/10.1021/jf071737b

Mark JE (1999) In polymer data handbook. Oxford University Press

Mahcene Z, Khelil A, Hasni S, Bozkurt F, Goudjil MB, Tornuk F (2020) Home-made cheese preservation using sodium alginate based on edible film incorporating essential oils. J Food Sci Technol 58:2406–2419. https://doi.org/10.1007/s13197-020-04753-3

Maqbool M, Ali A, Alderson PG, Mohamed MTM, Siddiqui Y, Zahid N (2011) Postharvest application of gum arabic and essential oils for controlling anthracnose and quality of banana and papaya during cold storage. Postharvest Biol Technol 62(1):71–76. https://doi.org/10.1016/j.postharvbio.2011.04.002

Martínez-Ortiz MA, Palma-Rodríguez HM, Montalvo-González E, Sáyago-Ayerdi SG, Utrilla-Coello R, Vargas-Torres A (2019) Effect of using microencapsulated ascorbic acid in coatings based on resistant starch chayotextle on the quality of guava fruit. Sci Horticult 256 (October):108604. https://doi.org/10.1016/j.scienta.2019.108604

Martins JT, Cerqueira MA, Souza BWS, do Carmo Avides M, Vicente AA (2010) Shelf life extension of ricotta cheese using coatings of galactomannans from nonconventional sources incorporating nisin against *Listeria monocytogenes*. J Agric Food Chem 58(3):1884–1891. https://doi.org/10.1021/jf902774z

McHugh T, Habig J-FA, Krochta JM (1994) Plasticized whey protein edible films: water vapor permeability properties. J Food Sci 59(2):416–419. https://doi.org/10.1111/j.1365-2621.1994.tb06980.x

Meindrawan B, Ofe O, Susanto CS, Ayman A, Mangindaan D, Kasih TP (2020) Glucomannan–Beeswax–Chitosan antimicrobial edible coating to maintain the storage quality of Salak fruit (*Salacca Zalacca*). Macromol Symp 391(1):1900164. https://doi.org/10.1002/masy.201900164

Menezes J (n.d.) Chapter 5—report on edible films and coatings, p 36

Min S, Harris LJ, Krochta JM (2006) Inhibition of Salmonella enterica and Escherichia coli O157:H7 on roasted Turkey by edible whey protein coatings incorporating the lactoperoxidase system. J Food Protect 69(4):784–793. https://doi.org/10.4315/0362-028X-69.4.784

Mohamed C, Elise N'g, Etienne TV, Loiseau G, Montet D (2020a) Antifungal activity of edible coating made from chitosan and lactoperoxidase system against *Phomopsis* sp. RP257 and

*Pestalotiopsis* sp. isolated from mango. J Food Safety 40(3):e12785. https://doi.org/10.1111/jfs.12785

Mohamed SAA, El-Sakhawy M, El-Sakhawy MA-M (2020b) Polysaccharides, protein and lipid -based natural edible films in food packaging: a review. Carbohydr Polym 238(June):116178. https://doi.org/10.1016/j.carbpol.2020.116178

Naseri HR, Beigmohammadi F, Mohammadi R, Sadeghi E (2020) Production and characterization of edible film based on gelatin–chitosan containing *Ferulago Angulate* essential oil and its application in the prolongation of the shelf life of Turkey meat. J Food Process Preserv 44(8): e14558. https://doi.org/10.1111/jfpp.14558

Nisar T, Wang Z-C, Yang X, Tian Y, Iqbal M, Guo Y (2018) Characterization of citrus pectin films integrated with clove bud essential oil: physical, thermal, barrier, antioxidant and antibacterial properties. Int J Biol Macromol 106(January):670–680. https://doi.org/10.1016/j.ijbiomac.2017.08.068

Oliveira MA, Bastos MSR, Magalhães HCR, Garruti DS, Benevides SD, Furtado RF, Egito AS (2017) α, β-citral from Cymbopogon citratus on cellulosic film: release potential and quality of coalho cheese. LWT—Food Sci Technol 85(November):246–251. https://doi.org/10.1016/j.lwt.2017.07.029

Ordoñez R, Contreras C, González-Martínez C, Chiralt A (2021) Edible coatings controlling mass loss and Penicillium roqueforti growth during cheese ripening. J Food Eng 290 (February):110174. https://doi.org/10.1016/j.jfoodeng.2020.110174

Oussalah M, Caillet S, Salmiéri S, Saucier L, Lacroix M (2004) Antimicrobial and antioxidant effects of milk protein-based film containing essential oils for the preservation of whole beef muscle. J Agric Food Chem 52(18):5598–5605. https://doi.org/10.1021/jf049389q

Ozdemir M, Floros JD (2008) Optimization of edible whey protein films containing preservatives for water vapor permeability, water solubility and sensory characteristics. J Food Eng 86(2):215–224. https://doi.org/10.1016/j.jfoodeng.2007.09.028

Padua GW, Wang Q (2002) Formation and properties of corn zein films and coatings, p 25

Park S-I, Daeschel MA, Zhao Y (2004) Functional properties of antimicrobial lysozyme-chitosan composite films. J Food Sci 69(8):M215–M221. https://doi.org/10.1111/j.1365-2621.2004.tb09890.x

Peralta-Ruiz Y, Tovar CG, Sinning-Mangonez A, Bermont D, Cordero AP, Paparella A, Chaves-López C (2020) *Colletotrichum gloesporioides* inhibition using chitosan-ruta graveolens L essential oil coatings: studies in vitro and in situ on carica papaya fruit. Int J Food Microbiol 326 (August):108649. https://doi.org/10.1016/j.ijfoodmicro.2020.108649

Pérez MJ, Moreno MA, Martínez-Abad A, Cattaneo F, Zampini C, Isla MI, López-Rubio A, Fabra MJ (2020) Interest of black carob extract for the development of active biopolymer films for cheese preservation. Food Hydrocoll 113:106436. https://doi.org/10.1016/j.foodhyd.2020.106436

Pérez-Gago MB, Krochta JM (1999) Water vapor permeability of whey protein emulsion films as affected by pH. J Food Sci 64(4):695–698. https://doi.org/10.1111/j.1365-2621.1999.tb15112.x

Pinzon MI, Sanchez LT, Garcia OR, Gutierrez R, Luna JC, Villa CC (2020) Increasing shelf life of strawberries ( *Fragaria* Ssp) by using a banana starch-chitosan-aloe vera gel composite edible coating. Int J Food Sci Technol 55(1):92–98. https://doi.org/10.1111/ijfs.14254

Ponce AG, Roura SI, del Valle CE, Moreira MR (2008) Antimicrobial and antioxidant activities of edible coatings enriched with natural plant extracts: in vitro and in vivo studies. Postharvest Biol Technol 49(2):294–300. https://doi.org/10.1016/j.postharvbio.2008.02.013

Porta R (2013) Edible coating as packaging strategy to extend the shelflife of fresh-cut fruits and vegetables. J Biotechnol Biomater 3(4):e124. https://doi.org/10.4172/2155-952X.1000e124

Pourmolaie H, Asl AK, Ahmadi M, Zomorodi S, Raeisi SN (2018) The effect of guar and tragacanth gums as edible coatings in cheddar cheese during ripening. J Food Safety 38(6): e12529. https://doi.org/10.1111/jfs.12529

Pranoto Y, Rakshit SK, Salokhe VM (2005) Enhancing antimicrobial activity of chitosan films by incorporating garlic oil, potassium sorbate and nisin. LWT—Food Sci Technol 38(8):859–865. https://doi.org/10.1016/j.lwt.2004.09.014

Quezada-Gallo J-A (2009) Delivery of food additives and antimicrobials using edible films and coatings. In: Huber KC, Embuscado ME (eds) Edible films and coatings for food applications. Springer New York, New York, pp 315–333. https://doi.org/10.1007/978-0-387-92824-1_11

Raeisi M, Tabaraei A, Hashemi M, Behnampour N (2016) Effect of sodium alginate coating incorporated with nisin, cinnamomum zeylanicum, and rosemary essential oils on microbial quality of chicken meat and fate of *Listeria monocytogenes* during refrigeration. Int J Food Microbiol 238(December):139–145. https://doi.org/10.1016/j.ijfoodmicro.2016.08.042

Ramírez-Guerra HE, Castillo-Yañez FJ, Montaño-Cota EA, Ruíz-Cruz S, Márquez-Ríos E, Canizales-Rodríguez DF, Torres-Arreola W, Montoya-Camacho N, Ocaño-Higuera VM (2018) Protective effect of an edible tomato plant extract/chitosan coating on the quality and shelf life of Sierra fish fillets. J Chem 2018:1–6. https://doi.org/10.1155/2018/2436045

Rashid F, Ahmed Z, Hussain S, Kausar T, Nadeem M, Ainee A, Mehmood T (2020) Optimization of fenugreek and flax polysaccharides-based edible coating formulation to preserve the quality and storability of apple after harvesting. J Food Process Preserv 44(10):e14812. https://doi.org/10.1111/jfpp.14812

Rasouli M, Saba MK, Ramezanian A (2019) Inhibitory effect of salicylic acid and aloe vera gel edible coating on microbial load and chilling injury of orange fruit. Sci Horticult 247 (March):27–34. https://doi.org/10.1016/j.scienta.2018.12.004

Ravishankar S, Zhu L, Olsen CW, McHugh TH, Friedman M (2009) Edible apple film wraps containing plant antimicrobials inactivate foodborne pathogens on meat and poultry products. J Food Sci 74(8):M440–M445. https://doi.org/10.1111/j.1750-3841.2009.01320.x

Raybaudi-Massilia R, Mosqueda-Melgar J, Soliva-Fortuny R, Martín-Belloso O (2016) Combinational edible antimicrobial films and coatings. In: Antimicrobial food packaging. Elsevier, pp 633–46. https://doi.org/10.1016/B978-0-12-800723-5.00052-8

Razavi SMA, Amini AM, Zahedi Y (2015) Characterisation of a new biodegradable edible film based on sage seed gum: influence of plasticiser type and concentration. Food Hydrocoll 43 (January):290–298. https://doi.org/10.1016/j.foodhyd.2014.05.028

Rompothi O, Pradipasena P, Tananuwong K, Somwangthanaroj A, Janjarasskul T (2017) Development of nonwater soluble, ductile mung bean starch based edible film with oxygen barrier and heat sealability. Carbohydr Polym 157:748–756. https://doi.org/10.1016/j.carbpol.2016.09.007

Rojas-Graü MA, Avena-Bustillos RJ, Olsen C, Friedman M, Henika PR, Martín-Belloso O, Pan Z, McHugh TH (2007a) Effects of plant essential oils and oil compounds on mechanical, barrier and antimicrobial properties of alginate–apple puree edible films. J Food Eng 81(3):634–641. https://doi.org/10.1016/j.jfoodeng.2007.01.007

Rojas-Graü MA, Raybaudi-Massilia RM, Soliva-Fortuny RC, Avena-Bustillos RJ, McHugh TH, Martín-Belloso O (2007b) Apple puree-alginate edible coating as carrier of antimicrobial agents to prolong shelflife of fresh-cut apples. Postharvest Biol Technol 45(2):254–264. https://doi.org/10.1016/j.postharvbio.2007.01.017

Rojas-Graü MA, Soliva-Fortuny R, Martín-Belloso O (2009) Edible coatings to incorporate active ingredients to fresh-cut fruits: a review. Trends Food Sci Technol 20(10):438–447. https://doi.org/10.1016/j.tifs.2009.05.002

Saleem MS, Ejaz S, Anjum MA, Nawaz A, Naz S, Hussain S, Ali S, Canan İ (2020) Postharvest application of gum arabic edible coating delays ripening and maintains quality of persimmon fruits during storage. J Food Process Preserv 44(8):e14583. https://doi.org/10.1111/jfpp.14583

Salgado PR, Ortiz CM, Musso YS, Di Giorgio L, Mauri AN (2015) Edible films and coatings containing bioactives. Curr Opin Food Sci 5(October):86–92. https://doi.org/10.1016/j.cofs.2015.09.004

Sanches R, Augusto M, Camelo-Silva C, da Silva Carvalho C, de Mello JR, Barroso NG, da Silva Barros EL, Silva PP, Pertuzatti PB (2021) Active packaging with starch, red cabbage extract and sweet whey: characterization and application in meat. LWT 135(January):110275. https://doi.org/10.1016/j.lwt.2020.110275

Sánchez-González L, Vargas M, González-Martínez C, Chiralt A, Cháfer M (2011) Use of essential oils in bioactive edible coatings: a review. Food Eng Rev 3(1):1–16. https://doi.org/10.1007/s12393-010-9031-3

Saremnezhad S, Azizi MH, Barzegar M, Abbasi S, Ahmadi E (2011) Properties of a new edible film made of faba bean protein isolate, p 12

Saricaoglu FT, Tural S, Gul O, Turhan S (2018) High pressure homogenization of mechanically deboned chicken meat protein suspensions to improve mechanical and barrier properties of edible films. Food Hydrocoll 84(November):135–145. https://doi.org/10.1016/j.foodhyd.2018.05.058

Sathivel S (2005) Chitosan and protein coatings affect yield, moisture loss, and lipid oxidation of pink salmon (Oncorhynchus Gorbuscha) fillets during frozen storage. J Food Sci 70(8):e455–e459. https://doi.org/10.1111/j.1365-2621.2005.tb11514.x

Saxena A, Sharma L, Maity T (2020) Enrichment of edible coatings and films with plant extracts or essential oils for the preservation of fruits and vegetables. In: Biopolymer-based formulations. Elsevier, pp 859–80. https://doi.org/10.1016/B978-0-12-816897-4.00034-5

Seydim AC, Sarikus G (2006) Antimicrobial activity of whey protein based edible films incorporated with oregano, rosemary and garlic essential oils. Food Res Int 39(5):639–644. https://doi.org/10.1016/j.foodres.2006.01.013

Shahbazi Y (2018) Application of carboxymethyl cellulose and chitosan coatings containing *Mentha spicata* essential oil in fresh strawberries. Int J Biol Macromol 112(June):264–272. https://doi.org/10.1016/j.ijbiomac.2018.01.186

Sharma S, Prasad RN, Tiwari S, Chaurasia SNS, Shekhar S, Singh J (2021) Effect of chitosan coating on postharvest quality and enzymatic activity of eggplant (*Solanum Melongena* L.) cultivars. J Food Process Preserv 45:e15098. https://doi.org/10.1111/jfpp.15098

Shin S-H, Chang Y, Lacroix M, Han J (2017) Control of microbial growth and lipid oxidation on beef product using an apple peel-based edible coating treatment. LWT 84(October):183–188. https://doi.org/10.1016/j.lwt.2017.05.054

Sothornvit R, Krochta JM (2005) Plasticizers in edible films and coatings. In: Innovations in food packaging. Elsevier, pp 403–33. https://doi.org/10.1016/B978-012311632-1/50055-3

Souza VGL, Pires JRA, Vieira ÉT, Coelhoso IM, Duarte MP, Fernando AL (2019) Activity of chitosan-montmorillonite bionanocomposites incorporated with rosemary essential oil: from in vitro assays to application in fresh poultry meat. Food Hydrocoll 89(April):241–252. https://doi.org/10.1016/j.foodhyd.2018.10.049

Tahergorabi R, Matak KE, Jaczynski J (2015) Fish protein isolate: development of functional foods with nutraceutical ingredients. J Funct Foods 18(October):746–756. https://doi.org/10.1016/j.jff.2014.05.006

Umaraw P, Verma AK (2017) Comprehensive review on application of edible film on meat and meat products: an eco-friendly approach. Crit Rev Food Sci Nutr 57(6):1270–1279. https://doi.org/10.1080/10408398.2014.986563

Ustunol Z (2009) Edible films and coatings for meat and poultry. In: Huber KC, Embuscado ME (eds) Edible films and coatings for food applications. Springer New York, New York, pp 245–268. https://doi.org/10.1007/978-0-387-92824-1_8

Vachon C, Yu H-L, Yefsah R, Alain R, St-Gelais D, Lacroix M (2000) Mechanical and structural properties of milk protein edible films cross-linked by heating and γ-irradiation. J Agric Food Chem 48(8):3202–3209. https://doi.org/10.1021/jf991055r

Valizadeh S, Naseri M, Babaei S, Hosseini SMH (2020) Shelf life extension of fish patty using biopolymer-coated active paper sheets. Food Packag Shelf Life 26(December):100603. https://doi.org/10.1016/j.fpsl.2020.100603

Venkatachalam K, Lekjing S (2020) A chitosan-based edible film with clove essential oil and nisin for improving the quality and shelf life of pork patties in cold storage. RSC Adv 10(30):17777–17786. https://doi.org/10.1039/D0RA02986F

Vieira MGA, da Silva MA, dos Santos LO, Beppu MM (2011) Natural-based plasticizers and biopolymer films: a review. Eur Polymer J 47(3):254–263. https://doi.org/10.1016/j.eurpolymj.2010.12.011

Vital ACP, Guerrero A, Kempinski EMBC, de Oliveira Monteschio J, Sary C, Ramos TR, del Mar Campo M, do Prado IN (2018) Consumer profile and acceptability of cooked beef steaks with edible and active coating containing oregano and rosemary essential oils. Meat Sci 143 (September):153–158. https://doi.org/10.1016/j.meatsci.2018.04.035

Wang SY, Gao H (2013) Effect of chitosan-based edible coating on antioxidants, antioxidant enzyme system, and postharvest fruit quality of strawberries (Fragaria x Aranassa Duch.). LWT—Food Sci Technol 52(2):71–79. https://doi.org/10.1016/j.lwt.2012.05.003

Wihodo M, Moraru CI (2013) Physical and chemical methods used to enhance the structure and mechanical properties of protein films: a review. J Food Eng 114(3):292–302. https://doi.org/10.1016/j.jfoodeng.2012.08.021

Wu S (2019) Extending shelflife of fresh-cut potato with cactus Opuntia Dillenii polysaccharide-based edible coatings. Int J Biol Macromol 130(June):640–644. https://doi.org/10.1016/j.ijbiomac.2019.03.022

Wu Y, Curtis LW, Hamouz F, Susan LC, Schnepf M (2002) Development and application of multicomponent edible coatings and films: a review. Adv Food Nutr Res 44:347–394. https://doi.org/10.1016/S1043-4526(02)44007-7

Xiong Y, Chen M, Warner RD, Fang Z (2020) Incorporating nisin and grape seed extract in chitosan-gelatine edible coating and its effect on cold storage of fresh pork. Food Control 110 (April):107018. https://doi.org/10.1016/j.foodcont.2019.107018

Yangilar F (2016) Effect of the fish oil fortified chitosan edible film on microbiological, chemical composition and sensory properties of Göbek Kashar cheese during ripening time. Korean J Food Sci Animal Resour 36(3):377–388. https://doi.org/10.5851/kosfa.2016.36.3.377

Yangilar F (2017) Effects of natamycin edible films fortified with essential oils on the safety and quality parameters of Kashar cheese. J Food Safety 37(2):e12306. https://doi.org/10.1111/jfs.12306

Yano T (1990) Physical properties and microbiology of foods. In: Food packaging. Elsevier, pp 3–24. https://doi.org/10.1016/B978-0-08-092395-6.50005-2

Youssef AM, Assem FM, Abdel-Aziz ME, Elaaser M, Ibrahim OA, Mahmoud M, Abd El-Salam MH (2019) Development of bionanocomposite materials and its use in coating of ras cheese. Food Chem 270(January):467–475. https://doi.org/10.1016/j.foodchem.2018.07.114

Yousuf B, Qadri OS (2020) Preservation of fresh-cut fruits and vegetables by edible coatings. In: Fresh-cut fruits and vegetables. Elsevier, pp 225–42. https://doi.org/10.1016/B978-0-12-816184-5.00011-2

Yuan L, Wu F, Zhang Z, Peng Y, Xiao Y, Chen J (2021) Effect of potato starch-based antibacterial composite films with thyme oil microemulsion or microcapsule on shelf life of chilled meat. LWT 139(March):110462. https://doi.org/10.1016/j.lwt.2020.110462

Zahedi Y (2019) Edible/biodegradable films and coatings from natural hydrocolloids. In: Razavi SMA (ed) Emerging natural hydrocolloids. Wiley, Chichester, pp 571–599. https://doi.org/10.1002/9781119418511.ch23

Zhang Y, Zhou L, Zhang C, Show PL, Anning D, JingChao F, Ashokkumar V (2020) Preparation and characterization of curdlan/polyvinyl alcohol/ thyme essential oil blending film and its application to chilled meat preservation. Carbohydr Polym 247(November):116670. https://doi.org/10.1016/j.carbpol.2020.116670

# Role of Sensors to Improve Food/Beverage Packaging

# 9

Mariusz Tichoniuk

**Abstract**

Traditionally, food and beverage packaging focus on the protection of its content and information on them within the supply chain. The development of modern packaging materials and technologies provides an opportunity to improve the communication on the product state and its surrounding. Sensors dedicated to food and/or beverage packaging could be simple time-temperature indicators or chemosensors sensitive to food spoilage products. A gas indicator implemented in the packaging could reveal its leakage or the development of undesirable processes in the packaged product. A wide variety of electrochemical sensors (e.g. temperature or humidity sensors) can work independently or are combined with advanced data carriers such as radio-frequency identification (RFID) tags. The application of sensors in various parts of the food/beverage supply chain could improve its controllability and reduce the scale of food waste more efficiently. The consumer will also appreciate a simple indicator informing about the freshness of food or the correct temperature of the drink.

**Keywords**

Food/beverages packaging · Sensors · Smart packaging · Intelligent packaging · Indicators · Data carriers

M. Tichoniuk (✉)
Department of Industrial Products and Packaging Quality, Poznań University of Economics and Business, Poznań, Poland
e-mail: mariusz.tichoniuk@ue.poznan.pl



A. K. Shukla (ed.), *Food Packaging: The Smarter Way*,
https://doi.org/10.1007/978-981-16-7196-8_9

## 9.1 Introduction

Food and beverages are a very diverse group of products that provide consumers with the required nutrient compounds, biologically active components in the diet, but also the pleasure of consuming our favourite dishes. Current trends in food and beverage consumption are in the strong influence of customer expectation for convenient, ready to eat and/or low processed products with unchangeable quality and extended shelf-life (Han et al. 2018; Rai et al. 2019). Most food products are a complex mixture of ingredients that are sensitive to external physicochemical factors, microbial contamination and biochemical processes occurring in the products themselves, and require protection against factors that could lower their quality and affect safety (Fuertes et al. 2016; Sohail et al. 2018; Witjaksono et al. 2018). Food, by definition, must be safe, but consumers demand also information about its composition, use-by date, method of preparation, and many other data associated with food storage and consumption. Traditionally, the need for food packaging stems from its practical applications of holding goods together and protecting them within the supply chain until the food or beverages reaches the end-user (Lydekaityte and Tambo 2019). According to FAO reports, about 14% of food products are lost in the food supply chain before they reach the final consumer (FAO 2019). In addition to the protective function, the packaging in its original use is intended to facilitate handling the packed product in storage and transportation and to communicate about the properties of their content (Vanderroost et al. 2014; Müller and Schmid 2019). The protection against adverse physical conditions, chemical and/or microbial contamination is an obvious result of the barrier capabilities of packaging construction and materials, which is a passive, conventional way of the product preservation against external factors occurring in the supply chain (Schaefer and Cheung 2018; Drago et al. 2020). The information contained on packaging provides important knowledge about the origin of the product, its weight/volume, ingredients, nutritional value, expiration date, precautions for use as well as the conditions of transport, storage and utilization, if necessary (Fuertes et al. 2016). Packaging should improve the convenience of food product consumption and distribution by the adaptation to customer's lifestyle and physiological capabilities (e.g. by the selected package size and easier food preparation) and to transportation/storage technologies (e.g. using standard packaging shape and dimensions, introduction of useful holders and fixings) (Müller and Schmid 2019). The same is referred as to the containment of food products which should ensure ease of their transportation and handling in the supply chain and the right portion of products for the final consumers (Schaefer and Cheung 2018).

The development of innovative materials and packaging technologies enabled the extension of the information functionality of conventional packaging by the introduction of intelligent systems (Mustafa and Andreescu 2018). The intelligent packaging for food products is recognized as material and article that monitor the condition of packaged food or the environment surrounding it (Commision Regulation 2009 (EC) No 450/2009). The intelligent food packaging systems are designed to monitor and signalize the changes of packed products and its surrounding in the

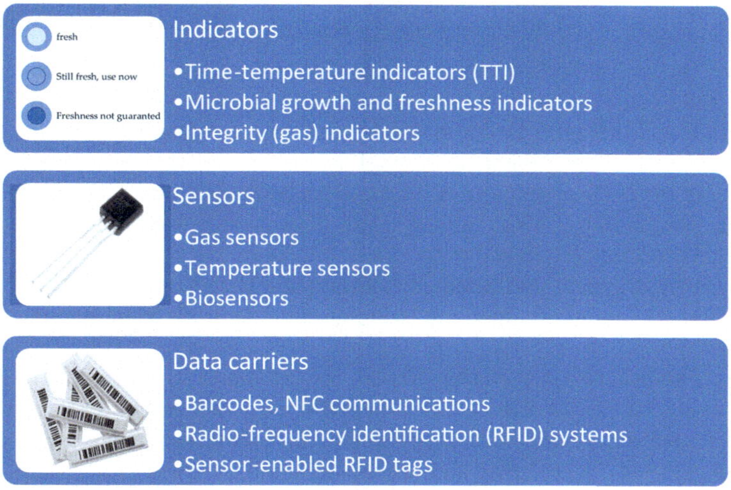


**Fig. 9.1** General types of intelligent packaging systems applied for food and beverages packaging (Source: Own work based on Ghaani et al. 2016, Sohail et al. 2018, Tichoniuk 2019, Drago et al. 2020)

food supply chain, which can facilitate decision-making procedures focused on preserving food quality, and indirectly extend its shelf life and/or improve overall safety (Müller and Schmid 2019). The subject of the real-time monitoring can be internal and external characteristics of the packaging environment (e.g. temperature, humidity), food freshness or microbiological condition, the integrity of packaging or location of the packaged product (Sohail et al. 2018; Chowdhury and Morey 2019). The functionality of intelligent packaging is based on the recognizing element that interacts physically and/or chemically with the monitored environment or product and signalizes the change of observed phenomena. This mechanism relies usually on the application of one of the specific sensors, indicators, or other data transmission items, that enables the collection and/or transfer of information through the packaging (Fig. 9.1). The upgrade of conventional packaging to the intelligent one often relies on attaching an external recognition element (for the product surrounding monitoring or data transmission) or internal sensor or indicator (for monitoring of packed product condition) (Fang et al. 2017; Mustafa and Andreescu 2018). A more detailed classification of intelligent packaging solutions is presented in Fig. 9.1, but it should be remembered that this is not an absolute distinction of types and they may be classified differently according to various sources. This applies in particular to the distinction between indicators and sensors, which both recognize the target compounds or physical phenomena, and the assignment to one or the other category is more related to the interpretation of the results provided.

Sensors and indicators are constructed to provide information relating to the product quality connected with the temperature measurements (time-temperature indicator, temperature sensor) or the presence of specific chemical compounds (gas sensor/indicator, microbial growth or freshness indicators, biosensors) (Ghaani et al. 2016; Sohail et al. 2018; Tichoniuk 2019). Data carriers implemented in intelligent packaging (barcodes, NFC communication, radio-frequency identification systems) are designed to support the management of supply chain logistics (Tichoniuk 2019; Drago et al. 2020) (Fig. 9.1). More reliable and effective communication can also serve to significant improvement of product traceability in the supply chain. A combination of RFID systems with sensors, that monitor the condition of products and their surroundings, could influence the reduction of food waste during food storage and distribution (Ghaani et al. 2016). Intelligent packaging with the function of data transfer might be implemented in Automatic Data Collection (ADC) system and support IT solutions such as Warehouse (Logistic) Management System (Yu et al. 2020).

Sensors, indicators and/or data carriers implemented in intelligent packaging support many aspects of food product preparation and distribution in the supply chain. The real-time information about the packed food or beverages and its surrounding is used to monitor and adjust the conditions within the supply chain and to make decisions affording the extension of food shelf-life, improve safety, ensure quality, provide information and warn against the threat for the final consumer (Drago et al. 2020). Besides the identification of target compounds or physical phenomena, intelligent packaging has the potential to support the effectiveness of different parts of the supply chain and reduce food waste (Yu et al. 2020).

According to the Deloitte report on the market potential of smart packaging, the innovative packaging systems could be successfully applied in three main areas (Deloitte 2018):

1. Inventory and product life cycle management
2. Product integrity assessment
3. Supporting consumer experience

Intelligent packaging with gas sensor, food freshness indicator or adopted for this purpose time-temperature indicator can be used by each member of food supply chain to the estimation of real end-of-life point for food and beverage considering the actual conditions within the distribution chain (regardless of the initially specified product expiration date) (Müller and Schmid 2019). A widespread introduction of intelligent systems for the supply chain would make it possible to change so-called fixed shelf life (FSL) estimation for food and beverages product into dynamic shelf life (DSL) attitude. The adjustable product expiration date (with DSL approach) could reduce food waste as consumers will be able to check the actual condition of food/beverages and their suitability to consumption (Herbon et al. 2012; Albrecht et al. 2019).

## 9.2 Sensors: Types and Construction Principles

The development of intelligent food and beverage packaging is strictly associated with the incorporation of sensors or indicators into its construction, which enables the assessment of packed product quality or the condition of its surrounding (Kuswandi and Moradi 2019). The sensor interacts with an internal target (e.g. food component, volatile metabolite in the packaging headspace) and/or external factor (e.g. temperature, a component of the atmosphere) (Realini and Marcos 2014). The main challenge in the sensor construction is to establish a reliable relationship between the result of the interaction and information about the state of the packed product and/or its surrounding (Fuertes et al. 2016; Poyatos-Racionero et al. 2018).

From the technical point of view, the sensor consists of two basic elements (Fig. 9.2)—chemical or biological receptor, that reacts specifically with the target analyte, and a transducer converting the analytical response into a recognizable signal, which can provide quantitative and/or qualitative output of the sensor activity (Nayak et al. 2009; Mustafa and Andreescu 2018). Biosensing systems rely on the recognition systems of biomolecules like enzymes, nucleic acids, immune system components (antigen, antibodies) or other bioactive compounds (Wang et al. 2012; Chauhan et al. 2019). The sensor transducer can be based on many different physicochemical and electronic solutions, which convert the observed target reaction into a useful analytical signal. Depending on the nature of this reaction, the transducer may use colorimetric arrays (Yang et al. 2011), electrochemical (Pwavodi et al. 2021), optical (Viter et al. 2017) or mass-based signal transmission (Debabhuti et al. 2021). Biosensors are usually designed for use in medical research or environmental protection, but the newest intelligent packaging is slowly starting to adopt these analytical devices (Mustafa and Andreescu 2018). Biosensing devices are

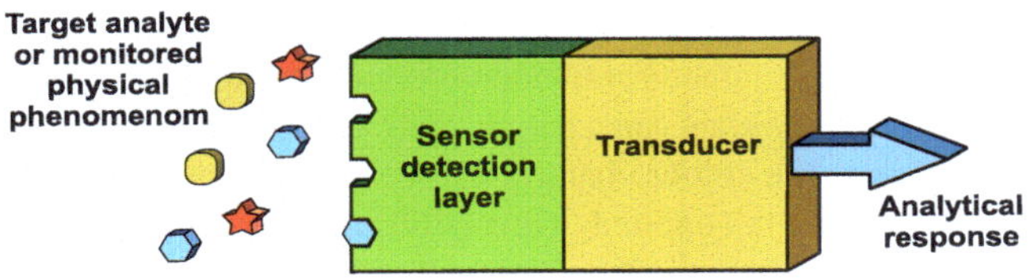


**Types of receptors:**
1. Chemical
2. Electromagnetic
3. Mechanic
4. Optical
5. Thermal
6. Acoustic
7. Biologically active (biosensors)

**Types of transducers:**
1. Electrochemical
   (*e.g.* conductometric, potentiometric)
2. Optical
   (*e.g.* Surface Plasmon Responce, UV-Vis absorbance)
3. Thermal
   (*e.g.* heat sensitive change in polymer film)
4. Mass-based
   (*e.g.* piezoelectric, magnetoelastic)

**Fig. 9.2** Types of receptors and transducers in sensors construction (Source: Own work based on Sohail et al. 2018, Chauhan et al. 2019)

commonly developed in laboratories, and their creators' aim is to achieve popularity at least equal to biosensors used to monitor blood glucose levels (glucometers currently account for 85% of commonly available biosensors) (Pereda et al. 2017).

A more direct assessment of the condition of the packed food or beverages is provided by the indicators, which signalize any change in packed product or its surrounding (e.g. pH, temperature) usually by visual changes (e.g. by changing the colour) (Sohail et al. 2018). The indicators have usually simplified recognition and transduction system in comparison to the sensors. Indicators provide direct and noticeable to the naked eye change the appearance of the their distinctive elements in the presence of the target phenomenon. The greatest challenge in the design of a usable intelligent packaging system is the identification of a reliable relationship between the measurable indicator and changes in food product quality and/or safety being under observation (Müller and Schmid 2019). The most popular indicators applied in intelligent food packaging are time-temperature indicators (TTI) mainly because of the simplicity of its construction, low cost, ease of calibration and interpretation of results compared to other indicators and sensors (Han et al. 2018). Packaging tightness indicators (gas indicators) and food freshness indicators (including ripening indicators and microbial growth indicators) are also becoming more and more popular (Drago et al. 2020).

## 9.3 Physical Sensors/Indicators

Temperature is one of the most important physical agents affecting the condition of food and beverages (especially refrigerated products and frozen food) (Fuertes et al. 2016). Deviations from the determined temperature profile lead to noticeable food changes or significantly increase the likelihood of these changes, which should be recognized in prerequisite kinetic study and modelling of food quality loss at different temperature levels (Kim et al. 2016). The simplest physical sensors applied in food and beverage packaging are designed for temperature monitoring base on the application of thermochromic inks which can change their colour as a result of exposure to elevated temperature. The colour changes can be reversible or irreversible depending on the thermochromic ink applied. Irreversible changes remain stable when the packed product reaches the specified temperature and the packaging element does not revert to its previous appearance until excessive cooling or overheating subsides (Vanderroost et al. 2014). The intelligent packaging can indicate undesirable exposure of the packed product to too high or too low temperature, which can be a desirable application of the indicator in the food supply chain. The reversible thermochromic inks systems are more desirable by the final food or beverages consumers. The temperature colour-changing imprints on the packaging may indicate the desired temperature of consumption of a given product (e.g. a chilled drink—Fig. 9.3a) or remind about the necessity to refrigerate a product that is sensitive to prolonged exposure to elevated temperatures (Fig. 9.3b).

The monitoring of thermal conditions of food storage and/or transportation can be easily carried out with time-temperature indicators (TTI) (Fig. 9.4) which are the

**Fig. 9.3** Examples of food and beverages packaging imprinted with thermochromic ink indicating (**a**) optimal cooling temperature or (**b**) a warning of increased temperature of the packed product (Source: Image provided by Chromatic Technologies Inc.)

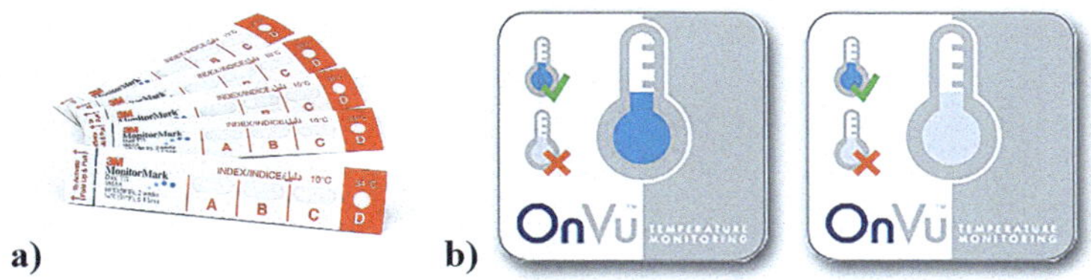


**Fig. 9.4** Examples of commercially available time-temperature indicators: (**a**) Monitor Mark™. (**b**) OnVu™ labels (Source: Images provided by 3M Company (**a**) and Freshpoint (**b**))

most popular intelligent packaging items (Müller and Schmid 2019). TTI indicators activity relies on different types of reactions, the intensity of which is proportional to the temperature increase above the assumed optimal value. The mechanisms can have mechanical (e.g. 3M Monitor Mark™), chemical (e.g. On Vu™, Fresh Check®), electrochemical (e.g. VITSLAB®), enzymatic (e.g. CheckPoint®) or microbiological (e.g. e0®, TopCryo®) background but the main goal is to provide irreversible and visible reaction of TTI indicator dependent on the monitored temperature of the packed product or its environment (Pereira Jr. et al. 2015; Wang et al. 2015). Table 9.1 presents commercially available TTI indicators and a short description of its operation modes.

Time-temperature indicators are designed to have a convenient and simple form of an element that can be easily inserted into conventional food packaging such as labels or strips. TTI indicators work usually in one of three modes: indicating the transgression of determining temperature (critical temperature indicators), measuring the time of exceeding the indicated temperature (partial history indicators), or recording the complete temperature profile in a given period (full history indicator) (Müller and Schmid 2019). The indicator should respond when the adverse thermal

**Table 9.1** Commercially available time-temperature indicators applicable for food and beverages intelligent packaging

| TTI indicator (company, country) | Type of action | Need of activation | Colour change (optical response) | Application |
|---|---|---|---|---|
| CheckPoint® types M, L (Vitsab Int. AB, Sweden) | Enzymatic | Yes | Tricolour: green to yellow to red | Meat, fish, dairy products |
| (e0)® (Cryolog, France) | Microbial | No | Green to red | Cold chain |
| FreshCheck® (TEMPTIME Corp. USA) | Polymeric | No | Colourless to blue | All kinds of fresh products |
| On Vu™ (Freshpoint, Switzerland) | Photochromic | Yes | Dark blue to colourless | Meat, fish, dairy products |
| Monitor Mark™ and Freeze Watch (3M Comp. USA) | Diffusion-reaction | Yes | Diffusion of coloured path/ material | Bakery products, beverage, meat |

Own work on the basis (Park et al. 2015; Sohail et al. 2018; Drago et al. 2020)

**Fig. 9.5** CheckPoint® label L5-8 (Smart TTI Seafood Label) indicating thermal exposure over the level recommended by U.S. Food and Drug Administration (Source: Image provided by Vitsab International)

condition occurs and provide information about the scale of the risk of food quality and safety loss (Fig. 9.5).

The usefulness of colorimetric indicators for all members of the food supply chain depends on the technological applicability of the indicator on the packaging and the readability of its signalling (especially for the final consumer) (Schaefer and Cheung 2018). A very good example to consider these two aspects of TTI application on food packaging is On Vu™ time-temperature indicator that can be activated by UV radiation at any stage of the supply chain. On the other hand, its response to the adverse thermal condition is fading the dark navy blue interior of the pictogram to grey colouring, which gives only the approximate time of unfavourable temperature influence on the monitored product (Branka 2012). The thermochromic inks used in physical sensors for intelligent packaging should provide a more distinguishable change of the indicator appearance, which will be easy to recognize in a situation threatening food safety and quality.

## 9.4 Chemical Sensors/Indicators

The atmosphere inside packaging has a tremendous influence on the quality of packed food and/or beverages, and some changes in its chemical composition indicate often undesirable processes in packed product or leakage from inside or outside of the packaging materials (Sohail et al. 2018). The chemical sensors are constructed to detect target analytes in the packaging atmosphere or close contact with packed food products (Table 9.2). This mechanism of action is exploited by integrity and food freshness indicators/sensors (Fuertes et al. 2016). The first type of intelligent packaging is based mainly on oxygen or carbon dioxide detection. When it is applied as an integrity indicator, it contains a dye compound that changes its colour in case of alternation of the chemical composition inside packaging (Meng et al. 2014).

The integrity (gas) sensor or indicator is very usable for food packaging with the modified atmosphere inside, when the chemical content of packaged product head-space prolongs its shelf life, and damage to the integrity of the packaging reduces the effectiveness of the applied packaging system. One of the most popular integrity indicators that is available on the market is Ageless Eye® label (provided by Mitsubishi Gas Chemical company) (Fig. 9.6). The indicator enables the detection of an oxygen level increase inside the packaging but it should be implemented together with oxygen scavengers to prevent the influence on the reliability of the indicator by residual oxygen, which could be present in the packaging or could be released from the packaged product (Tichoniuk 2019).

The release of metabolites from packed food could indicate a significant decrease in its quality and/or safety due to the development of undesirable microflora or as a result of internal physicochemical and/or enzymatic changes in the product. Freshness indicators and sensors are designed to recognize metabolites released into the packaging interior during food spoilage or ripening (Fig. 9.7). The target analytes in the recognition process are usually carbon dioxide, ethyl alcohol, esters, organic acids, sulphur derivatives or volatile nitrogen compounds (Sohail et al. 2018; Tichoniuk 2019). The freshness indicators are usually simple colorimetric arrays

**Table 9.2** Examples of target compounds and food products monitored by freshness indicators

| Target compounds (metabolites) | Food product | Freshness indicator/sensor |
|---|---|---|
| Biogenic amines | Fish, seafood, meat | Colour-changing indicator with pH-sensitive dye/electrochemical sensor for enzyme redox reaction |
| Carbon dioxide | Fermented food, meat | pH-sensitive with colorimetric response/electrochemical sensor, e.g. with silicon-based polymer recognition layer |
| Glucose/ Lactic acid | Fermented food, meat | pH-sensitive with colorimetric response/electrochemical sensor for redox reaction |
| Oxygen | Meat, fruits, vegetables | Oxygen-sensitive indicator with pH-sensitive dye/optical sensor by fluorescence |

Own work on the basis of Sohail et al. (2018)

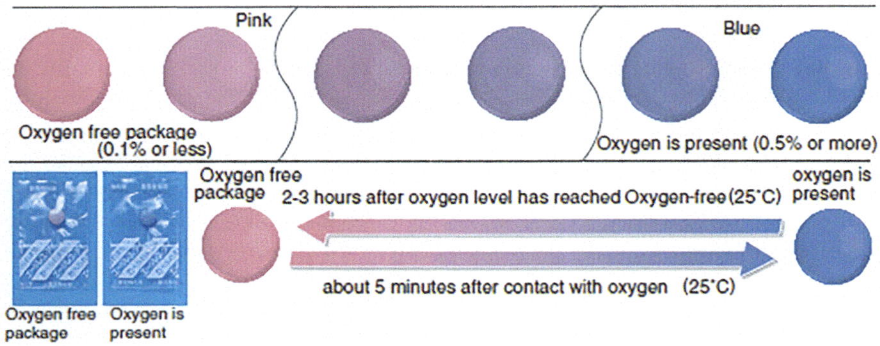


**Fig. 9.6** Ageless Eye® gas (oxygen) indicator (Source: Image provided by MITSUBISHI GAS CHEMICAL)

**Fig. 9.7** RipeSense fruits ripening indicator (Source: Image provided by Ripesense Limited)

that change their colour in the presence of target chemical compounds and signal change of product quality or loss of freshness (Fuertes et al. 2016).

The intelligent packaging systems able to monitor food freshness and/or ripeness are intensively developed but some of them are already available on the market (Table 9.3). Most of the published research results focused on the application of freshness indicator for non-destructive monitoring of food spoilage metabolites in the packaging headspace during the storage of packaged perishable foods

**Table 9.3** Examples of commercially available chemical indicators applicable in food and beverages intelligent packaging

| Type of indicator | Commercial name | Company (Country) | Area of application |
|---|---|---|---|
| Integrity (gas) indicator | Ageless Eye® | Mitsubishi Gas Chemical (Japan) | All packed food products—specially applicable in aseptic and modified atmosphere packaging systems |
| Integrity (gas) indicator | Novas® | Insignia Technologies (United Kingdom) | |
| Freshness indicator | Fresh Tag® | COX Technologies (USA) | Perishable food products, with volatile metabolic products, e.g. meat, fish and seafood, dairy products, vegetables and fruits |
| Freshness indicator | Raflatac | VTT and UPM Raflatac (Finland) | |
| Freshness indicator | RipeSense | RipeSense (New Zealand) | |
| Freshness indicator | SensorQ® | DSM NV and Food Quality Sensor International Inc. (Denmark) | |

Source: own work based on (Fuertes et al. 2016; Ahmed et al. 2018; Poyatos-Racionero et al. 2018)

(e.g. vegetables, fruits, seafood or meat) (Ahmed et al. 2018; Tichoniuk et al. 2021). The colorimetric freshness indicators are very promising elements of intelligent packaging, but their effectiveness and credibility of the information provided depend on many factors such as positioning inside the package, indicator's activation procedure, the chemical content of packaging atmosphere, or temperature changes within the whole food supply chain (Tichoniuk et al. 2017). The influence of the above-mentioned factors has to be considered both in the laboratory study on new freshness indicators and in the evaluation of commercially available indicators. Table 9.3 presents the indicators dedicated to food freshness and tightness of their packaging that are available on the market and are recognizable elements of intelligent packaging. Freshness indicators could be a very good tool for the introduction of dynamically adjustable expiration dates in the so-called dynamic shelf life (DSL) approach in the food supply chain. However, due to technological limitations, they give way to TTI indicators, which can be much easier implemented on a mass scale in intelligent packaging (Herbon et al. 2012; Albrecht et al. 2019). The application of time-temperature indicators in the food supply chain based on DSL attitude can reduce food waste by retailers and help to maintain food quality and safety for the final consumers (Buisman et al. 2019).

## 9.5 Electronic Sensors and Data Carriers

Food and beverages distributors appreciate the ability to trace their products along the supply chain thanks to intelligent packaging designed for efficient data transfer. Data carriers are not designed to monitor the condition of a packed product (or its surrounding), but they are intended to facilitate the flow of information in the supply chain (Müller and Schmid 2019). This group of packaging systems includes different variations of barcodes one-dimensional and two-dimensional (e.g. Quick Response (QR) codes), radio-frequency identification (RFID) systems, and their combinations with sensors such as temperature or mechanical shock sensor (Fang et al. 2017; Yu et al. 2020). Barcodes are now the global standard for capturing and communicating product information throughout the supply chain (GS1 2020). A typical one-dimensional barcode consists of a set of black and white bars and a corresponding numeric notation of encoded information. Two-dimensional (2D) barcodes have a square or rectangular shape and contain black and white spots creating the whole pattern and encoding required information. All these codes are printed on the surface of the packages and are used in optical Automatic Data Collection (ADC) systems, but intelligent packaging technology enables data transfer without visual contact with the label using electromagnetic waves in RFID technology (Yu et al. 2020).

Radio-frequency identification (RFID) systems are emerging intelligent packaging technology usable in real-time data collecting, storage and transmission, which do not require direct contact and visibility of carriers (labels or tags) (Poyatos-Racionero et al. 2018). RFID systems exploit electromagnetic fields to collect and transfer information about the packed product for its automated identification and traceability (Abad et al. 2009; Yu et al. 2020). Radio-frequency tags can be divided into three main groups depending on applied power supply, the ability of communication and implementation of additional features (e.g. temperature measurement). The most advanced are active RFID systems, less developed are semi-passive or semi-active labels, and the simplest construction have passive RFID tags (Poyatos-Racionero et al. 2018). The radio-frequency identification tags contain an integrated electronic circuit built into an antenna for the transmission of data stored in the chip to a reader, which enables a different type of packaging to be monitored at the same time and collect various information. The tag can be coupled to a primary food packaging, box, transportation container, or pallet and therefore can be identified and tracked (Cerqueira et al. 2018; Sohail et al. 2018). Table 9.4 presents examples of commercially available RFID systems combined with sensing arrays designed for food packaging. Such sensors could be used to monitor the freshness and current state of perishable food products in the supply chain to minimize the risk of delivering spoiled food to the final consumer.

A market simulation performed for the inventory management of perishable food products in a dynamic pricing system and with the application of RFID tags with TTI indicators confirmed the usability and economic justification of the application of such intelligent packaging in the supply chain (Herbon et al. 2012). According to the researches the introduction of RFID systems combined with the time-temperature

**Table 9.4** Examples of commercially available RFID systems combined with sensors for food packaging

| Commercial name | Company | RFID system |
|---|---|---|
| Easy2log® | CAEN RFID Srl | Time-temperature sensor tag |
| CS8304 | Convergence Systems Ltd. | Time-temperature sensor tag |
| TempTRIP | TempTRIP LLC | Time-temperature sensor tag |
| Intelligentbox | MondiPlc | Box with integrated time-temperature sensor tag |
| Intelligentfishbox | CraemerGroup GmbH | Box with integrated time-temperature sensor tag |

Source: own work based on (Fuertes et al. 2016; Poyatos-Racionero et al. 2018)

sensor is profitable (at all price discrimination levels) provided the penalties for the sale of damaged items are high and the cost of using intelligent packaging is relatively low. On the other hand, the increase of packaging cost by over 30% of the value of a packaged item and relatively low penalties for selling bad quality products may make the introduction of intelligent packaging unprofitable. However, in the market where customers' awareness of product quality is strong, using TTI-RFID automatic devices can be very profitable in connection with pricing differentiation. Radio-frequency identification systems were also introduced into food quality and safety assurance approaches, for example, for food cold chain monitoring (Badia-Melis et al. 2015), food freshness (Eom et al. 2014), or food product stored in modified atmosphere packaging (Martínez-Olmos et al. 2013).

The most desirable intelligent packaging for food and beverages, considered by both consumer and supplier, includes sensors and indicators used to expand its informative functionality of packaging (Schaefer and Cheung 2018). Figure 9.8 presents QLIKTAG project of food packaging featured with barcodes, Quick Response (QR) codes, and authentication tags to provide all participants of the supply chain additional information about the product. Additionally, Internet of Thing (IOT) technologies could be also connected with intelligent packaging in the food supply chain to support product traceability, monitoring food products' condition, and live-sharing the collected data with the chain participants. (Witjaksono et al. 2018). All mentioned smart technologies could include simple indicators or sensors such as temperature or integrity ones especially usable for monitoring product quality and safety (Fuertes et al. 2016).

## 9.6 Summary

Real-time control of food and beverages condition and evaluation of its surrounding could be facilitated with intelligent packaging systems. Indicators and/or sensors, included in this packaging group, are capable of real-time monitoring selected parameters influencing the quality and safety of food products such as temperature

**Fig. 9.8** Example of intelligent (smart) packaging designed for food products and equipped with different identification tags and data carrier labels (Source: Image provided by Qliktag Software Inc.)

or chemical composition of the atmosphere inside the packaging. They could be also adjusted to recognize symptoms of food ripening or spoilage, e.g. detecting pH change or confirming the presence of volatile metabolites being the results of undesirable changes in food and beverages quality. However, intelligent packaging systems require validation of its signalling about the changes in monitored product or its surrounding regarding the influence of external factors on the indicator/sensor activity.

The electronic data carriers based on Radio Frequency Identification (RFID) technologies are already introduced in many areas of the supply chain, which supports the flow of information during product storage and distribution. RFID systems could facilitate real-time product monitoring, more efficient inventory and precise product traceability, which support the maintenance of food product quality and safety. RFID tags combined with electronic sensors such as temperature recorders or gas sensors could be very useful in the supply chain, especially for perishable food products. Additionally, Internet of Thing (IOT) elements introduced into intelligent packaging in a form of QR codes or product authentication tags, for example, significantly extends the possibility of obtaining additional information about food products for all members of the food supply chain (and especially for the final consumers).

**Reviewers** **Mehran Moradi** (Department of Food Hygiene and Quality Control, Urmia University, Iran).

**Alfredo Ernesto Di Noia** (Department of Economics, Management and Territory, Universitàdeglistudi di Foggia, Italy).

**Anna Masek** (Institute of Polymer and Dye Technology, Lodz University of Technology, Poland).

## References

Abad E, Palacio F, Nuin M, de Zárate AG, Juarros A, Gómez JM, Marco S (2009) RFID smart tag for traceability and cold chain monitoring of foods: demonstration in an intercontinental fresh fish logistic chain. J Food Eng 93:394–399. https://doi.org/10.1016/j.jfoodeng.2009.02.004

Ahmed I, Lin H, Zou L, Li Z, Brody AL, Qazi IM, Lv L, Pavase TR, Khan MU, Khan S, Sun L (2018) An overview of smart packaging technologies for monitoring safety and quality of meat and meat products. Packag Technol Sci 31:449–471. https://doi.org/10.1002/pts.2380

Albrecht A, Ibald R, Raab V, Reichstein W, Haarer D, Kreyenschmidt J (2019) Implementation of time temperature indicators to improve temperature monitoring and support dynamic shelf life in meat supply chains. J Packag Technol Res 4:22–32. https://doi.org/10.1007/s41783-019-00080-x

Badia-Melis R, Ruiz-Garcia L, Garcia-Hierro J, Villalba JIR (2015) Refrigerated fruit storage monitoring combining two different wireless sensing technologies: RFID and WSN. Sensors 15:4781–4795. https://doi.org/10.3390/s150304781

Branka L (2012) Chromogenic inks for packaging labels application—conference materials—new possibilities for print media and packaging—combining print with digital, Helsinki, Finland, 15–16 October 2012. https://www.bib.irb.hr/599093. Accessed 15 June 2019

Buisman ME, Haijema R, Bloemhof-Ruwaard JM (2019) Discounting and dynamic shelf life to reduce fresh food waste at retailers. Int J Prod Econ 209:274–284. https://doi.org/10.1016/j.ijpe.2017.07.016

Cerqueira MA, Nurmi M, Gregor-Svetec D (2018) Intelligent packaging. Industry leaflet. http://www.actinpak.eu/leaflets-intelligent-packaging/. Accessed 15 June 2020

Chauhan N, Jain U, Soni S 2019 Sensors for food quality monitoring. In: Pudake RN, Chauhan N, Kole C (Eds) Nanoscience in sustainable agriculture. Springer, pp 601–626. https://doi.org/10.1007/978-3-319-97852-9_23

Chowdhury EU, Morey A (2019) Intelligent packaging for poultry industry. J Appl Poult Res 28: 791–800. https://doi.org/10.3382/japr/pfz098

Commision Regulation (EC) No 450/2009 of 29 May 2009 on active and intelligent materials and articles intended to come into contact with food (L 135/3 30.5.2009)

Debabhuti N, Neogia S, Mukherjee S, Dhara A, Sharma P, Vekariya RL, Sarkar MP, Tudu B, Bhattacharyya N, Bandyopadhyay R, Muddassir M (2021) Development of QCM sensor to detect α-terpinyl acetate in cardamom. Sensors Actuators A Phys 319:11251. https://doi.org/10.1016/j.sna.2020.112521

Deloitte (2018) Capturing value from the smart packaging revolution. https://www2.deloitte.com/content/dam/insights/us/articles/4353_Smart-packaging/4353_Smart-packaging.pdf. Accessed 20 Dec 2020

Drago E, Campardelli R, Pettinato M, Perego P (2020) Innovations in smart packaging concepts for food. Foods 9(1628):1–34. https://doi.org/10.3390/foods9111628
Eom KH, Hyun KH, Lin S, Kim JW (2014) The meat freshness monitoring system using the smart RFID tag. Int J Distrib Sensor Netw 1–9, 591812. https://doi.org/10.1155/2014/591812
Fang Z, Zhao Y, Warner RD, Johnson SK (2017) Active and intelligent packaging in meat industry. Trends Food Sci Technol 61:60–71. https://doi.org/10.1016/j.tifs.2017.01.002
FAO (2019) The state of food and agriculture 2019. Moving forward on food loss and waste reduction. Rome: FAO. http://www.fao.org/3/ca6030en/ca6030en.pdf. Accessed 1 Feb 2021
Fuertes G, Soto I, Carrasco R, Vargas M, Sabattin J, Lagos C, 2016 Intelligent packaging systems: sensors and nanosensors to monitor food quality and safety. J Sensors, 1–8, 4046061. https://doi.org/10.1155/2016/4046061
Ghaani M, Cozzolino CA, Castelli G, Farris S (2016) An overview of the intelligent packaging technologies in the food sector. Trends Food Sci Technol 51:1–11. https://doi.org/10.1016/j.tifs.2016.02.008
GS1 (2020) GS 1 barcodes. www.gs1.org/standards/barcodes. Accessed 25 Oct 2020
Han JW, Ruiz-Garcia L, Qian J-P, Yang X-T (2018) Food packaging: a comprehensive review and future trends. Compr Rev Food Sci Food Saf 17:860–877. https://doi.org/10.1016/B978-0-12-394601-0.00001-1
Herbon A, Levner E, Cheng E (2012) Perishable inventory management and dynamic pricing using TTI technologies. Int J Innov Manag Technol 3(3):262–266. http://www.ijimt.org/papers/235-D0111.pdf
Kim JU, Ghafoor K, Ahn J, Shin S, Lee SH, Shahbaz HM, Shin H-H, Kim S, Park J (2016) Kinetic modeling and characterization of a diffusion-based time-temperature indicator (TTI) for monitoring microbial quality of non-pasteurized angelica juice. LWT—Food Sci Technol 67:143–150. https://doi.org/10.1016/j.lwt.2015.11.034
Kuswandi B, Moradi M (2019) Sensor trends in beverages packaging. In: Grumezescu AM, Holban AM (Eds) The science of beverages, vol. 16. Trends in beverages packaging. Woodhead Publishing, pp 279–302. https://doi.org/10.1016/B978-0-12-816683-3.00010-4
Lydekaityte J, Tambo T (2019) Smart packaging: definitions, models and packaging as an intermediator between digital and physical product management. Int Rev Retail Distrib Consumer Res, 1–34. https://doi.org/10.1080/09593969.2020.1724555
Martínez-Olmos A, Fernández-Salmerón J, Lopez-Ruiz N, Torres AR, Capitan-Vallvey LF, Palma AJ (2013) Screen printed flexible radiofrequency identification tag for oxygen monitoring. Anal Chem 85(22):11098–11105. https://doi.org/10.1021/ac4028802
Meng X, Kim S, Puligundla P, Ko S (2014) Carbon dioxide and oxygen gas sensor-possible application for monitoring quality, freshness, and safety of agricultural and food products with emphasis on importance of analytical signals and their transformation. J Korean Soc Appl Biol Chem 57(6):723–733. https://doi.org/10.1007/s13765-014-4180-3
Müller P, Schmid M (2019) Intelligent packaging in the food sector: a brief overview. Foods 8(16): 1–12. https://doi.org/10.3390/foods8010016
Mustafa F, Andreescu S (2018) Chemical and biological sensors for food-quality monitoring and smart packaging. Foods 7(168):1–20. https://doi.org/10.3390/foods7100168
Nayak M, Kotian A, Marathe S, Chakravortty D (2009) Detection of microorganisms using biosensors. A smarter way towards detection techniques. Biosens Bioelectron 25(4):661–667. https://doi.org/10.1016/j.bios.2009.08.037
Park YW, Kim SM, Lee JY, Jang W (2015) Application of biosensors in smart packaging. Mol Cell Toxicol 11:277–285. https://doi.org/10.1007/s13273-015-0027-1
Pereda M, Marcovich NE, Ansorena MR (2017) Nanotechnology in food packaging applications: barrier materials, antimicrobial agents, sensors, and safety assessment. In: Myrian et al (eds) Handbook of ecomaterials. Springer, pp 1–22. https://doi.org/10.1007/978-3-319-48281-1_30-1
Pereira VA Jr, de Arruda IZQ, Stefani R (2015) Active chitosan/PVA films with anthocyanins from *Brassica oleraceae* (Red cabbage) as time-temperature indicators for application in intelligent food packaging. Food Hydrocoll 43:180–188. https://doi.org/10.1016/j.foodhyd.2014.05.014

Poyatos-Racionero E, Ros-Lis JV, Vivancos J-L, Martínez-Máñez R (2018) Recent advances on intelligent packaging as tools to reduce food waste. J Clean Prod 172:3398–3409. https://doi.org/10.1016/j.jclepro.2017.11.075

Pwavodi PC, Ozyurt VH, Asir S, Ozsoz M (2021) Electrochemical sensor for determination of various phenolic compounds in wine samples using $Fe_3O_4$ nanoparticles modified carbon paste electrode. Micromachines 12:312. https://doi.org/10.3390/mi12030312

Rai M, Ingle AP, Gupta I, Pandit R, Paralikar P, Gade A, Chaud MV, dos Santos CA (2019) Smart nanopackaging for the enhancement of food shelf life. Environ Chem Lett 17(1):277–290. https://doi.org/10.1007/s10311-018-0794-8

Realini CE, Marcos B (2014) Active and intelligent packaging systems for a modern society. Meat Sci 98:404–419. https://doi.org/10.1016/j.meatsci.2014.06.031

Schaefer D, Cheung WM (2018) Smart packaging: opportunities and challenges. Procedia CIRP 72: 1022–1027. https://doi.org/10.1016/j.procir.2018.03.240

Sohail M, Sun D-W, Zhu Z (2018) Recent developments in intelligent packaging for enhancing food quality and safety. Crit Rev Food Sci Nutr 58(10):2650–2662. https://doi.org/10.1080/10408398.2018.1449731

Tichoniuk M (2019) Innovative packaging improving food quality and extending its shelf life. Polish J Commodity Sci 1(58):21–35

Tichoniuk M, Radomska N, Cierpiszewski R (2017) Application of food freshness indicators in intelligent packaging. In: Śmigielska H (ed) Current trends in commodity science: food safety and analysis of bioactive substances. PUEB, Poznań, pp 189–202

Tichoniuk M, Biegańska M, Cierpiszewski R (2021) Intelligent packaging. In: Galanakis CM (ed) Sustainable food processing and engineering challenges. Cambridge, Academic, pp 279–313. https://doi.org/10.1016/B978-0-12-822714-5.00008-5

Vanderroost M, Ragaert P, Devlieghere F, De Meulenaer B (2014) Intelligent food packaging: the next generation. Trends Food Sci Technol 39(1):47–62. https://doi.org/10.1016/j.tifs.2014.06.009

Viter R, Tereshchenko A, Smyntyn V, Ogorodnichuk J, Starodub N, Yakimova R, Khranovskyy V, Ramanavicius A (2017) Toward development of optical biosensors based on photoluminescence of TiO2 nanoparticles for the detection of *Salmonella*. Sensors Actuators B Chem 252:95–102. https://doi.org/10.1016/j.snb.2017.05.139

Wang Y, Ye Z, Ying Y (2012) New trends in impedimetric biosensors for the detection of foodborne pathogenic bacteria. Sensors 12(3):3449–3471. https://doi.org/10.3390/s120303449

Wang S, Liu X, Yang M, Zhang Y, Xiang K, Tang R (2015) Review of time temperature indicators as quality monitors in food packaging. Packag Technol Sci 28:839–867. https://doi.org/10.1002/pts.2148

Witjaksono G, Rabih AAS, bt Yahya N, Alva S (2018) IOT for agriculture: food quality and safety. IOP Conf Series Mater Sci Eng 343:012023, 1–8. https://doi.org/10.1088/1757-899X/343/1/012023

Yang C, Wang Y, Marty J-L, Yang X (2011) Aptamer-based colorimetric biosensing of Ochratoxin A usingunmodified gold nanoparticles indicator. Biosens Bioelectron 26:2724–2727. https://doi.org/10.1016/j.bios.2010.09.032

Yu Z, Jung D, Park S, Hu Y, Huang K, Rasco BR Wang S, Ronholm J, Lu X, Chen J 2020 Smart traceability for food safety. Crit Rev Food Sci Nutr, 1–12. https://doi.org/10.1080/10408398.2020.1830262

# Functional Nanomaterials for Food Packaging Applications

# 10

Vivek Kumar, Deepika Umrao, and Anjali Srivastava

**Abstract**

The utilization of nanotechnology in the establishment of innovative food packaging materials has had a significantly remarkable enhancement in the previous years, and researches are expected to have an essential impact on the food packaging market in the coming years. Nanotechnology is the emerging engineering field of functional system at the molecular level and has the potential to play a significant role in agriculture and food security, molecular and cellular biology sensors for pathogen identification, environmental protection and allows designers to change the packaging materials on the molecular level to enhance their desired properties such as mechanical strength, temperature and moisture stability, long durability, gas barrier, and flexibility. Due to the revolution in the food packaging materials, the traditional packaging is being replaced by the active packaging, biochemical improved packaging, physically improved and smart packaging to enhance food quality and safety with the significant use of nanotechnology. The focused application of nanotechnology is utilized in food packaging because of their better balance in processing, characteristics, and overall production cost. Nowadays the country regulations and the market are claiming for sustainable higher performance of the food packages, such as longer shelf life, maximum ways of preserving the food in better conditions and shows lower environmental impact. The most promising approaches to overcome defined problems with success have been significantly achieved by preparing nanotechnology-based food packaging material, for instance, preparing nanocomposites that are based on inorganic nanoparticles and polymer matrix dispersed as reinforcement. The techniques used with high aspect ratio of nanoparticles allow adding new functionalities to traditional packaging materials,

V. Kumar (✉) · D. Umrao · A. Srivastava
Harcourt Butler Technical University, Kanpur, Uttar Pradesh, India


A. K. Shukla (ed.), *Food Packaging: The Smarter Way*,
https://doi.org/10.1007/978-981-16-7196-8_10

for example, antimicrobial activity, control emitter substances, and enhanced gas and water vapor barrier characteristics. The main classification of nanotechnology packaging materials and their significant functions and applications are reviewed in this chapter with safety concerns and future prospective.

## 10.1 Introduction

The major losses in the food industry are wastage of more than 1.3 billion metric tons of consumable food every year. The more focused reason behind the loss is low-quality post-harvesting techniques, unsuitable storage facilities, transport facilities, and also last but not the least wastage of food in the marketplaces as well as by consumer uses (Reynolds 2007). The food production rate is increasing on one side but on the other side to focus on solving the problem of food crises due to food wastage, environmental adverse conditions as well as increase in greater population rate. Researchers and scientifically proven studies showed that food wastage is mostly born due to microbial contamination; this will lead toward deterioration in shelf life of food products and also increases the high risk of foodborne diseases (Sperber 2009). The problem of food commodities storages and decrease in shelf life of food products were minimized by the use of "nano-packaging for food" or utilization of nanotechnology (Ekrami et al. 2014). Nano-technology is the combination of development, fabrication, and characterization of material or the structure measures length in the nanometers. If a macro-sized particle lowers down to a nano-size particle, the producing nanoparticles have entirely different chemical and physical property from the original macro-size particles (Ravichandran 2009). Nanoparticles are the owner of unique characteristics of physical, chemical, biological, and large surface area volume ratio. This newborn technology helps in improving better health, quality of life, wealth as well as also shows significantly positive impact on the environmental issues (Dasgupta et al. 2015). Besides the tremendous development of nanomaterial in the field of agro-food, sewage treatment, medicines, and electronics, it lacks behind in the field of food packaging. However, since various nanomaterials developed with many functional properties can be used to improve the quality of food packaging, hence even in its infancy, the nonomaterials are being developed and applied increasingly in the food packaging industry nowadays (Kuswandi 2017).

The utilization of functional nanomaterials for food packaging could also be helpful in developing properties which could modulate the release of antimicrobials, antioxidants, enzymes as well as flavors nutraceuticals to enhance the shelf life and overall quality of food (Berekaa 2015). Nanotechnology helps in manufacturing nanomaterials in a form of film which works by refusing permeability of unwanted gases such as carbon dioxide or oxygen, which affect the shelf life of food products (Silvestre et al. 2011). Previously plastic polymers of non-biodegradable nature were mostly used for food packaging but these are a threat to human beings, animals, and the environment too. The biomaterials are edible and biodegradable in nature and can be used in food packaging to maintain food product freshness and food safety during distribution, storages, and consumption period. Nanotechnology used for food packaging is based on nanocomposite material to reduce the wastes of the food packaging (Kuswandi 2017). The nanocomposite materials are helpful in biodegradation film, edible, environment friendly, easy to coating, and significantly manage freshness and shelf life of food (Sorrentino et al. 2007).

The nanotechnology sector emerges very fast but in nano-food packaging it is still playing a role of infancy. The present chapter focuses on acknowledging the recent methodologies in the development of nano-food packaging. Covered topics include the functional nanomaterials used for the improvement of food packaging and classification, physical improved packaging consists of nano coatings, nano surface biocides, nano-laminates for enhancing physical, mechanical, barrier, and stability, biochemical improved packaging consists of various bio-based nanomaterials for enhancing biodegradability and eco-friendly, improved packaging with active functions such as UV absorbing film, scavenger film to protect from oxygen, antimicrobials, improved packaging with smart functions such as nanosensors for freshness identification, oxygen detection, techniques to analyze functional nanomaterials, environmental and safety concerns and the chapter ended with future prospective and conclusion.

## 10.2 Classification of Nanomaterials-Based Food Packaging

The utilization of nanomaterials in food packaging is divided into two main classes: firstly, nano-object materials and second one is nanostructured materials. The application of nano-object materials is as fillers in the nano-plates and nanofibers, for example: carbon nanotubes, metallic nanoparticles. In the case of nanostructured material, the nanomaterials are utilized into matrix of polymer by the dispersion of nanostructured materials in the form of nanocomposites as shown in Fig. 10.1.

Food packaging can be improved by incorporation of nanomaterials and they can be categorized into four main classes on the basis of the functions of nanomaterials (Table 10.1). Physically improved packaging; the nanomaterials improve the physical properties of packaging such as temperature and moisture stability, mechanical strength, durability and gas barrier and flexibility properties. Biochemical improved packaging; the biochemical characteristics of packaging such as biodegradability, eco-friendly, edible and low waste generation. Improvement of packaging by active functions of nanomaterials as antimicrobial properties also includes antioxidant and UV absorbance. These properties were intentionally introduced into and show a positive effect on the packed food regarding taste, freshness, and shelf life. Smart functions of nanomaterials are used to improve food packaging by utilization of nanosensors to examine and monitor the condition of the food in terms of freshness, oxygen and pathogen level, etc.

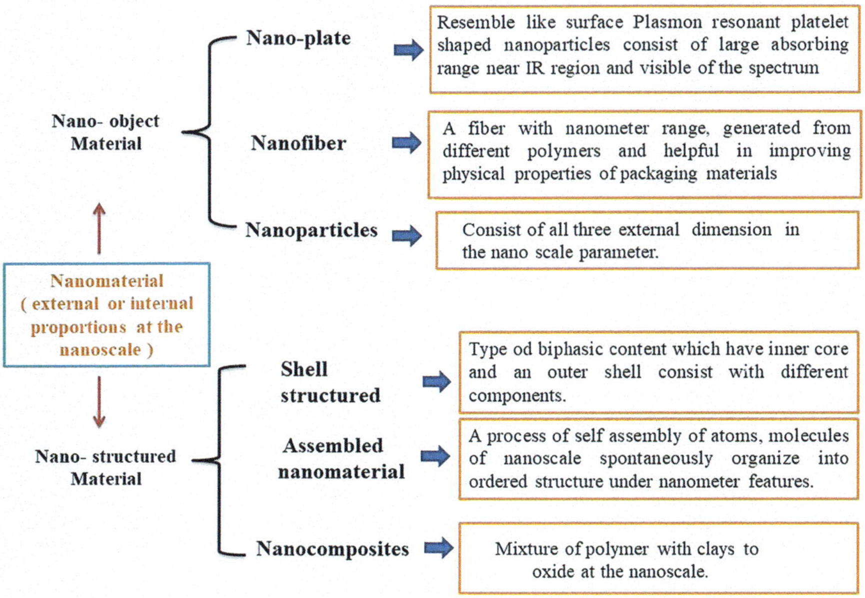


**Fig. 10.1** Classification of nano-object materials and nanostructured materials

### 10.2.1 Physically Improved Nanomaterials-based Food Packaging

Incorporation of nanomaterial to enhance the physical and mechanical properties of food packaging is known as physically improved packaging. To manufacture nano-food packaging material, mixing of nanomaterials into the polymer matrix is an important phenomenon and in addition aids to increase physical characteristics such as gas barriers, temperature resistance, mechanical strength, flexibility, and humidity resistance of food packaging (Vasile 2018). Nanotechnology methods for improving physical and mechanical properties of food packaging are shown in Fig. 10.2.

### 10.2.2 Nano-Coatings

It is the most common method of application nanomaterials in the form of thin layer or film on the food surface. Nano-coatings are measured in nanoscale thickness ranges from less than 1–100 nm. It is used in covering the surface and also imparting particular physical and chemical functions. Nano coating does not play a role in improving surface topography and is not able to fill cracks or gaps in the surface like paint. Nano coating is formulated by placing one or multiple molecular layers on the surface of basic packaging materials such as glass, metals, ceramic, and polymer (Brody 2006). This film coating acts as a barrier for gases, moisture, and lipids. Edible coating films are directly applied to the food either by liquid film-forming

**Table 10.1** Classification of food packaging based on the functions of nanomaterials

| Nanomaterials | Physically improved packaging (Brody 2007) | Biochemical improved packaging (Mangiacapra et al. 2006) | Active functions improved packaging (Asadi and Mousavi 2006) | Smart functions improved packaging (Ahuja et al. 2007) |
|---|---|---|---|---|
| Properties | Utilization of nanomaterial to improve the physical properties of food packaging such as temperature and moisture stability, mechanical power, long durability, gas barrier, and flexibility | Utilization of nanomaterial to improve biochemical properties such as biocompatibility, eco-friendly, edible, biodegradation, and low waste generation | Utilization of nanomaterials with antimicrobial, antioxidant, and UV absorbance properties to improve taste, freshness, and shelf life of packed foods | Utilization of nanosensors to examine and monitor the condition of the food, for example, oxygen indicator, pathogen, and freshness indicator |
| Type | Nano-coating, nano-laminates, clay nanoparticles, and nanocrystals | Tactoid (phase-separated micro composite), intercalated nanocomposites, exfoliated polymer clay | Antimicrobial function, oxygen scavenging film, UV absorbing films | Freshness nanosensors, oxygen nanosensors, Biosensors for tracking and for active tags |
| Material | Carbon, metallic, metal oxide, nano-clay nanoparticles | Different types of polymers such as bio-nanocomposites | Silver nanoparticles, metal oxides, and gold nanoparticles | Different types of biosensors |
| Application | Clay nanoparticles composites are used up to 5% (w/w), these nanomaterials help in 80–90% reduction of oxygen and carbon dioxide permeation | Biodegradable coating film prepared from polysaccharides such as chitosan, starch, pectin, alginate, pullulan, cellulose. These coating materials have good biodegradability, biocompatibility, low waste, and eco-friendly | Prevention of fat oxidation by utilizing metallic iron powder, activated iron, and enhancement of fruit and vegetable shelf life by incorporation of lead-impregnated zeolite, polyvinyl chloride film having ZnO nanoparticles | Toxin (mycotoxin) detection by transducer element tri-layer oxide ($SiO_2$ 10 nm/$Si_3O_4$ 10 nm/$SiO_2$ 10 nm)<br>For microbes detection done by single-walled carbon nanotubes |

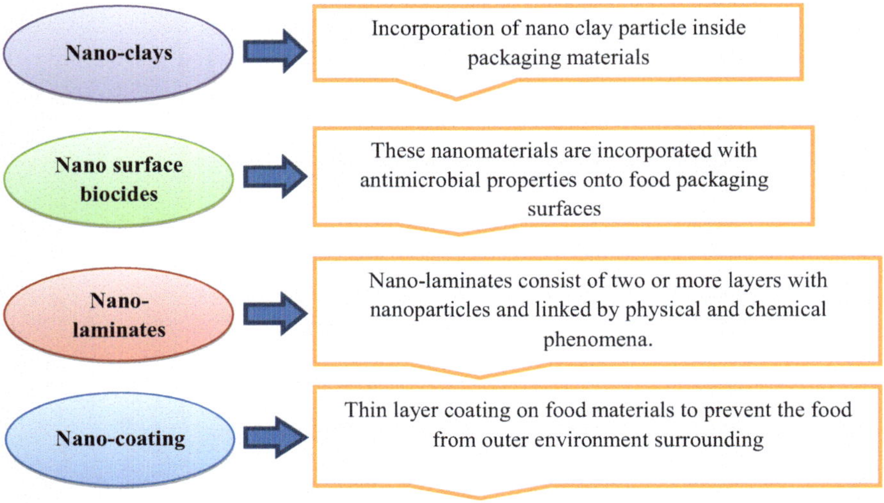


**Fig. 10.2** Nanotechnology methods for improving physical and mechanical properties of food packaging

solution or molten compounds (Baldwin et al. 1996). Hydrocolloids, proteins, and lipids are commonly used as an edible coating material for films formation.

### 10.2.3 Nano Surface Biocide

Specifically utilize nano silver as metallic. For instance: Silver, silver nitrate, zinc oxide, titanium oxide, magnesium oxide, etc. These nanomaterials are incorporated with antimicrobial properties onto food packaging surfaces. They increased shelf life of product, provide significant league quality of mechanical and machinability characteristics. It is the most commonly used packaging for fresh foods and fat-rich products (Guilbert et al. 1997).

### 10.2.4 Nano-Laminates

Layer by layer deposition of nanomaterials was significantly utilized for edible coating. Layer by layer coating application of active agents was employed between the layers or within the individual polyelectrolytes. Polyelectrolytes with opposite charges interact and deposit by mutual attraction between polymeric activated surfaces and polyelectrolytes having covalent entities. For instance, polyelectrolytes are: bio-based polyelectrolytes (proteins, polysaccharides), lipids (surfactants, phospholipids), colloidal particles (micelles, vesicles, droplets) (Kuswandi and Moradi 2019). Layer by layer phenomena allow integration of active compounds into the films (some instances of active compounds are: antimicrobials, anti-browning agents, enzymes, odor, and flavors) and these agents assist in enhancing shelf life and quality of laminated food commodities (Galus et al. 2020).

Nano-laminates are formulated by dipping and washing. Nano-laminate properties depend on many features such as structure, thickness, and composition, and in addition these characteristics also depend on total dipping steps, varieties of absorbing agents used in dipping solutions and accounting environmental factors which play important role in formulation of nano-laminates (for example, temperature, ionic strength, pH and dielectric constant, etc.). Covalently linked layer by layer films enhanced the modulus and nano-films stability (Kuswandi and Moradi 2019).

### 10.2.5 Nano-Clays

Nano-clays are defined as clays having nano-size and are significantly employed to enhance the physical properties of food packaging materials. It is utilized in nanocomposites (formation of layered silicates). The presence of silicates in polymer formulation assists in improving diffusive path with the tortuosity (for a molecule to penetrate through it and help in providing better barrier properties) (Bharadwaj 2001).

Nano-clay enhances the polymers mechanical strength and its cooperation with biopolymers enhances the utilization of functional nanomaterial in food packaging. Nano-clays are utilized in a variety of food commodities to enhance the shelf life and stability of food packaging, some instances are processed meats, cereals, boil in bag foods, meat, etc. (Brody 2007). Nanocomposites combine with clays leads in the transport of diffusing molecules are blocked by impenetrable particles/clay and due to this interfacial zones consist of different permeability characteristics as in comparison to basic polymer formed as shown in Fig. 10.3. Consequently, the nonlinear pathway enhances the gas diffusion length, which leads to the prolongation of shelf life of rapidly spoiled foods (Adame and Beall 2009).

The general goal of improved food packaging material is achieved by incorporation of functional nanomaterials into polymer to enhance the mechanical and physical properties. The nano-clay particles are incorporated in different proportions into polymers to enhance the gas barrier properties of the packaging materials for the beverages and oil industry (Table 10.2).

## 10.3 Biochemical Improved Nanomaterials-Based Food Packaging

Bio-based functional nanomaterials are biodegradable films that are significantly used in food packaging to control the moisture and oxygen transfer as a result it enhances the shelf life, maintain the nutritional values, sensory quality, and safety of foods (Siracusa et al. 2008). As in comparison between plastic and bio-based nanomaterial food packaging, the bio-based nanomaterial seems to be more in favor of eco-friendly nature and in addition to provide best protection to food from

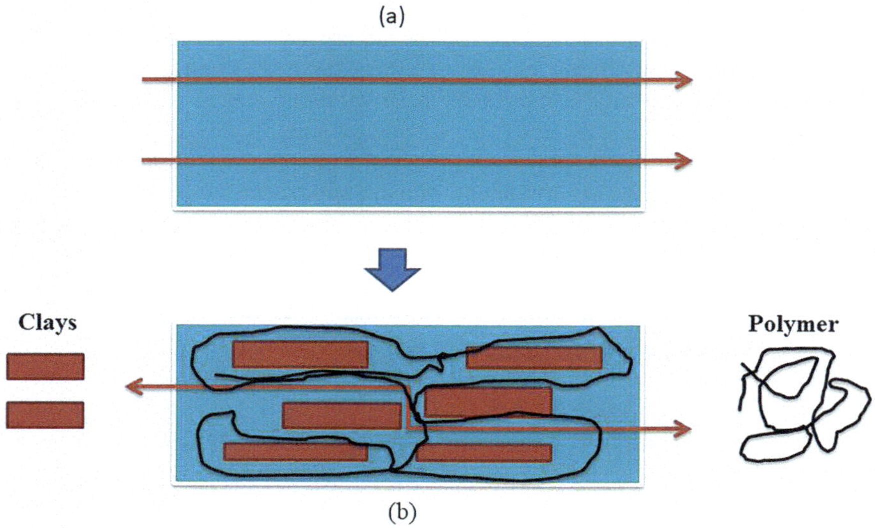


**Fig. 10.3** Incorporation of clay nanoparticles into a film matrix

**Table 10.2** Effect of nano-clay proportions on food packaging

| Nano-clay percentage | Reduction in oxygen permeation | Remarks | References |
|---|---|---|---|
| 5% (w/w) | Reduction of oxygen permeation of 80–90% | The properties of the nanocomposites films were significantly improved as in comparison to the material without added clay nanoparticles | Brody (2007) |
| 3% (w/w) | Oxygen penetration is reduced up to 59% and water vapor barrier by 90% | Excess clay loading in ethylene-vinyl alcohol copolymer results in reduction of tensile strength and optical transparency due to clay agglomerates formation | Kim and Cha (2014) |
| 4% (w/w) | Results in increase in oxygen barrier properties | Utilization of (PS) polystyrene to improve oxygen barrier features | Arora et al. (2011) |

external environment by preventing it from microorganism growth, gas penetrations, humidity, and to avoid further deterioration of food (Dasgupta et al. 2015). Presenting some instances of bio-based materials that are significantly applicable in functional nanomaterial food packaging systems. The maximum use of polyester (PET; polyethylene terephthalate) for food packaging is prepared from bio-based extracted raw materials (Chandra and Rustgi 1998). In the current scenario it is formulated by using petroleum classified materials and PLA (polylactide) is also biodegradable and its incorporation with nanomaterial enhances the mechanical and barrier properties of food packaging. PEF (polyethylenefuroate) is formulated from 2,5-furan dicarboxylic acid (FDCA) and ethylene glycol both are derived from

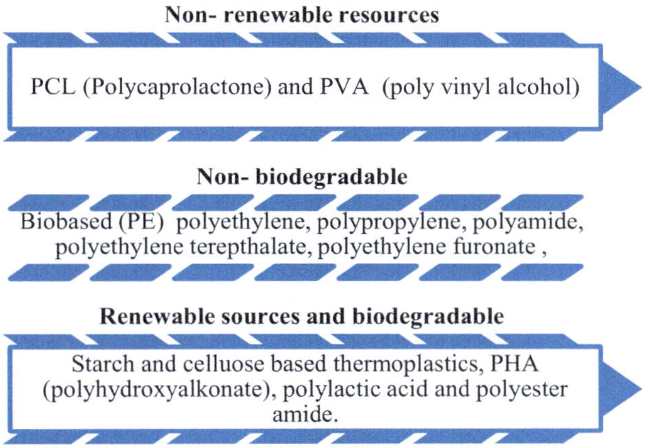


**Fig. 10.4** Classification of bio-based material for food packaging applications

renewable resources. The (PTF; polytrimethylenefurandicarboxylate) is made from furan dicarboxylic methyl ester (FDME) and has biodegradable properties. They are successfully utilized for manufacturing of conventional PET, glass, and aluminum bottles. Many of these nanomaterials are not fully eco-friendly but having best sustainability for packaging material (Doi et al. 2002).

Bio-based nanomaterials for food packaging applications are shown in Fig. 10.4 that are easily decomposable in comparison to plastic nanomaterials. Requirement of natural situations such as temperature, oxygen availability, moisture, and biodegradation process of bio-based nanomaterials is a nontoxic phenomenon for the environment (Bhardwaj et al. 2006).

The further classification of biodegradable plastic polymers based on their origin: (a) Biomass-based biodegradable polymer: direct extracted from biomass (polysaccharides, polypeptides, polynucleotide, and protein). (b) Mixed biomass and petrochemical: fabrication of polymer with bio monomers (for instance: polylactic acid/bio-polyester). (c) Production of polymers by microorganisms or GMOs (genetically modified organisms): for example; bacterial cellulose, xanthane, curdin, and pullan (Bhardwaj et al. 2006). The use of nanotechnology in the preparation of biopolymers leads to improve physical and chemical properties of biopolymers and also decreases the overall cost of packaging. The most common nanomaterials used for food packaging are PLA derivates, polyhydroxybutyrate (PHB), chitosan, protein, and cellulose.

### 10.3.1 Starch-Based Functional Natural Nanomaterials

The convincing characteristics of bio-nanocomposites called "starch-based" having renewable capacity and perfectly fit in a low-cost productive process. Starch-based food packaging can be created to manufacture a film having appropriate mechanical strength. Starch is successfully utilized into a thermoplastic nanomaterial by using extrusion and improves both mechanical and thermal power (Wang et al. 2010). During extrusion process of thermoplastic starch production, plasticizers were applied to lower intramolecular hydrogen bonds therefore helpful in enhancing the stability of packaging material (Abdulmola et al. 1996). The important sources of starch for bioplastic materials are corn, rice, barley, oat, wheat, soy, and potato sources. Starch nature is hygroscopic, therefore their absorbent pads were significantly applied in meat effusions (Park et al. 2002). Starch-based functional films are employed for packaging of both perishable foods and less moisture or dry products (Yoon and Deng 2006).

Due to the hydrophilicity nature of thermoplastic starch (TPS), sometimes it is not appropriate because of the fluctuations in moisture content during processing (Chen and Evans 2005). To overcome this problem nano-filler clay is recommended to improve the TPS property and also enhance the stability of food packaging (Park et al. 2003). The stability of TPS further improved by employing sodium montmorillonite in a concentration of less than 5%. In addition to stability, it also improves decomposition temperature, tensile strength and modulus and reduce relative diffusion coefficient of water vapor (Wilhelm et al. 2003). In recent time, the most common bio-nanocomposites used for food packaging material are starch and its derivates, such as PLA, polyhydroxybutyrate (PHB), and aliphatic polyester, chitosan, proteins, and cellulose.

#### 10.3.1.1 PLA (Polylactic Acid)-Based Functional Nanomaterials

Polylactic acid is produced by chemical processes under a category of bio polyester. PLA is a biodegradable polymer synthesis from phenomena called thermoplastic aliphatic polyester (Chen and Evans 2005). PLA has great capacity in producing a wide range of renewable food packaging materials. The fermentation process is required to manufacture PLA from renewable resources (corn starch) and further polymerization via lactic acid polymerization (Murariu et al. 2008). PLA has good biodegradability characteristics and outstanding mechanical properties (Martino et al. 2006).

The PLA physical and chemical characteristics are based on the ratio of L-PLA and D-PLA. The L-PLA type consists of a high melting point and high crystalline point but a mixture of L- and D-PLA results in an amorphous polymer for the glass transition (Murariu et al. 2008). Some drawbacks are related to PLA, its low reactive side chain constituents decrease its use as packaging element, slow degradation, lower hardiness as well as hydrophobicity. To enhance the mechanical and physical characteristics of PLA, chemical modification is applied by using clay nanocomposite elements (Bandyopadhyay et al. 1999). The clay incorporation with PLA to develop nano-composites of PLA/clay helps in speeding up the process of degradation. Other nano-composites that are also blended with PLAs are silicates (PLA mix with montmorillonites and flurocteroit clays) (Ogata et al. 1997). Moreover, PLA with polycaprolactone is also formed by melt-mixing via advanced kaolinite. All these blended nano-composites PLA has shown significantly great mechanical, gas barrier, and thermal properties in comparison to other polymers without clay (Tang et al. 2008).

PLA incorporation with zinc oxide and silver was employed in different food simulants. Food simulants are part of food elements which reflect the characteristics of food stuffs and in addition this will come in contact with the component of food constituents (Maiti and Batt 2003). Six types of food stimulants are used in migration testing. Stimulant A: 10% ethanol (v/v), Stimulant B: 3% acetic acid (w/v), Stimulant C: 20% ethanol (v/v), Stimulant D1: 50% ethanol, Stimulant D2: Vegetable oil and Stimulant E; poly (2,6-diphenyl-p-phenylene oxide) has particle size 69–80 mesh, pore size (200 nm) (Aguzzi et al. 2007). Food stimulants A, B, and C are utilized for food commodities that have hydrophilic properties (aqueous, alcoholic, and acidic foods) and Stimulants D1 and D2 are employed for food that has lipophilic properties and in addition covers both dairy products and non-dairy products. Stimulant E was utilized for testing specific dry foods. PLA/ZnO; Cu/Ag shows great results in antimicrobial activity in food packaging and also providing great mechanical and thermal properties, barrier to UV rays, water vapor, carbon dioxide, and oxygen (Cabedo et al. 2006).

#### 10.3.1.2 PHB (Polyhydroxybutyrate)-Based Functional Nanomaterials

PHB is an eco-friendly polymeric functional material and formulated by utilizing microorganisms (*Bacillus megaterium*) (Oliva et al. 2007). This functional polymer is utilized as molecule energy storage in the microbial cellular structure. PHB has good biodegradability and biocompatibility characteristics and this makes it fitter for the food packaging industry (Lorz et al. 2000). It is mainly applied for perishable food commodities like fresh meat, dairy products and beverages. In comparison to PP (polypropylene) it has great physical properties and outstanding biodegradable capacity for utilization in food packaging (Lenz and Marchessault 2005).

The microalga *Spirulina species* is involved in synthesizing PHB as a bioproduct and has similar biodegradability rate, mechanical, and thermal characteristics (Weber 2000). Moreover, as this is the byproduct of microalgae, phenolic compounds have antibacterial, antioxidant, and antifungal properties. PHB incorporated with nanofibrils has phenolic compounds and hydrophobicity properties enable PHB nanofibrils to protect food from relative moisture from the surrounding. These nanofibers showed an inhibitory effect on *Staphylococcus aureus* with an inhibitory zone of approx. 7.5 mm. PHB nanofibers produced by the electrospinning process have great barrier to the outer environment due to nonmetric dimensions (Sudesh et al. 2000).

#### 10.3.1.3 Polycaprolactone-Based Functional Nanomaterials

Polycaprolactone is a linear polymer with the semi-crystalline model and high crystalline point and in addition low modulus and high elongation property (Lim et al. 2003). Its biochemical properties are useful for food packaging as well as in biomedical applications and agriculture fields. Due to low melting of polycaprolactone, it is essential to incorporate it with other polymers to increase its stability (Pitt et al. 1981). An incorporation of layer of silicates with the polycaprolactone has developed the nanocomposites which have improved physical properties (Jha et al. 2019).

### 10.3.2 Cellulose-Based Functional Nanomaterials

Cellulose-based nanomaterials are obtained from lignocellulosic biomass, this has become popular due to its biodegradable and ecological nature. The varieties of nanocellulose materials are employed in food packaging such as cellulose nanofibrils, cellulose nanocrystals, nanocellulose-based hybrid materials, and bacterial nanocellulose. Cellulose nanomaterials had proven good results in reduction in manufacturing costs and also lead as eco-friendly materials. Cellulose nanoparticles are mostly used for food packaging applications, various materials such as in a matrix of alginate, chitosan, PLA, polycaprolactone, pectin, etc. (Khalil et al. 2016). These materials had been incorporated with cellulose nanoparticles by employing different techniques. The nanoscale structure and high surface area of cellulose aids in cellulosic nanocomposites having significantly high mechanical, biodegradation, optical, and barrier characteristics. The effect of different concentration of cellulose nanoparticles on the functionality of packaging has been shown in Table 10.3.

The films made from bacterial cellulose produced from cashew apple juice, and added with lignin (0–15 wt%) and cellulose nanocrystals (0–8 wt%), showed enhanced tensile properties and decreased water vapor permeability (Sa et al. 2020). In addition, cellulose nanocrystals (CNC) combined with super magnetic iron oxide NPs (nanoparticles) were used for the production of high-performance nanocomposites coating films. Soy protein isolate nanocomposites incorporated with CNC and zinc oxide NPs were prepared by different methods and showed good tensile strength, water vapor barrier, water-resistant ability, oxygen barrier, and thermal stability (Xiao et al. 2020).

**Table 10.3** Effect of concentration of cellulose nanoparticles on mechanical properties of food packaging

| Cellulose nanoparticles concentration (% (w/w)) | Mechanical properties | Function remarks | References |
|---|---|---|---|
| 5 | Tensile characteristics | Enhanced by 42% | Qasim et al. (2020) |
| 5 | Water vapor permeability | Reduced by 28% | Qasim et al. (2020) |
| 1 | Oxygen permeability | Reduced by 21% | Qasim et al. (2020) |
| 8 | Tensile and water vapor permeability | Tensile property increases but water vapor permeability decreases | Rana et al. (2021) |

### 10.3.3 Chitosan-Based Functional Nanomaterials

Chitosan is the second most abundantly found polysaccharide in the world and obtained from many renewable sources. Consequently, it is available at low cost and more commercially feasible. The focused drawback of chitosan utilization as packaging material as in comparison to the non-biodegradable polymers produced from petroleum results in poor mechanical, barrier, and thermal properties (Cazón and Vázquez 2019). The utilization of nanotechnology features aids in improving properties of the chitosan matrix by the combination of newer formulated nanoparticles. Chitosan films are significantly utilized for food packaging materials to sustain the quality of preserved food commodities. Chitosan has high antimicrobial activity which works against a wide variety of pathogenic and food spoilage microorganisms such as fungi, gram-positive and gram-negative bacteria (Kong et al. 2010; Kravanja et al. 2019). The two phenomena used by the chitosan as antimicrobial against: (a) Chitosan successfully bind to the negatively charged bacteria cell wall and leads to the distribution of the cell membrane and also altered the permeability and consequently, inhibiting the DNA replication that results to cell death (Nagy et al. 2011) and (b) chitosan worked as chelating agents that have capability to bound the trace metal and the toxins formulation resulting in inhibiting the microbial growth (Yilmaz Atay 2020). The incorporation of other molecules into the matrix of chitosan-based polymer is proposed for the formulation of new composites film which has significantly high mechanical properties and improved microbial characteristics as depicted in Table 10.4.

The major drawbacks of chitosan usage as packaging material in comparison to other commonly used petroleum-based nonbiodegradable polymers are reflected in its poor mechanical, thermal, and barrier properties (Radhakrishnan et al. 2015). The functional properties of chitosan can be improved by the use of nanotechnology with the incorporation of newer nanoparticles, carbohydrates, proteins, and other polymers shown in Fig. 10.5 (Zubair et al. 2020). Therefore, presently, chitosan films are increasingly used as packaging materials to maintain the quality of preserved foods.

## 10.4 Active Functions Improved Nanomaterial-Based Food Packaging

In active function nano packaging phenomena, nanomaterials come in direct contact with contained food to improve the environment for better protection and enhance shelf life of food commodities (For instance: silver nanoparticles significantly works as an antimicrobial role) (Chaudhry et al. 2008). Moreover, nano-magnesium oxide, nano-titanium dioxide, nano-copper oxide, and carbon nanotubes are also reported as an antimicrobial agent used in food packaging. Commercial production of antimicrobial packaging as oxygen scavengers has been established by Kodak Company (Asadi and Mousavi 2006). Polyethylene films with active enzymes as

**Table 10.4** Effect of incorporation of nanomaterials into the chitosan composites

| Incorporation of nanomaterials into the chitosan composites | Function remarks | References |
|---|---|---|
| Incorporation of titanium oxide and silver having concentration of (0.5%) into chitosan composites | Enhances the water solubility of the film, highest antibacterial property and shows the lowest light transmission of 54.6%, tensile strength was also declined | Lin et al. (2020) |
| Addition of ZnO and gallic acid into chitosan composites films | Improve the mechanical and physical properties of chitosan composites such as water vapor and oxygen barrier, water solubility, and UV-vis light transmission and also proved to be significantly showing best results in antibacterial and antioxidant activity | Yadav et al. (2021) |
| Silver NPs/chitosan/PVA | The composite nanolayer is significantly utilized in packaging material for fresh meat with enhanced bioactivity and long shelf life of meat products. The combination of chitosan with nanoparticles shows antimicrobial activity *E. coli, S. entericaserovar Typhimurium, S aureus,* and *Listeria innocua* | Arkoun et al. (2017) |
| ZnO NPs incorporation into the linseed oil in chitosan/potato-based polymer | Applicable for determination of storage quality of fresh meat and improve the tensile strength and transparency of the film | Wang et al. (2020) |
| Chitosan gelatin incorporated with nanoencapsulated tarragon essential oil | Applicable for preserving the pork slices. Moreover chitosan incorporation with a variety of essential oils (cinnamomum, assai pulp, thyme, lemon) showed oxidation and microbial growth retardation in a variety of food products | Zhang et al. (2020) |

oxygen scavenging function are also reported in the literature and they showed to inhibit microbial growth and ensures the safety of food products (Lopez-Rubio et al. 2006).

Active functional packaging has more advantages over conventional packaging, in that nanomaterials are coated or laminated as antimicrobial agents in order to ensure microbial safety of food products (Lotfi et al. 2018). Antimicrobial films solve the problem of controlling unwanted microbial growth and also reduce health threat microorganisms, mostly those causing contamination on the surface of food products during a post-processing step. The improvement in the functional nanomaterial in the form of active microbial inhibiting functions is a breakthrough process in the food packaging industry for maintaining food packaging sustainability (Kuswandi and Moradi 2019).

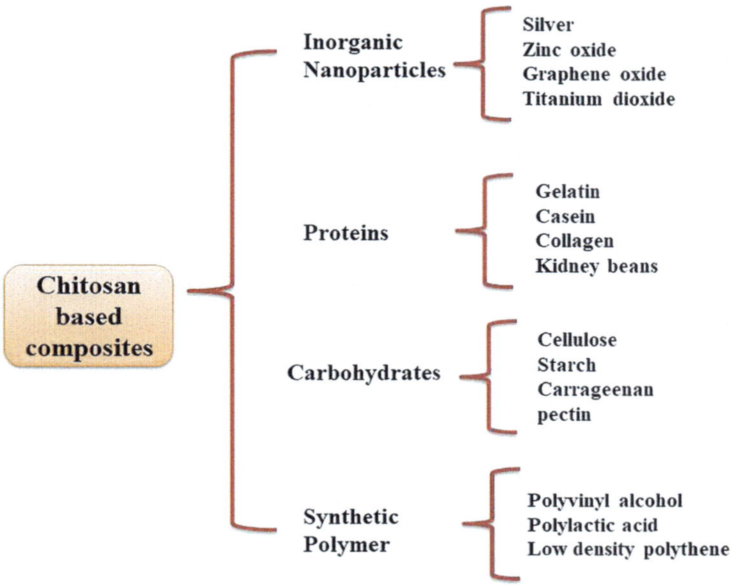


**Fig. 10.5** Classification of chitosan-based composites

### 10.4.1 Antimicrobial Active Packaging

The active antimicrobial function in the packaging mainly consists of the incorporation of antimicrobial agents like silver nanoparticles or silver coatings into packaging materials. These packaging films can minimize the growth of spoilage and pathogenic microorganism. The choice of antimicrobial agents is based on the characteristics of microorganisms such as cell wall composition, oxygen requirements, growth stage, acid/osmosis resistance, optimal growth temperatures, etc. (Malhotra et al. 2015). Different species of microorganisms such as *Salmonella, Staphylococcus, Listeria, Bacillus, Escherichia, Pseudomonas, Lactobacillus, Rhizopus, Aspergillus, Candida*, etc. are responsible for food spoilage (Vilela et al. 2018).

The effective antibacterial characteristics are understood by many molecular mechanisms for instance: production of ROS (reactive oxygen species) having ability to disturb enzyme activity, disturb DNA synthesis, and damage cell organelles as shown in Fig. 10.6. Interactions between cell membrane and nanoparticles are carried by zeta potential present in nanoparticles and this results in cell membrane breakdown influencing the transmission channel through plasma membrane and finally causing cellular death (Gu et al. 2003). Surface of nanoparticles have positive charge and bacterial cell surface comprise with negative charge, this opposite charge responsible for an interaction between enzyme present in bacterial cell and nanoparticles. Therefore, causes cellular death because

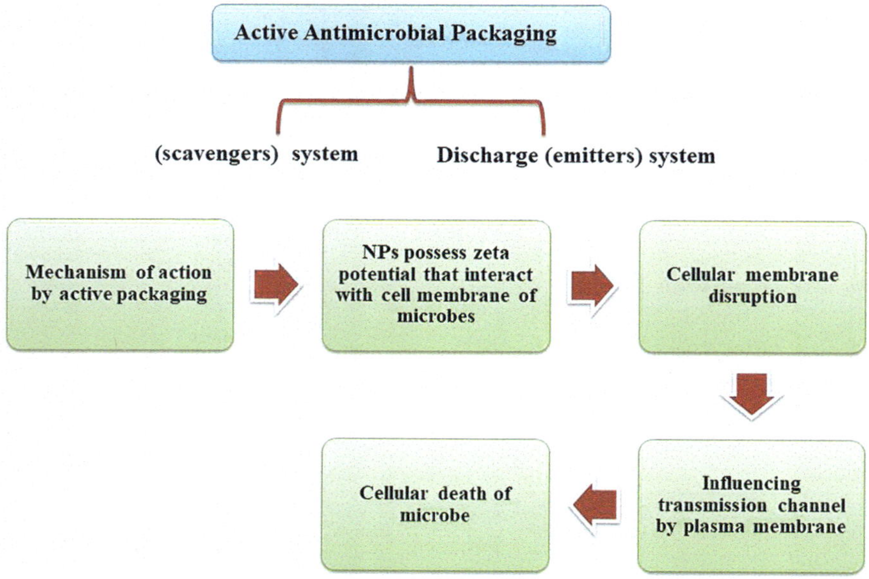


**Fig. 10.6** Antimicrobial mechanism of functional nanoparticles

respiratory enzyme present in bacterial cell membrane are destroyed by an efflux-influx of ion inside the cell (Suski et al. 2012). Moreover, signaling chain activates with the interaction of nanoparticles and bacterial cell membrane, for instance ROS production and oxidation of respiratory enzymes that leads to damage of DNA.

The powerhouse of the cell called mitochondria, during ATP synthesis, reduction of molecular oxygen by an ETC (electron-proton transmission cycle), leads the formation of ROS species (superoxide's anions free radicals, hydroxyl, hydrogen peroxide) found to be a threat for microbial species (Yin et al. 2012). It is well discovered that production of ROS results in per-oxidation of cell matters, oxidative stress, disturb communication channels and also generation of protein radicals. Many reported studies proved that nanoparticles such as titanium dioxide devastate the growth of *S. mutants* due to the cohesion of nanoparticles and titanium oxide (Rezaei et al. 2020).

*Paracoccusdenitrificans*: inhibited by CuO nanoparticles by hampering the enzyme activity of *nitrate* and *nitrite reductase (*Evans et al. 2004*). E coli*: silver nanoparticles hinder the *E. coli* growth, and furthermore ethylene decomposes by silver nanoparticles leads to increased vegetables and fruit shelf life (Shi et al. 2004). Silver nanoparticles are most popular for toxicity to microorganisms, having low volatility and large temperature stability. Titanium oxide, carbon nanotubes, nisin, and chitosan are the other common antimicrobial functional nanomaterials that are applied in food packaging (Chiang et al. 2012).

### 10.4.2 UV Absorbing Film

Nanomaterial film of nanocrystallineTitania (titanium oxide) utilized as UV absorbing films. The absorbance of t oxide is improved by metal doping and leads to improve its photocatalytic properties under UV radiations. Titanium oxide-coated film treated with UV is most effective to inactivate fecal *E. coli* present in water (Gelover et al. 2006). Doping of titanium oxide with silver nanoparticles and incorporated with PVC (nanocomposites) having great antibacterial characteristics (Cheng et al. 2006).

### 10.4.3 Oxygen Scavenger Film

Oxygen causes deterioration of many food commodities in a direct or indirect way as it leads to oxidation by direct reactions. Therefore, it is responsible for browning effect of fruits and rancidity effect in vegetable oils. The most common reason for food degradation is indirect reaction by aerobic microorganisms. The problem of food spoilage is solved by integrating the oxygen scavengers into the food packaging materials. So that oxygen scavengers can reduce the oxygen availability and enhances the shelf life of food (Kuswandi and Moradi 2019). Oxygen hunter films are formulated by incorporation of titanium oxide into polymers. These are mainly beneficial for more oxygen-sensitive food products (Xiao-e et al. 2004).

## 10.5 Smart Functions Improved Nanomaterial-Based Food Packaging

Nanoparticles have a kind of smart functioning in detecting chemical, biochemical, or microbial growth inside the food. Specific microbes and gases nanosensors have been developed for detection of food spoilage (Kuswandi et al. 2011). In this scenario, nanomaterials are employed as reactive materials in food packaging to indicate and communicate the inner condition of packed food commodities, known as nanosensors. Nanosensors are employed to respond as per changes in internal and external parameters of food products and moreover to assure the quality and safety of food (Kuswandi and Moradi 2019).Ongoing research showed that polymeric functional nanomaterials significantly prove to improve the packaging with smart functions (for instance: oxygen determining nanosensors, nanosensors to detect freshness in food, utilization of nano-devices to track the food product and their identification). Nanosensors are capable to detect spoilage of food by change in color parameters due to extreme variation in temperature and humidity with time (Bouwmeester et al. 2009). Inbuilt nanosensors in plastic packaging could detect gases produced by spoiled food and change the package color itself as alarm to the customer about the food quality. Food manufacturing industries predict the shelf life of particular food commodities on the bases of calculating distribution time and storage temperature (Liao et al. 2005). Carbon nanotubes (CNTs) are employed as nanosensors, making the detection process comparatively rapid and precise as compared to the conventional detection process like high-performance liquid chromatography (Nachay 2007). Carbon nanotubes have high-throughput screening capacity, are easy to handle and low cost, less consumption of power, and nice feasibility in recycling. Multi-walled CNTs are significantly utilized in the detection of toxic proteins, microbial growth of specific microorganisms, and food degradation. In the running research scenario, many industries (Nestle, British Airways, and Monoprix) employ nanosensors, useful in the identification of color change under spoilage of food products (Pehanich 2006).

### 10.5.1 Nanosensors for Freshness Identification

The surfaces of packaging material are laminated with nanosensors for freshness indicator. Electrochemically polymerized conducting polymers such as polyacetylene or polypyrrole are widely used for smart packaging due to excellent change between oxidized over reduced form that is the basis of most of the smart functions (Ahuja et al. 2007). Fish spoilage and microbial growth indication were observed by use of polyaniline film as a nanosensor by color change due to volatile amines produced during spoilage of fish (Kuswandi et al. 2012). This nanosensor aids in indicating spoilage of fish in a constant as well as in fluctuating temperature. The spoilage of food is also done by gas penetration and this is indicated by metal oxides because microbial growth produces gases that are determined by metal gas nanosensor. Nanosensors consist of polymer nanocomposites that have conducting features and are immobilized in a polymeric matrix. Polymer nanocomposites help in detecting gases that evolve from microbial growth (Zhang et al. 2020). The common food-borne pathogens such as *Bacillus cereus, vibrio parahaemolyticus*, and *Salmonella* species are detected by nanosensors based on nanocomposite of black carbon and polyaniline (Arshak et al. 2007). The pattern of chicken freshness was determined by employing metal oxide sensor and responses were further analyzed using neural network (Galdikas et al. 2000).

### 10.5.2 Nanosensors to Detect Oxygen

Oxygen provides favorable conditions for aerobic microorganisms to enhance their growth and lead to spoilage of food products (Kuswandi and Moradi 2019). Therefore, food manufacturing industries make sure the absence of oxygen in the food package by vacuumed or degassed the package with nitrogen. Nanosensor for oxygen-sensitive foods has been developed by nanocomposite film of titanium oxide and methylene blue. Under ultraviolet radiation, applied nanosensors start bleaching and finally resulted in colorlessness until oxygen is exposed and again blue color appears (Lee et al. 2005).

### 10.5.3 Nanomaterial Embedded Devices for Tracking and Anti-counterfeiting

Smart functions of nanomaterials are significantly aid in developing tracking devices and anti-counterfeiting to detect the safety of food. The Nanotech Corporations have been involved in the production of nanomaterial-based tracking devices (for instance: Bio silicon) incorporated in the food package to detect the pathogen growth and also easy for customers to identify safe food (Scrinis and Lyons 2007). Nano barcodes are used in each food product, and this is readable by using a microscope for anti-counterfeiting (Roberts 2007). Nowaday's nano barcodes are commercially manufactured by electroplating of inert metal materials like gold, silver, and platinum.

### 10.5.4 Nanomaterials for Active Tags: RFID

Radiofrequency identification (RFID) is the active electronic information-based systems. It is working on radiofrequency, and used to transfer obtained data from a tag (attached to a packaging material) to automatically trace and identify the object containing particular food products. RFID improves the barcode and manual system of tacking (Abad et al. 2007). Due to the presence of radiofrequency range, long reading ranges have strong capacity to penetrate in adverse conditions like temperature and pressure. Nanotechnology provides nanosensors to work with cheap RFID active tags, called nano labeled RFID tags. The characteristics of RFID tags are: flexible, small printed on external layers of food packaging, and also support low-cost production.

## 10.6 Techniques to Analyze Functional Nanomaterials

The focused characteristics of nanoparticles are size and shape. To evaluate the surface chemistry of NPs researchers measured the size distribution, degree of aggregation, surface area, and charge (Mourdikoudis et al. 2018). The techniques for characterization of nanoparticles should be fully proven regarding the availability, cost, selectivity, non-destructive nature, affinity to certain material, and simplicity. Techniques needed to understand and used for characterization of NPs have been discussed in Table 10.5. TEM (transmission electron microscopy), SEM (scanning electron microscope), and AFM (atomic force microscopy) are the best instances of microscopic phenomena and used to characterize nanomaterials in terms of size, structure, shape, dispersion, and rate of coagulations state (Pitt et al. 1981). Alone, this technique is not sufficient for complete verification of nanomaterials and it requires other spectroscopic techniques such as X-ray diffraction (XRD) and UV-visible spectroscopy for more accurately characterization of nanomaterials (Jha et al. 2019). Spectroscopic techniques help in the characterization of structure as well as in the analysis of nanomaterials. The XRD is especially employed for the

**Table 10.5** Techniques for characterization of nanoparticles

| Nanoparticle characteristics | Techniques to characterize NPs |
|---|---|
| Shape | TEM, AFM, HRTEM, FMR |
| Size distribution | DCS, DLS, SAXS, NTA, FMR |
| Chemical state-oxidation state | XAS, EELS, XPS |
| Growth kinetics | SAXS, NMR, TEM, liquid TEM |
| Ligand | XPS, FTIR, NMR, SIMS, FMR |
| Size | TEM, XRD, NTA, SEM |
| Surface area, specific surface area | BET, liquid NMR |
| Surface charge | Zeta potential, EPM |
| Concentration | ICP-MS, UV-VIS, RMM-MEMS, PTA, DCS |
| Agglomeration | DLS, DCS, UV-VIS, SEM |
| Optical properties | UV-VIS, NIR, EELS-STEM |
| Magnetic properties | MFM, FMR |

characterization of montmorillonite exfoliation in nanomaterial used in food packaging (Tortora et al. 2002). Quantification of functional nanomaterials in food packaging is also a big evaluating question due to safety reasons. For this scenario migrated nanoparticles can be probed by inductively coupled plasma mass spectroscopy (ICP-MS), inductively coupled plasma atomic emission spectroscopy (ICP-AES), and inductively coupled optical emission spectrometry (ICP-AES).

By utilizing mentioned techniques, nanoparticles up to the size of 0.1–1.0 ppm can be detected (Huang et al. 2015). The determination of heavy metals (Pb, Cd, Cr, Ni, Zn) in food packaging was identified by employing ICP-MS, consequently researchers found the concentration of Pb: 0.023 μg/L, Cd: 0.030 μg/L, Cr: 0.025 μg/L, Ni: 0.012 μg/L, Zn: 0.017 μg/L (Azlin-Hasim et al. 2016). A titanium metal was identified in food packaging by ICP-AES and the concentration was 5.0 mg/kg, the linear dynamic range 100–5000 μg/L and the recovery was 94.7–100.1% observed (Li et al. 2013).

An experiment was conducted on LDPE film coated with three layers of silver precursor at different concentrations 0.5, 2.0, and 5% and migration of silver NPs under different parameters were determined by ICP-AES. The results showed that the mean migration levels ranged between 0.01 and 1.75 mg/L and further the presence of NPs also confirmed by the SEM (Hannon et al. 2016).

## 10.7 Safety Aspects of Nanomaterials

Food contact articles (FCAs) is a category of functional nanomaterials used in food packaging (Arora and Padua 2010). In FCAs all foods are touched during all stages of production process and transportation. FCAs are combinations of FCMs (food contact materials) such as plastics, papers, metals, board, and glass, and adhesive materials used in ink for printing and coating and in addition functional nanomaterials (Huang et al. 2015). Safety concerns must be fulfilled by determining all food contact chemicals (FCCs) more precisely and including functional nanomaterials that migrate into food commodities and to which consumers are exposed (Alexandre and Dubois 2000).

Food safety is related to human health and environmental health too. The study of migration of nanomaterials present in packaging into food commodities is to assess their potential hazard to environment and human (Duncan 2011). The European legislation reported an overall migration limit of 10 mg per dm square surface area with respect to all substances which have capacity to migrate from nano packaging material to food products (Commission regulation (EU) No. 10/2011) (Cushen et al. 2012). Such as, 1 liter cubic food packaging material with 1 kg of food containing product equal to a migration of 60 mg of packaging material per kg of food product. This criterion is not the same for all food stuffs, it is varied case by case (Foschi and Bonoli 2019). The silver nanoparticle migration from various kinds of nanocomposites into food products has been identified by analytical techniques (Echegoyen and Nerín 2013). In acidic rich food the level of silver migration could happen at a high level and in addition reported that heating is favorable to induce migration and microwave heating showed more pronounce effect on migration. The silver nanoparticle migration can occur by two different mechanisms, firstly by silver nanomaterial getting detached from the composite material and secondly, by silver particles undergoing an oxidation process (Cushen et al. 2013). The silver and copper migration from functional nanomaterials that have antimicrobial action in food packaging is also reported (von Goetz et al. 2013).

The important parameters that would modulate the migration process are nanofillers present in nanocomposite materials, contact time, and particle size. Researchers are more focused on less migration of silver ions into food stuffs; this model benefits the industry regarding decreasing cost and time in migration process studies (Cushen et al. 2014). An incorporation of PLA (polylactic acid) nanocomposites and organo-modified clay utilized an FCM and found that migration of ions into water was less than 10 mg per dm square of the functional nanomaterials (Maisanaba et al. 2014a). Moreover, rats were also exposed for 90 days to the extracts of migration in drinking water and results showed that migration extracts from nanoparticles did not show toxicity, for instance, inflammation or oxidative stress. This defined evidence proved that nanomaterial migration into food stuff was high as in comparison to migration into water (Maisanaba et al. 2014b). Functional improved nanomaterials are needed to investigate more and make sure of the safety of nanomaterials application in the nanoparticle-based food packaging industry (Maisanaba et al. 2014a).

Three important regulations are undertaken to make sure that nanomaterials used in food packaging have high standards for safety of food, safety of consumer, and safety of environment (Jorda-Beneyto et al. 2014). Moreover, these all are essential requirements for successful utilization of functional nanomaterial for the food packaging industry and also having special investigation knowledge regarding; nanomaterials toxicological effects, migration of ions into foodstuffs and their exposure level in working place and consumers (Houtman et al. 2014). If these three regulations were followed properly with all guidelines, nanomaterial for food packaging would become safer, healthier, and tastier as well as more nutritious with eco-friendly nature (Maisanaba et al. 2014a).

## 10.8 Conclusion and Future Potential

The current investigation results showed that nanotechnology provides a number of opportunities in functional nanomaterial in food packaging ranging from bio base to smart packaging. The important beneficial parameters that make nanomaterials more suitable for food packaging are: significantly stronger role, lighter, more durability, and also enhances shelf life of food commodities. Moreover, positive effects on health and low production cost. To develop new food packaging features, nanotechnology proves to increase food characteristics (healthier, tastier, delicious, and full with more nutritious value). The nanosensors incorporation as intelligent packaging provides the food state inside packaging material with visual information. Nanosensors are advantageous to consumers by alerting them when food material inside the packaging material is already spoiled. The conventional food packing system is changed by smart functional nanomaterial food packaging. The future packaging of food materials is safer, healthier, and less harmful to the environment. Nanomaterial food packaging is beneficial for preservation of perishable food products. By utilizing smart and active nanomaterial for food packaging, it could support the food packaging with better mechanical properties; provide stronger barriers from the external environment and better thermal properties. For example, nanomaterial utilized food packaging avoids bacterial invasion and growth for safety of food.

The conventional method of food packaging by biodegradable and natural polymers is not sufficient to provide better safety and healthier food commodities to the consumers due to their poor mechanical and barrier properties. Incorporation of nanofillers (clay) into biopolymers provides a path to improve in general performances (mechanical, thermal, and barrier properties). Despite all positives of nanomaterial function in food packaging, their proper use for packaging may develop safety concerns to human health due to their different physicochemical properties from their macroscale properties. More infestations are needed to understand the impact of nanomaterial or nanoparticles on human health and in addition research should also be focused on identification, quantification, and characterization of functional nanomaterials. Migration of nanomaterial ions into food commodities is a crucial point to understand functional nanomaterial for food packaging. Nanotechnology in the food packaging industry is in its infant stage, more research and investigations are needed to make functional nanomaterial more feasible for food packaging regarding safety, health, and environmental concerns.

## References

Abad E, Zampolli S, Marco S (2007) Flexible tag microlab development: gas sensors integration in RFID flexible tags for food logistic. Sensors Actuators B Chem 127(1):2–7

Abdulmola NA, Hember MWN, Richardson RK, Morris ER (1996) Effect of xanthan on the small-deformation rheology of crosslinked and uncrosslinked waxy maize starch. Carbohydr Polym 31(1–2):65–78

Adame D, Beall GW (2009) Direct measurement of the constrained polymer region in polyamide/clay nanocomposites and the implications for gas diffusion. Appl Clay Sci 42(3–4):545–552

Aguzzi C, Cerezo P, Viseras C, Caramella C (2007) Use of clays as drug delivery systems: possibilities and limitations. Appl Clay Sci 36(1–3):22–36

Ahuja T, Mir IA, Kumar D (2007) Biomolecular immobilization on conducting polymers for biosensing applications. Biomaterials 28(5):791–805

Alexandre M, Dubois P (2000) Polymer-layered silicate nanocomposites: preparation, properties and uses of a new class of materials. Mater Sci Eng R Rep 28(1–2):1–63

Arkoun M, Daigle F, Heuzey MC, Ajji A (2017) Mechanism of action of electrospun chitosan-based nanofibers against meat spoilage and pathogenic bacteria. Molecules 22:585

Arora A, Padua GW (2010) Nanocomposites in food packaging. J Food Sci 75(1):R43–R49

Arora A, Choudhary V, Sharma DK (2011) Effect of clay content and clay/surfactant on the mechanical, thermal and barrier properties of polystyrene/organoclay nanocomposites. J Polym Res 18(4):843–857

Arshak K, Adley C, Moore E, Cunniffe C, Campion M, Harris J (2007) Characterisation of polymer nanocomposite sensors for quantification of bacterial cultures. Sensors Actuators B Chem 126(1):226–231

Asadi G, Mousavi SM (2006) Application of nanotechnology in food packaging. In: 13th world congress of food science & technology, pp 739–739

Azlin-Hasim S, Cruz-Romero MC, Cummins E, Kerry JP, Morris MA (2016) The potential use of a layer-by-layer strategy to develop LDPE antimicrobial films coated with silver nanoparticles for packaging applications. J Colloid Interface Sci 461:239–248

Baldwin EA, Nisperos MO, Chen X, Hagenmaier RD (1996) Improving storage life of cut apple and potato with edible coating. Postharvest Biol Technol 9(2):151–163

Bandyopadhyay S, Chen R, Giannelis EP (1999) Biodegradable organic inorganic hybrids based on poly (L-lactic acid). Polym Mater Sci Eng 81:159–160

Berekaa MM (2015) Nanotechnology in food industry; advances in food processing, packaging and food safety. Int J Curr Microbiol App Sci 4(5):345–357

Bharadwaj RK (2001) Modeling the barrier properties of polymer-layered silicate nanocomposites. Macromolecules 34(26):9189–9192

Bhardwaj R, Mohanty AK, Drzal LT, Pourboghrat F, Misra M (2006) Renewable resource-based green composites from recycled cellulose fiber and poly (3-hydroxybutyrate-co-3-hydroxyvalerate) bioplastic. Biomacromolecules 7(6):2044–2051

Bouwmeester H, Dekkers S, Noordam MY, Hagens WI, Bulder AS, de Heer C (2009) Review of health safety aspects of nanotechnologies in food production. Regul Toxicol Pharmacol 53(1): 52–62

Brody AL (2006) Nano and food packaging technologies converge. Food Technol (Chicago) 60(3): 92–94

Brody AL (2007) Case studies on nanotechnologies for food packaging. Food Technol

Cabedo, L., Luis Feijoo, J., Pilar Villanueva, M., Lagarón, J.M. and Giménez, E., 2006. Optimization of biodegradable nanocomposites based on aPLA/PCL blends for food packaging applications. In Macromolecular symposia (233, 1, pp. 191-197). Weinheim: Wiley VCH Verlag

Cazón P, Vázquez M (2019) Applications of chitosan as food packaging materials. In: Sustainable agriculture reviews, vol 36. Springer, Cham, pp 81–123

Chandra RUSTGI, Rustgi R (1998) Biodegradable polymers. Prog Polym Sci 23(7):1273–1335

Chaudhry Q, Scotter M, Blackburn J, Ross B, Boxall A, Castle L, Aitken R, Watkins R (2008) Applications and implications of nanotechnologies for the food sector. Food Addit Contamin 25(3):241–258

Chen B, Evans JR (2005) Thermoplastic starch–clay nanocomposites and their characteristics. Carbohydr Polym 61(4):455–463

Cheng Q, Li C, Pavlinek V, Saha P, Wang H (2006) Surface-modified antibacterial TiO2/Ag+ nanoparticles: preparation and properties. Appl Surf Sci 252(12):4154–4160

Chiang HM, Xia Q, Zou X, Wang C, Wang S, Miller BJ, Howard PC, Yin JJ, Beland FA, Yu H, Fu PP (2012) Nanoscale ZnO induces cytotoxicity and DNA damage in human cell lines and rat primary neuronal cells. J Nanosci Nanotechnol 12(3):2126–2135

Cushen M, Kerry J, Morris M, Cruz-Romero M, Cummins E (2012) Nanotechnologies in the food industry–recent developments, risks and regulation. Trends Food Sci Technol 24(1):30–46

Cushen M, Kerry J, Morris M, Cruz-Romero M, Cummins E (2013) Migration and exposure assessment of silver from a PVC nanocomposite. Food Chem 139(1–4):389–397

Cushen M, Kerry J, Morris M, Cruz-Romero M, Cummins E (2014) Evaluation and simulation of silver and copper nanoparticle migration from polyethylene nanocomposites to food and an associated exposure assessment. J Agric Food Chem 62(6):1403–1411

Dasgupta N, Ranjan S, Mundekkad D, Ramalingam C, Shanker R, Kumar A (2015) Nanotechnology in agro-food: from field to plate. Food Res Int 69:381–400

Doi Y, Steinbuchel A, Chen GQ (2002) Biopolymers, polyesters III—applications and commercial products, vol 4. Wiley-VCH, Weinheim

Duncan TV (2011) Applications of nanotechnology in food packaging and food safety: barrier materials, antimicrobials and sensors. J Colloid Interface Sci 363(1):1–24

Echegoyen Y, Nerín C (2013) Nanoparticle release from nano-silver antimicrobial food containers. Food Chem Toxicol 62:16–22

Ekrami M, EmamJomeh Z, Mirzakhani M (2014) Physical and mechanical properties of biodegradable edible film obtained from Salep. Iran J Biosyst Eng 45(1):45–51

Evans MD, Dizdaroglu M, Cooke MS (2004) Oxidative DNA damage and disease: induction, repair and significance. Mutat Res 567(1):1–61

Foschi E, Bonoli A (2019) The commitment of packaging industry in the framework of the European strategy for plastics in a circular economy. Admin Sci 9(1):18

Galdikas A, Mironas A, Senuliene V, Šetkus A, Zelenin D (2000) Response time based output of metal oxide gas sensors applied to evaluation of meat freshness with neural signal analysis. Sensors Actuators B Chem 69:258–265

Galus S, Arik Kibar EA, Gniewosz M, Kra'sniewska K (2020) Novel materials in the preparation of edible films and coatings—A review. Coatings 10:674

Gelover S, Gómez LA, Reyes K, Leal MT (2006) A practical demonstration of water disinfection using TiO2 films and sunlight. Water Res 40(17):3274–3280

Gu H, Ho PL, Tong E, Wang L, Xu B (2003) Presenting vancomycin on nanoparticles to enhance antimicrobial activities. Nano Lett 3(9):1261–1263

Guilbert S, Cuq B, Gontard N (1997) Recent innovations in edible and/or biodegradable packaging materials. Food Addit Contam 14(6–7):741–751

Hannon JC, Kerry JP, Cruz-Romero M, Azlin-Hasim S, Morris M, Cummins E (2016) Assessment of the migration potential of nanosilver from nanoparticle-coated low-density polyethylene food packaging into food simulants. Food Addit Contamin A 33(1):167–178

Houtman J, Maisanaba S, Puerto M, Gutiérrez-Praena D, Jordá M, Aucejo S, Jos A (2014) Toxicity assessment of organomodified clays used in food contact materials on human target cell lines. Appl Clay Sci 90:150–158

Huang JY, Li X, Zhou W (2015) Safety assessment of nanocomposite for food packaging application. Trends Food Sci Technol 45(2):187–199

Jha K, Kataria R, Verma J, Pradhan S (2019) Potential biodegradable matrices and fiber treatment for green composites: a review. AIMS Mater Sci 6(1):119–138

Jorda-Beneyto M, Ortuño N, Devis A, Aucejo S, Puerto M, Gutiérrez-Praena D, Houtman J, Pichardo S, Maisanaba S, Jos A (2014) Use of nanoclay platelets in food packaging materials: technical and cytotoxicity approach. Food Addit Contam A 31(3):354–363

Khalil HA, Davoudpour Y, Saurabh CK, Hossain MS, Adnan AS, Dungani R, Paridah MT, Sarker MZI, Fazita MN, Syakir MI, Haafiz MKM (2016) A review on nanocellulosicfibres as new material for sustainable packaging: process and applications. Renew Sust Energ Rev 64:823–836

Kim SW, Cha SH (2014) Thermal, mechanical, and gas barrier properties of ethylene–vinyl alcohol copolymer based nanocomposites for food packaging films: effects of nanoclay loading. J Appl Polym Sci, 131(11)

Kong M, Chen XG, Xing K, Park HJ (2010) Antimicrobial properties of chitosan and mode of action: a state of the art review. Int J Food Microbiol 144(1):51–63

Kravanja G, Primožič M, Knez Ž, Leitgeb M (2019) Chitosan-based (nano) materials for novel biomedical applications. Molecules 24(10):1960

Kuswandi B (2017) Environmental friendly food nano-packaging. Environ Chem Lett 15(2): 205–221

Kuswandi B, Moradi M (2019) Improvement of food packaging based on functional nanomaterial. In: Nanotechnology: applications in energy, drug and food. Springer, Cham, pp 309–344

Kuswandi B, Wicaksono Y, Jayus AA, Heng LY, Ahmad M (2011) Smart packaging: sensors for monitoring of food quality and safety. Sens Instrumen Food Qual Safety 5:137–146

Kuswandi BJ, Restanty A, Abdullah A, Heng LY, Ahmad M (2012) A novel colorimetric food package label for fish spoilage based on polyaniline film. Food Contam 25:184

Lee SK, Sheridan M, Mills A (2005) Novel UV-activated colorimetric oxygen indicator. Chem Mater 17(10):2744–2751

Lenz RW, Marchessault RH (2005) Bacterial polyesters: biosynthesis, biodegradable plastics and biotechnology. Biomacromolecules 6(1):1–8

Li W, Zamani R, Rivera Gil P, Pelaz B, Ibáñez M, Cadavid D, Shavel A, Alvarez-Puebla RA, Parak WJ, Arbiol J, Cabot A (2013) CuTe nanocrystals: shape and size control, plasmonic properties, and use as SERS probes and photothermal agents. J Am Chem Soc 135(19):7098–7101

Liao F, Chen C, Subramanian V (2005) Organic TFTs as gas sensors for electronic nose applications. Sensors Actuators B Chem 107(2):849–855

Lim ST, Hyun YH, Lee CH, Choi HJ (2003) Preparation and characterization of microbial biodegradable poly (3-hydroxybutyrate)/organoclay nanocomposite. J Mater Sci Lett 22(4): 299–302

Lin D, Yang Y, Wang J, Yan W, Wu Z, Chen H, Zhang Q, Wu D, Qin W, Tu Z (2020) Preparation and characterization of TiO2-Ag loaded fish gelatin-chitosan antibacterial composite film for food packaging. Int J Biol Macromol 154:123–133

Lopez-Rubio A, Gavara R, Lagaron JM (2006) Bioactive packaging: turning foods into healthier foods through biomaterials. Trends Food Sci Technol 17:567–575

Lorz PM, Towae FK, Enke W, Jäckh R, Bhargava N, Hillesheim W (2000) Phthalic acid and derivatives. In: Ullmann's encyclopedia of industrial chemistry

Lotfi M, Tajik H, Moradi M, Forough M, Divsalar E, Kuswandi B (2018) Nanostructured chitosan/monolaurin film: preparation, characterization and antimicrobial activity against listeria monocytogenes on ultrafiltered white cheese. LWT 92:576–583

Maisanaba S, Pichardo S, Jordá-Beneyto M, Aucejo S, Cameán AM, Jos Á (2014a) Cytotoxicity and mutagenicity studies on migration extracts from nanocomposites with potential use in food packaging. Food Chem Toxicol 66:366–372

Maisanaba S, Gutierrez-Praena D, Puerto M, Llana-Ruiz-Cabello M, Pichardo S, Moyano R, Blanco A, Jorda-Beneyto M, Jos A (2014b) In vivo toxicity evaluation of the migration extract of an organomodified clay–poly (lactic) acid nanocomposite. J Toxic Environ Health A 77(13): 731–746

Maiti P, Batt C (2003) Renewable plastics: synthesis and properties of PHB nanocomposites. Polym Mater Sci Eng 88:58–59

Malhotra B, Keshwani A, Kharkwal H (2015) Antimicrobial food packaging: potential and pitfalls. Front Microbiol 6:611

Mangiacapra P, Gorrasi G, Sorrentino A, Vittoria V (2006) Biodegradable nanocomposites obtained by ball milling of pectin and montmorillonites. Carbohydr Polym 64(4):516–523

Martino V, Ruseckaite R, Jiménez A (2006) Thermal and mechanical characterization of plasticized poly (L-lactide-co-D, L-lactide) films for food packaging. J Therm Anal Calorim 86(3):707–712

Mourdikoudis S, Pallares RM, Thanh NT (2018) Characterization techniques for nanoparticles: comparison and complementarity upon studying nanoparticle properties. Nanoscale 10(27): 12871–12934

Murariu M, Da Silva Ferreira A, Alexandre M, Dubois P (2008) Polylactide (PLA) designed with desired end use properties: 1. PLA compositions with low molecular weight ester like plasticizers and related performances. Polym Adv Technol 19(6):636–646

Nachay K (2007) Analyzing nanotechnology. Food Technol 61(1):34–36

Nagy A, Harrison A, Sabbani S, Munson RS Jr, Dutta PK, Waldman WJ (2011) Silver nanoparticles embedded in zeolite membranes: release of silver ions and mechanism of antibacterial action. Int J Nanomedicine 6:1833

Ogata N, Jimenez G, Kawai H, Ogihara T (1997) Structure and thermal/mechanical properties of poly (l-lactide) clay blend. J Polym Sci B Polym Phys 35(2):389–396

Oliva J, Payá P, Cámara MÁ, Barba A (2007) Removal of famoxadone, fluquinconazole and trifloxystrobin residues in red wines: effects of clarification and filtration processes. J Environ Sci Health 42(7):775–781

Park HM, Li X, Jin CZ, Park CY, Cho WJ, Ha CS (2002) Preparation and properties of biodegradable thermoplastic starch/clay hybrids. Macromol Mater Eng 287(8):553–558

Park HM, Lee WK, Park CY, Cho WJ, Ha CS (2003) Environmentally friendly polymer hybrids part I mechanical, thermal, and barrier properties of thermoplastic starch/clay nanocomposites. J Mater Sci 38(5):909–915

Pehanich M (2006) Small gains in processing, packaging. Food Proc 11:46–48

Pitt CG, Chasalow FI, Hibionada YM, Klimas DM, Schindler A (1981) Aliphatic polyesters. I. The degradation of poly (€-caprolactone) in vivo. J Appl Polym Sci 26(11):3779–3787

Qasim U, Osman AI, Ala'a H, Farrell C, Al-Abri M, Ali M, Vo DVN, Jamil F, Rooney DW (2020) Renewable cellulosic nanocomposites for food packaging to avoid fossil fuel plastic pollution: a review. Environ Chem Lett 19:613–641

Radhakrishnan Y, Gopal G, Lakshmanan CC, Nandakumar KS (2015) Chitosan nanoparticles for generating novel systems for better applications: a review. Mol Genet Med 9:1–10

Rana AK, Frollini E, Thakur VK (2021) Cellulose nanocrystals: pretreatments, preparation strategies, and surface functionalization. Int J Biol Macromol 182:1554–1581

Ravichandran R (2009) Nanoparticles in drug delivery: potential green nano biomedicine applications. Int J Green Nanotechnol Biomed 1(2):B108–B130

Reynolds G (2007) FDA recommends nanotechnology research, but not labelling food production. Daily.com News

Rezaei M, Pirsa S, Chavoshizadeh S (2020) Photocatalytic/antimicrobial active film based on wheat gluten/ZnO nanoparticles. J Inorg Organomet Polym Mater 30(7):2654–2665

Roberts R (2007) The role of nanotechnology in brand protection. Packag Dig

Sa NMSM, Mattos ALA, Silva LMA, Brito ES, Rosa MF, Azeredo HMC (2020) From cashew byproducts to biodegradable active materials: bacterial cellulose-lignin-cellulose nanocrystal nanocomposite films. Int J Biol Macromol 161:1337–1345

Scrinis G, Lyons K (2007) The emerging nanocorporate paradigm: nanotechnology and the transformation of nature, food and agrifood systems. Int J Sociol Food Agric 15(2):22–44

Shi H, Hudson LG, Liu KJ (2004) Oxidative stress and apoptosis in metal ion-induced carcinogenesis. Free Radic Biol Med 37(5):582–593

Silvestre C, Duraccio D, Cimmino S (2011) Food packaging based on polymer nanomaterials. Prog Polym Sci 36(12):1766–1782

Siracusa V, Rocculi P, Romani S, Dalla Rosa M (2008) Biodegradable polymers for food packaging: a review. Trends Food Sci Technol 19(12):634–643

Sorrentino A, Gorrasi G, Vittoria V (2007) Potential perspectives of bio-nanocomposites for food packaging applications. Trends Food Sci Technol 18(2):84–95

Sperber WH (2009) Introduction to the microbiological spoilage of foods and beverages. In: Compendium of the microbiological spoilage of foods and beverages. Springer, New York, pp 1–40

Sudesh K, Abe H, Doi Y (2000) Synthesis, structure and properties of polyhydroxyalkanoates: biological polyesters. Prog Polym Sci 25(10):1503–1555

Suski JM, Lebiedzinska M, Bonora M, Pinton P, Duszynski J, Wieckowski MR (2012) Relation between mitochondrial membrane potential and ROS formation. Methods Mol Biol 810:183–205

Tang X, Alavi S, Herald TJ (2008) Barrier and mechanical properties of starch-clay nanocomposite films. Cereal Chem 85(3):433–439

Tortora M, Vittoria V, Galli G, Ritrovati S, Chiellini E (2002) Transport properties of modified montmorillonite-poly (ε-caprolactone) nanocomposites. Macromol Mater Eng 287(4):243–249

Vasile C (2018) Polymeric nanocomposites and nano coatings for food packaging: a review. Materials 11(10):1834

Vilela C, Kurek M, Hayouka Z, Röcker B, Yildirim S, Antunes MDC, Nilsen-Nygaard J, Pettersen MK, Freire CSR (2018) A concise guide to active agents for active food packaging. Trends Food Sci Technol 80:212–222

von Goetz N, Fabricius L, Glaus R, Weitbrecht V, Günther D, Hungerbühler K (2013) Migration of silver from commercial plastic food containers and implications for consumer exposure assessment. Food Addit Contam A 30(3):612–620

Wang L, Kwok SK, Ip WH (2010) A radio frequency identification and sensor-based system for the transportation of food. J Food Eng 101(1):120–129

Wang C, Chang T, Dong S, Zhang D, Ma C, Chen S, Li H (2020) Biopolymer films based on chitosan/potato protein/linseed oil/ZnO NPs to maintain the storage quality of raw meat. Food Chem 332:127375

Weber CJ (2000) Biobased packaging materials for the food industry. The Royal Veterinary and Agricultural University, pp 1–69

Wilhelm HM, Sierakowski MR, Souza GP, Wypych F (2003) Starch films reinforced with mineral clay. Carbohydr Polym 52(2):101–110

Xiao Y, Liu Y, Kang S, Wang K, Xu H (2020) Development and evaluation of soy protein isolate-based antibacterial nanocomposite films containing cellulose nanocrystals and zinc oxide nanoparticles. Food Hydrocoll 106:105898

Xiao-e L, Green AN, Haque SA, Mills A, Durrant JR (2004) Light-driven oxygen scavenging by titania/polymer nanocomposite films. J Photochem Photobiol A Chem 162(2–3):253–259

Yadav S, Mehrotra GK, Dutta PK (2021) Chitosan based ZnO nanoparticles loaded gallic-acid films for active food packaging. Food Chem 334:127605

Yilmaz Atay H (2020) Antibacterial activity of chitosan-based systems. Funct Chitosan 457–489

Yin JJ, Liu J, Ehrenshaft M, Roberts JE, Fu PP, Mason RP, Zhao B (2012) Phototoxicity of nano titanium dioxides in HaCaT keratinocytes—generation of reactive oxygen species and cell damage. Toxicol Appl Pharmacol 263(1):81–88

Yoon SY, Deng Y (2006) Clay–starch composites and their application in papermaking. J Appl Polym Sci 100(2):1032–1038

Zhang H, Liang Y, Li X, Kang H (2020) Effect of chitosan-gelatin coating containing nano-encapsulated tarragon essential oil on the preservation of pork slices. Meat Sci 166:108137

Zubair M, Arshad M, Pradhan RA, Ullah A (2020) Chapter 20—Chitosan/chitin-based composites for food packaging applications. In: Gopi S, Thomas S, Pius A (eds) Handbook of chitin and chitosan. Elsevier, Amsterdam, pp 641–670. isbn:978-0-12-817966-6

# Rules and Regulations Related to Packaging 11

Amrita Bhanja and Monalisa Mishra

**Abstract**

There is a rapid growth in the food supply chain starting from the manufacturing to packaging and distribution. These processes in one or the other way can damage or hamper the product quality at any stage of handling. Therefore, proper packaging is the right way to protect the product from different stresses that would eventually extend the shelf life of the product satisfying the consumers need. This chapter basically deals with the rules and regulations related to the packaging as per the Ministry of Health and Family Welfare, regulated by the Food Safety and Standards Authority of India (FSSAI). It briefly provides information regarding the types of packaging materials suitable for a specific product while packaging. Along with that, the factors responsible for affecting the food quality and the innovative techniques of packaging are also discussed in this chapter.

**Keywords**

Packaging material · FSSAI · Regulations · Food quality · Quality degradation

## 11.1 Introduction

The growing interest of consumers towards processed or ready-to-eat food have been attracting the food industries and markets to produce the type of food that would store for several days without any quality degradation and wastage. With the

A. Bhanja · M. Mishra (✉)
Neural Developmental Biology Laboratory, Department of Life Science, National Institute of Technology, Rourkela, Odisha, India
e-mail: mishramo@nitrkl.ac.in





A. K. Shukla (ed.), *Food Packaging: The Smarter Way*,
https://doi.org/10.1007/978-981-16-7196-8_11

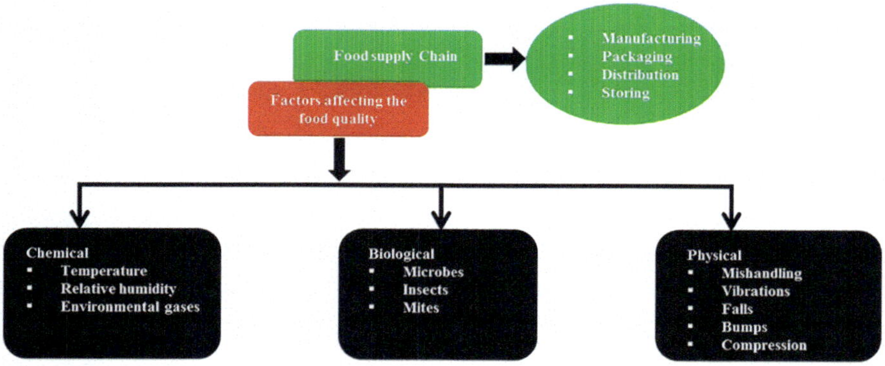


**Fig. 11.1** Schematic diagram showing the processes of food supply chain and the factors affecting the food quality during these processes

increasing demand for the processed food, it also targets the safety of the food throughout the supply chain since the wastage and quality loss of food is a very common and frequent problem starting from its manufacturing, transportation and distribution process. As per the information given by Food and Agriculture Organization (FAO), around 1.3 billion tons of food approximately are found to be wasted during the food supply process (Chen et al. 2020). Thus, to cope up with the consumer's demands, it is very much necessary to process and pack the food in a way so that it gets least affected by different environmental factors, mishandling, transportation, etc. Proper packaging not only protects the food from the outside damage, but also it is efficient enough to retain the nutritive quality extending the shelf life of the food. In other words, packaging is also done to neutralize the negative reactions that might affect the stability of the packed food (Gilbert 1985). Along with this, a proper packaging should contain all the relevant information like the manufactured and the date of expiry, the composition, and quantity of the contained food. Above that, it should be approved by the FSSAI (Lamberti and Escher 2007). The type of material plays a very crucial role, when it comes to the process of packaging. Plastic is the largely used packaging material as it not only provides the necessary safety to the food but also can be modified and incorporated with different other materials that would enhance the quality of the food. On the other hand, plastics are non-biodegradable substances that have become the main reason for the global waste problem. It has been badly affecting marine life along with animal and human health in a large ratio which has become a great matter of concern (Lu et al. 2019). Therefore, it is very much necessary to reduce the single use of plastic and promote ways that would balance the utilization of plastic as a packaging material along with its disposal measures. This chapter provides information about the recent rules and regulations formulated by the FSSAI regarding the packaging materials to be used for specific products (Fig. 11.1).

## 11.2 External Factors Influencing the Quality of the Food

The main purpose of packaging the food is to prevent its quality from getting deteriorated, extending its shelf-life. Proper packaging of the food helps to retain its quality by protecting it from the outside environment which includes several factors like biological changes (microbial contamination), damage caused from the transportation, any sort of theft and manipulation of the product during transportation, and climatic changes that could negatively influence the compositional and nutritional quality of the product (Yildirim et al. 2018). External factors that may hinder the quality of the food can be categorized into chemical, biological and physical factors.

### 11.2.1 Chemical Factors

Exposure to different environmental conditions, like temperature, relative humidity and environmental gases, plays a major role in degrading the quality of the unpacked or not properly packed food (Marsh and Bugusu 2007). Temperature plays a very crucial role in maintaining the quality of food. The effect of factors like temperature and relative humidity on fruit and vegetable has been explained (do Nascimento Nunes 2008). Direct contact of unpacked fruits and vegetables with the excess amount of atmospheric temperature may deteriorate their nutritional quality, texture and appearance. The deterioration of the quality increases constantly by two to three folds with the elevation of the temperature at every 10 °C (Qin et al. 2014). In order to minimize the quality deterioration, postharvest conditions such as temperature management, storage, along with adequate knowledge of the packaging material have to be given priority. Like temperature, the ambient relative humidity is also a matter of concern when it comes to the quality of unpacked food. The water vapour pressure between the surrounding and the product is affected by the humidity along with the temperature. To balance and minimize the level of water vapour pressure loss, the surface humidity has to be maintained in order to prevent the loss of moisture that takes place due to the process of transpiration. As a result, the moisture loss can be neutralized and maintained by controlling the movement of the surrounding air between the food and the environment by using different protective layers or appropriate packaging. The use of proper packaging materials like glass, plastics or film wrapping may prevent the moisture loss from the product by increasing the surface relative humidity. The increased amount of oxygen in the environment could damage the quality of the food by increasing the reactive oxygen species (ROS), damaging the cytoplasm and hindering metabolic activities (Choudhury et al. 2017). There is a high chance of rancidity (Li et al. 2014) and colour degradation (Teixeira et al. 2016) of the fruit due to the low oxygen and high carbon dioxide concentration respectively. Therefore, different products require different atmospheric conditions under modified atmospheric packaging (MAP) (Belay et al. 2019).

### 11.2.2 Biological Factors

Fermented foods are becoming more prone to microbial contamination as compared to fresh products due to mishandling, unhygienic environment and exposure to contaminated water. These fermented foods without appropriate packaging serve as a niche for the bacteria and fungi leading to different intoxications and foodborne illnesses (Roy et al. 2007). As it is previously been mentioned that unpackaged food are more affected by high temperature, likewise packaged food are highly prone to warm temperature. Even though the food is protected by packaging, there are several studies revealing the inefficiency of the poor quality packaging material. Protecting and maintaining the quality of the foodstuffs for longer period of time from being easily affected by insects, mites and microbes is a great matter of discussion as not all packaging are safe and appropriate for the product. Packaging should be done depending on the type and quality of food. Different products require different packaging. Packaging material used should be capable of avoiding infestations by mites and insects. Like unpackaged product, low-quality packaging or poor packaging is equally responsible for affecting the product. A small damage in the packaging material may allow the pest to enter the food followed by the quality deterioration. A chemical known as pheromones released by the pest helps them to increase their rate of population by approximately 20–30 folds every month if they get favourable conditions. This may lead to direct rejection of the whole product raising several legal consequences for the entire chain (Bell 2011).

### 11.2.3 Physical Factors

Physical factors are the damages that may take place during the handling, storing, transportation and distribution of food products. Such damages include vibrations, falls, bumps and compression that could take place during the process of transportation. Therefore, packaging is considered to be a very basic as well as an important parameter for maintaining the safety of the food products (Robertson 2012a).

## 11.3 Impact of Packaging

Starting from the manufacturing, handlings, storing of the fresh and processed food products till the distribution to the consumers, all the processes are carried out with proper standard operating procedures (SOPs) from the government. There are several amendments published by the FSSAI regarding the packaging of the specific food products. The recent amendment that was generated by the FSSAI regarding the rules and regulations of packaging was in the year 2018. The packaging material is chosen depending on the type and quality of food products in order to avoid the degradation of the food. Some of the materials that are commonly used for packaging are cardboard, glass, metal, plastic, etc. (Robertson 2012b). Airtight containers are used for packaging processed vegetables and fruits that would resist the oxygen

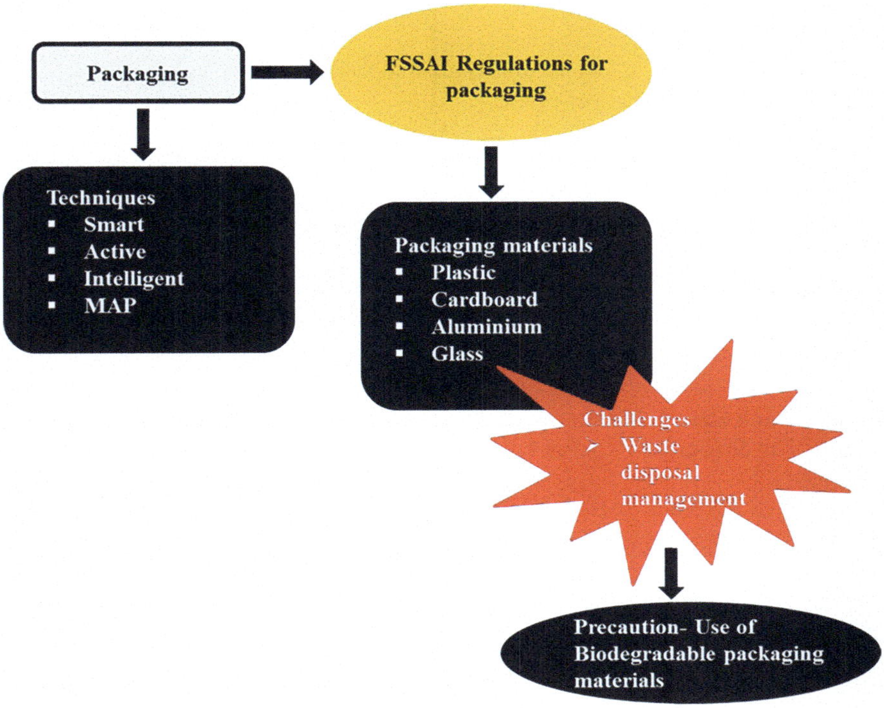


**Fig. 11.2** Schematic diagram showing the overall view of packaging

transmission responsible for product spoilage due to microbial contamination and lipid oxidation (Robertson 2009). The liquid products are packaged using glass jars and bottles whereas the solid stuffs are usually packed by cardboards and plastics. Almost 70% products are packed in the food industry of which around 48% of the total packaging is made from the paperboard (Opara 2013) (Fig. 11.2).

## 11.4 FSSAI Regulations for Packaging

All the rules, regulations and the decisions regarding the food are implemented by the FSSAI, Government of India. Recently the regulations vide no F.No. 1-95/Stds/Packaging/SP(L&C/A)/FSSAI-2017 under sub-section (1) of section 92 of Food Safety and Standards Act, 2006 (34 of 2006) regarding packaging was published and made available for the public in the year 2018 (FSSAI 2018). Some of the important regulations from the above-mentioned document have been mentioned here very briefly.

General requirements that must be followed by the food business authorities regarding the packaging material are that the packaging material should be (of):

- Food grade quality
- Suitable for the product
- Withstand mechanical, chemical and thermal stress
- Hygienic, clean and tamper-proof
- Compatible sealing material
- Not to re-use tin containers
- Printing inks (IS: 15495 conformed)
- Food products not to come in contact with the printed surface
- Not to use newspaper for storing and wrapping
- The layer in the multilayer packaging that would be coming in contact with the food should conform to the necessary requirements of the packaging material

## 11.5 Packaging Materials Suggested by FSSAI to Be Used for Specific Product

According to the rules of FSSAI vide no F.No. 1-95/Stds/Packaging/SP(L&C/A)/FSSAI-2017 published in the year 2018 (FSSAI 2018), there has been a list of packaging materials that are suggested to be used for specific food products. Some of them are being mentioned here very briefly (Table 11.1).

## 11.6 Innovative Techniques of Packaging

The innovation of different modern techniques has made the packaging system more valuable and informative, providing the necessary information regarding the quality of the food in their packaged form. Such types of packaging are provided with smart sensors and indicators to monitor the quality. Apart from this, the environment of the packaging materials is modified and incorporated with the required edible material coming in direct contact with the food resulting in the quality enhancement of the product. Some of the innovative packaging techniques are discussed below.

### 11.6.1 Smart Packaging

Smart packaging is a broad concept and can be categorized mainly into two types of packaging like active and intelligent packaging. The intelligent packaging works by monitoring the external and internal alteration of the product (Vanderroost et al. 2014). The ultimate purpose of smart packaging is not only to keep up the quality and the storage life of the food product, but also to reduce the overall food waste and loss by using several modified techniques, detectors and hardware elements that have been briefly described in active packaging and intelligent packaging. This trending system of packaging is being applied and used by food and pharmaceutical industries to improve and extend their food supply business in a convenient, safe,

**Table 11.1** FSSAI regulations regarding the use of packaging materials for specific products

| Sl. no | Food products | Materials used for packaging |
|---|---|---|
| 1 | Milk | • Glass bottle + metal caps/plastic caps (polypropylene (PP) or high-density polyethylene (HDPE))<br>• Plastic container of polyalkylene terephthalate (PET) + PP or HDPE caps<br>• Plastic container of HDPE/PP/polystyrene (PS) + PP or HDPE caps (Food Safety and Standards (Packaging) Regulations 2018).<br>• Flexible plastic pouch of polyethylene (PE) or PP-based multi-layered packaging<br>• Aseptic/flexible packaging material of paper board/ aluminium foil/polyethylene-based multi-layered structure<br>• Container made of tin plate (Food Safety and Standards (Packaging) Regulations 2018a)<br>• Plastic cups made of (PP or PS) + paper/peel-off lid<br>• Paper butter wrappers coated with wax<br>• Metal containers + plastic PP caps/metal/plastic lid<br>• Plastic pet container + plastic lid<br>• Thermoform cup/tray + paper/peel-off lids (Food Safety and Standards (Packaging) Regultions 2018)<br>• Packaging material made of paper and paper board with/without plastic film lamination, greaseproof paper<br>• Mud or clay pots |
| qx | Fats, oils and fat emulsions | • Container made of tin plate<br>• Glass bottle + metal caps /plastic caps (polypropylene (PP) or high-density polyethylene (HDPE))<br>• Plastic container/jar of HDPE<br>• Plastic bottle/jar (PET) + plastic caps<br>• Plastic pouch (multi-layered laminated or co-extruded structure<br>• Aseptic/flexible packaging material of paper board/ aluminium foil/polyethylene-based multi-layered structure<br>• Plastic laminated pouch (Food Safety and Standards (Packaging) Regulations 2018)<br>• Thermoform plastic jar + plastic caps |
| 3 | Fruit and vegetable products | • Glass bottle + metal caps/plastic caps (polypropylene (PP) or high-density polyethylene (HDPE))<br>• Can be made of aluminium with an easy open end<br>• Container made of tin plate<br>• Aseptic/flexible packaging material of paper board/ aluminium foil/polyethylene-based multi-layered structure<br>• Plastic jar of HDPE/co-extruded with plastic (PP/HDPE caps)<br>• Flexible plastic pouch of (PE/laminated structure)<br>• Plastic jar + metal caps<br>• PET/PP/PVC Punnets (Food Safety and Standards (Packaging) Regulations 2018a) |

(continued)

**Table 11.1** (continued)

| Sl. no | Food products | Materials used for packaging |
|---|---|---|
| 4 | Sweets and confectionary | • Metal containers + plastic PP caps/metal/plastic lid (multi-layered laminated heat sealed pouches)<br>• Composite containers prepared of paper board/ aluminium foil/plastic based films + plastic/metal lids (Food Safety and Standards (Packaging) Regulations 2018a)<br>• Foil wrap<br>• Plastic film based twist wraps (PET/PP/PVC)<br>• Thermoformed tray/punnet + lid<br>• Glass bottle + metal/plastic caps<br>• Plastic cups + film lid |
| 5 | Cereals and cereal products | • Tin container<br>• Aluminium foil-based laminated pouch in metal container<br>• Wrapper of wax-coated paper (Food Safety and Standards (Packaging) Regulations 2018b)<br>• Plastic heat-sealed multi-layered laminated pouch<br>• Plastic thermoform container + plastic lid<br>• Plastic multi-layered laminated Zipper pouch<br>• Thermoform trays + plastic lids/over wraps<br>• Glass bottle + metal caps<br>• PET/plastic-based rigid containers + metal/plastic caps ((PP) or (HDPE))<br>• Plastic/co-extruded film/PP/PE |
| 6 | Meat/meat products/poultry products | • Glass jars + plastic (PP)/(HDPE) caps<br>• Metal containers + metal lid (lacquered tin containers)<br>• Plastic flexible pouches in paper/paperboard carton<br>• Overwrapped plastic tray<br>• Aluminium foil wrap<br>• PET punnets/containers with plastic caps(Food Safety and Standards (Packaging) Regulations 2018) |
| 7 | Fish/fish products/seafood | • Glass jars + plastic caps (PP/HDPE)<br>• Metal containers + metal lid (lacquered tin containers)<br>• PET punnets/containers with plastic caps (Food Safety and Standards (Packaging) Regulations 2018b)<br>• Plastic heat-sealed multi-layered laminated pouch<br>• Overwrapped plastic tray |
| 8 | Honey and other sweetening agents | • Glass bottle + metal caps /plastic caps (polypropylene (PP) or high-density polyethylene (HDPE)) (Food Safety and Standards (Packaging) Regulations 2018)<br>• Plastic based thermoformed containers<br>• PET container + plastic caps<br>• Laminated tube of plastic |
| 9 | Salt/spices/condiments | • Glass bottle + metal lid/plastic caps (PP + HDPE)<br>• Plastic container + plastic cap (PET + HDPE)<br>• Plastic heat-sealed multi-layered laminated pouch |
| 10 | Beverages (excluding diary/ fruits/vegetables based) | • Plastic bottles made of PET/polycarbonate (PC) + plastic (PP/HDPE)/aluminium caps |

(continued)

**Table 11.1** (continued)

| Sl. no | Food products | Materials used for packaging |
|---|---|---|
| | | • Heat-sealed plastic pouch made of PE<br>• Glass bottles + metal/plastic caps<br>• Plastic pouch made of PE in corrugated fibreboard boxes<br>• Aluminium can have an easy open end<br>• Container made of tin plate (Food Safety and Standards (Packaging) Regulations 2018a)<br>• Wooden cask (for wines) |

and in a better way in order to minimize the large amount of loss (Janjarasskul and Suppakul 2018; Poyatos-Racionero et al. 2018).

### 11.6.2 Active Packaging (AP)

Active packaging is introduced as a well-planned, effective and chemical approach to protect and extend the quality and shelf life of the food (Qiu et al. 2019). The chemicals used in active packaging are not added directly to the food rather they are incorporated into the packaging materials. As per the definition given by the European regulation (EC) No 450/2009, active packaging is the packaging that uses "*deliberately incorporate components that would release or absorb substances into or from the packaged food or the environment surrounding the food*" (Commission E 2009). This type of packaging involves the incorporation of a chemical compound either absorbers or emitters to the packaging materials. One such example of the chemical that is commonly used is the oxygen absorber. The former scavenging systems (absorbers) absorbs the excess or unwanted compounds like moisture, ethylene, oxygen, carbon dioxide of the food to the environment that helps to extend the shelf life of the product. On the other hand, the emitters or the releasing systems help in adding the compounds like antimicrobial compounds, ethylene and antioxidants from the environment to the food ultimately balancing the packaging environment, extending the shelf life of the product (Yildirim et al. 2018). One such application of active packaging is that it was used to retain the colour of the chicken breast at 4 °C for 60 days (Stratakos et al. 2015).

### 11.6.3 Intelligent Packaging

As mentioned earlier, intelligent packaging is another form of smart packaging that uses various hardware and electronic devices to monitor the internal and external conditions of the packaged food product (Kruijf et al. 2002). It provides information about the products during their transportation and storage (Kerry et al. 2006). Through this system of packaging, it becomes easy to monitor the parameters like microbial contamination, temperature, and product purity. Along with this, it also gives information about the changes due to the pH and enzymatic reactions through the Vitsab time-temperature (TTI) indicator (Opara 2013). It also incorporates radio-frequency identification (RFID) devices that would provide data about the pH, food volatiles, gases, and conductivity with the use of chemical elements (Chen et al. 2020).

### 11.6.4 Modified Atmosphere Packaging (MAP)

The modified atmosphere packaging can be explained in a way where the atmosphere surrounding the food inside the packaging can be altered or modified as per the requirement so that it will not affect the shelf life along with preventing the quality of the food. It basically works on inhibiting the microbial contamination and stops the enzymatic spoilage. The packaging is made up of semi-permeable films in such a way that it tends to lower the exchange of the respiratory gases inside the package (Chitravathi et al. 2015). Along with carbon dioxide, which is known for its antibacterial ability (Sun et al. 2017), oxygen and nitrogen are also used. Specifically, these are the three gases that are used for the packaging for extending the shelf life of the food in MAP. The ratio and the composition of the gases completely rely on the type of food. Every food has its own environmental stability. Nitrogen is basically used as a filler gas. Several studies have been carried out that proved the efficiency of MAP. This method was successfully used for packaging smoked catfish without influencing its organoleptic qualities (Qiu et al. 2019).

## 11.7 Materials Used for Packaging

The selection, construction and designing of the packaging material play a key role in maintaining the storage life of the product. The material to be used for the product should be efficient enough to retain the quality, originality and the freshness of the product throughout the food supply process. Paper, aluminium, glass and metal are some of the traditionally used materials. Along with this, plastic has been the highly used packaging material that is being used in several forms.

### 11.7.1 Plastic

The use of plastic as a packaging material is the most utilized and given the topmost priority in the food industries. It is the highest utilized packaging material all over the world. Apart from the well-known disadvantages, there are a lot of advantages that makes it liable as a packaging material. Thermosets and thermoplastics are the two categories of plastics used for different purposes. Amongst them, thermoplastics are used for the packaging of food because of their flexible nature of getting easily moulded into different shapes and sizes depending on the type of food (Marsh and

Bugusu 2007). As compared to the other packaging materials like aluminium and metal, plastic provides more transparency that makes them easily visible to the consumers in spite of the packaging (Opara 2013). They are considered to be of lightweight, chemically inert, easy to print and handle as compared to the other packaging materials. The major concern that raises question regarding the frequent use of plastic includes health as they contain plasticizers and stabilizers and environmental problems as these are non-biodegradable substances (Marsh and Bugusu 2007). Its effective barrier property makes it microbiologically safe (Lamberti and Escher 2007).

#### 11.7.1.1 FSSAI Regulations Regarding Plastic as Packaging Material

Food products packed using plastic shall obey the Indian Standards specifications

- Packaging of the drinking water shall be done using transparent, tamper-proof bottles conforming to.
- IS: 9845—migration limit (60 mg/kg or 10 mg/dm$^2$)
- It also provides the migration limit (mg/kg) of the substances like barium, cobalt, copper, iron, lithium, manganese and zinc from the plastic materials
- IS: 9833—for pigments to be used in plastics
- Not to use recycled plastic products for packaging, carrying and storing of food (Table 11.2)

### 11.7.2 Paperboard

The recyclable, eco-friendly and inexpensive nature of cardboard makes it reliable for using it as a packaging material. In the food industries mostly cardboards are used in the form of cartons for packaging freshly produced fruits and vegetables. Sulphite and sulphate are used to convert cellulose fibres obtained from wood into valuable paper and cardboard (Robertson 2012b). Cardboards are made up of a wide range of plain paper. A normal plain paper is not utilized for packaging purpose as they are inefficient in preserving the food for a long time. Therefore when used for packaging, it is added with some materials like resins, wax to improve its strength. As compared to the paper, paperboard is of more strength (Marsh and Bugusu 2007).

**Table 11.2** FSSAI standards for specific plastic materials

| Indian Standards specifications (IS) | Material |
|---|---|
| IS: 10146 | Polyethylene (PE) |
| IS: 10151 | Polyvinyl chloride (PVC) |
| IS: 12252 | Polyalkylene terephthalate (PET) |
| IS: 10910 | Polypropylene (PP) |
| IS: 14971 | Food grade polycarbonate |
| IS: 10142 | Polystyrene |

#### 11.7.2.1 FSSAI Regulations Regarding Paper and Board as Packaging Materials

- Should maintain uniform thickness
- Free of grease marks, pinholes, cuts
- Should be contaminant free and be of food grade
- Should conform to Indian Standards Specification

### 11.7.3 Aluminium

Aluminium is one of the most highly available metallic constituent of the earth. It is silvery white and malleable in nature. It is expensive and used for preparing different seamless containers. According to Morris, 2011, it is mostly used for packaging sea food (Morris 2011), and soft drinks (Deshwal and Panjagari 2020). Mainly aluminium is used in the form of aluminium foil in the food industries for packaging as it prevents the light to pass through it and retains the original moisture, aroma and gases of the food. When it comes to the toxicity of aluminium, it is safe as a packaging material. It is of lightweight and can be moulded into different shapes as demanded by the nature of the food. The benefit of using aluminium as a packaging material is that it is recyclable, environmental friendly and obtained in the same quantity after being recycled(Lamberti and Escher 2007).

#### 11.7.3.1 FSSAI Regulations Metal as Packaging Material

Containers shall be considered unfit for human consumption, when the metal (packaging material) shows the following characteristics:

- Rusty
- Enamelled, chipped
- Not properly tinned copper or brass containers
- Should use appropriate grades of metal for packaging
- Shall conform to Indian standards specifications

### 11.7.4 Glass

Glass is another most frequently used packaging material. It is mostly used for packaging soft drinks, juices, pickles and many more. It tends to be a permanent kind of packaging material that is reusable. It is considered as a highly useful packaging material as it is chemically inert, recyclable and easily manufactured from raw materials (Le Bourhis 2008). Apart from the above-mentioned advantages, there is a disadvantage of requiring a lot of raw materials for its production like silica, dolomite, and also consumes a lot of energy (Patrascu et al. 2009; Kovacec et al. 2011).

#### 11.7.4.1 FSSAI Regulations Regarding Glass as Packaging Material

- Should be free from blisters, stones, chippings, mould marks and visible defects
- Should be of smooth surface without cracks, sharp edges and pinholes
- Sealing surface should be crack free

### 11.7.5 Incorporation of Edible Films in the Packaging Material

In order to prevent or lower the quality deterioration of fruits and vegetables caused due to the water activity and the ambient temperature, packaging incorporated with hydrocolloids could be a better option for using it as oxygen barriers (Robles-Sánchez et al. 2013). The use of edible packaging was also applied on several foods to safeguard their nutritional and organoleptic properties. These sorts of packaging help in neutralizing the antioxidant activity along with reducing the oxygen permeability. The use of such edible films in the food products works as an oxygen barrier that would eventually prevent or reduce the oxidation of lipids, flavours and colorants. Such oxygen barrier may also delay the process of respiration in the fresh produce especially fruits and vegetables. Different types of modifications like the use of synthetic nanofibres, carbon nanotubes and nano clays can also be implicated in these edible films as well as coatings that would improve the oxygen barrier properties. Polysaccharides and protein edible films can be used as good and effective oxygen barriers as they are known to carry antioxidants that would migrate towards the surface of the food. The oxygen permeability is influenced by different environmental properties as well as packaging conditions like temperature, film thickness, relative humidity and moisture content (Gómez-Estaca et al. 2009a, b). The oxygen permeability tends to decrease with the lower humidity as the structure of the film remains intact. There are studies showing the increase in the oxygen permeability due to the increase of the humidity when plastic films were used to coat the whey protein (Hong and Krochta 2006). Like humidity, increase in temperature also leads to the increase of oxygen permeability. This may reduce the extended shelf life of the food product. The use of antioxidants in edible coatings and film may reduce the process of rancidity, discoloration and degradation. Natural antioxidants extracted from the plants can also be incorporated to films that would result in the production of biopolymers having antioxidant activity. The oxygen permeability is directly influenced by the thickness of the film. According to (Bonilla et al. 2012), the oxidative reactions are affected by the factors like continuity and homogeneity of the films. Excess light exposure can lead to photo-oxidation of food. So packaging incorporated with low light permeability film like hydroxypropyl methycellulose can be used while packing products like fats and oils. The utilization of such films is helpful in reducing the process of photo oxidation (Akhtar et al. 2010). Apart from the above-mentioned coating, there are several other coatings like gellan, alginate and pectin-based edible coatings that were used to coat fresh-cut melon. These coatings were useful for reducing the level of dehydration. They were also effective in preventing ethylene production resulting in the reduction of carbon dioxide along with oxygen permeability. Using alginate and pectin coating reduced the wounding stress (Oms-Oliu et al. 2008). Chitin derived from the yeasts, insects, fungi and invertebrates after deacetylation forms chitosan (Dehghani et al. 2018) that can be used for film formation due to their unique nature of toughness and biodegradability (Sahraee et al. 2019).

## 11.8 Challenges

Despite several advantages regarding the use of the packaging material especially plastic, it has come up with a lot many challenges that are eventually and immensely affecting the earth in some or other ways. This global problem is not only affecting the terrestrial level but is also in a verge to deplete the marine life, rivers. The steadinesses of the measures are being overpowered by the large amount of plastic waste (Geyer et al. 2017). Most of the wastes are being collected from the source of packaging. Therefore, it is very much essential to implement proper regulations regarding its disposal that would not affect the environment and on the other hand would find some alternatives to such problems. The innovations of social-ecological systems might be an appropriate approach to overcome such problems (Sattlegger et al. 2020).

## 11.9 Conclusion and Future Trends

Packaging being an essential part of the food system plays a very chief role in maintaining the quality and also enhances the marketing possibility of the product. Therefore, in order to reach the consumer's demand, not only the organoleptic and the nutritive quality of the product should be checked, but also the packaging of the product must be given equal importance in order to gain the consumer's attention. Therefore, proper packaging protocols should be maintained. Along with the packaging, proper packaging waste disposal should also be taken care. The increasing environmental issues due to the packaging waste disposal are the biggest concerns that are needed to be resolved through legal procedures. The problems may not be solved by only the process of recycling or proper waste management, it might also require the production of more and more bio-based materials for the packaging purpose. Biodegradable packaging materials could be an alternative to the single-use plastic packaging materials. The raw materials of such type of packaging are mostly originated from agricultural resources that are renewable. These packaging materials are environmental friendly and are a better option for the countries that use landfill for managing the waste disposal. FSSAI should collaborate with different organizations working on waste management to resolve such types of challenges and also strictly implement and execute proper rules for disposing of the non-biodegradable packaging materials (Petersen et al. 1999).

## References


Akhtar MJ et al (2010) Control of salmon oil photo-oxidation during storage in HPMC packaging film: influence of film colour. Food Chem 120(2):395–401

Belay ZA, Caleb OJ, Opara UL (2019) Influence of initial gas modification on physicochemical quality attributes and molecular changes in fresh and fresh-cut fruit during modified atmosphere packaging. Food Packag Shelf Life 21:100359

Bell C (2011) Insect and mite penetration and contamination of packaged foods. In: Food and beverage stability and shelf life. Elsevier, pp 106–131

Bonilla J et al (2012) Edible films and coatings to prevent the detrimental effect of oxygen on food quality: possibilities and limitations. J Food Eng 110(2):208–213

Chen S et al (2020) The role of smart packaging system in food supply chain. J Food Sci 85 (3):517–525

Chitravathi K, Chauhan O, Raju P (2015) Influence of modified atmosphere packaging on shelf-life of green chillies (Capsicum annuum L.). Food Packag Shelf Life 4:1–9

Choudhury FK et al (2017) Reactive oxygen species, abiotic stress and stress combination. Plant J 90(5):856–867

Commission E., EU Guidance to the Commission Regulation (EC) No 450/2009 of 29 May 2009 on active and intelligent materials and articles intended to come into the contact with food (version 1.0). 2009

Dehghani S, Hosseini SV, Regenstein JM (2018) Edible films and coatings in seafood preservation: a review. Food Chem 240:505–513

Deshwal GK, Panjagari NR (2020) Review on metal packaging: materials, forms, food applications, safety and recyclability. J Food Sci Technol 57(7):2377–2392

do Nascimento Nunes MC (2008) Impact of environmental conditions on fruit and vegetable quality. Stewart Postharvest Rev 4(4):1–14

Food Safety and Standards (Packaging) Regulations (2018). https://fssai.gov.in

Food Safety and Standards (Packaging) Regulations, (2018a). https://foodsafetyhelpline.com

Food Safety and Standards (Packaging) Regulations (2018b). https://archive.fssai.gov.in/home

FSSAI, *Food Safety and Standards, (Packaging) regulations, 2018*, F.S.a.S.A.o. India, Editor. 2018, Gazette of India, Extraordinary, Part III, section 4: New Delhi

Geyer R, Jambeck JR, Law KL (2017) Production, use, and fate of all plastics ever made. Sci Adv 3 (7):e1700782

Gilbert S (1985) Food/package compatibility. Food technology (USA)

Gómez-Estaca J et al (2009a) Antioxidant properties of tuna-skin and bovine-hide gelatin films induced by the addition of oregano and rosemary extracts. Food Chem 112(1):18–25

Gómez-Estaca J et al (2009b) Incorporation of antioxidant borage extract into edible films based on sole skin gelatin or a commercial fish gelatin. J Food Eng 92(1):78–85

Hong S-I, Krochta JM (2006) Oxygen barrier performance of whey-protein-coated plastic films as affected by temperature, relative humidity, base film and protein type. J Food Eng 77 (3):739–745

Janjarasskul T, Suppakul P (2018) Active and intelligent packaging: the indication of quality and safety. Crit Rev Food Sci Nutr 58(5):808–831

Kerry J, O'grady M, Hogan S (2006) Past, current and potential utilisation of active and intelligent packaging systems for meat and muscle-based products: a review. Meat Sci 74(1):113–130

Kovacec M, Pilipovic A, Stefanic N (2011) Impact of glass cullet on the consumption of energy and environment in the production of glass packaging material. In: Recent researches in chemistry, biology, environment and culture. Monteux, Switzerland

Kruijf ND et al (2002) Active and intelligent packaging: applications and regulatory aspects. Food Addit Contaminants 19(S1):144–162

Lamberti M, Escher F (2007) Aluminium foil as a food packaging material in comparison with other materials. Food Rev Intl 23(4):407–433

Le Bourhis E (2008) Glass rheology. In: Glass: mechanics and technology. Wiley-VCH Verlag GmbH & Co. KGaA, Weinheim, pp 83–134
Li X et al (2014) Effects of ascorbic acid and high oxygen modified atmosphere packaging during storage of fresh-cut eggplants. Food Sci Technol Int 20(2):99–108
Lu L et al (2019) Interaction between microplastics and microorganism as well as gut microbiota: a consideration on environmental animal and human health. Sci Total Environ 667:94–100
Marsh K, Bugusu B (2007) Food packaging—roles, materials, and environmental issues. J Food Sci 72(3):R39–R55
Morris SA (2011) Food and package engineering. Wiley, Hoboken
Oms-Oliu G, Soliva-Fortuny R, Martín-Belloso O (2008) Using polysaccharide-based edible coatings to enhance quality and antioxidant properties of fresh-cut melon. LWT—Food Sci Technol 41(10):1862–1870
Opara UL (2013) A review on the role of packaging in securing food system: adding value to food products and reducing losses and waste. Afr J Agric Res 8(22):2621–2630
Patrascu R, Minciuc E, Darie G (2009) Complex analysis and evaluation of processes from the glass industry. In: Proceedings of the 4th IASME/WSEAS international conference on energy & enviroment (EE'09)
Petersen K et al (1999) Potential of biobased materials for food packaging. Trends Food Sci Technol 10(2):52–68
Poyatos-Racionero E et al (2018) Recent advances on intelligent packaging as tools to reduce food waste. J Clean Prod 172:3398–3409
Qin Y, Wang J, Wei C (2014) Joint pricing and inventory control for fresh produce and foods with quality and physical quantity deteriorating simultaneously. Int J Prod Econ 152:42–48
Qiu L et al (2019) Innovative technologies for producing and preserving intermediate moisture foods: a review. Food Res Int 116:90–102
Robertson GL (2009) Food packaging and shelf life: a practical guide. CRC Press, Boca Raton
Robertson GL (2012a) Introduction to food packaging. In: Food packaging: principles and practice, 3rd edn. CRC Press, Boca Raton, pp 1–8
Robertson G (2012b) Paper-based packaging of frozen foods. In: Handbook of frozen food processing and packaging, 2nd edn. CRC Press, Boca Raton, pp 743–758
Robles-Sánchez RM et al (2013) Influence of alginate-based edible coating as carrier of antibrowning agents on bioactive compounds and antioxidant activity in fresh-cut Kent mangoes. LWT—Food Sci Technol 50(1):240–246
Roy A, Moktan B, Sarkar PK (2007) Microbiological quality of legume-based traditional fermented foods marketed in West Bengal, India. Food Control 18(11):1405–1411
Sahraee S et al (2019) Protection of foods against oxidative deterioration using edible films and coatings: a review. Food Biosci 32:100451
Sattlegger L et al (2020) Plastic packaging, food supply, and everyday life: adopting a social practice perspective in social-ecological research. Nat Culture 15(2):146–172
Stratakos AC et al (2015) Synergism between high-pressure processing and active packaging against listeria monocytogenes in ready-to-eat chicken breast. Innovative Food Sci Emerg Technol 27:41–47
Sun B et al (2017) The combined efficacy of superchilling and high CO2 modified atmosphere packaging on shelf life and quality of swimming crab (Portunus trituberculatus). J Aquatic Food Prod Technol 26(6):655–664
Teixeira GH et al (2016) Quality of guava (Psidium guajava L. cv. Pedro Sato) fruit stored in low-O2 controlled atmospheres is negatively affected by increasing levels of CO2. Postharvest Biol Technol 111:62–68
Vanderroost M et al (2014) Intelligent food packaging: the next generation. Trends Food Sci Technol 39(1):47–62
Yildirim S et al (2018) Active packaging applications for food. Compr Rev Food Sci Food Saf 17 (1):165–199

The manufacturer's authorised representative in the EU is Springer Nature Customer Service Centre GmbH, Europaplatz 3, 69115 Heidelberg, Germany. If you have any concerns regarding our products, please contact ProductSafety@springernature.com

Printed and bound by CPI Group (UK) Ltd, Croydon, CR0 4YY
23/09/2026
02221541-0003